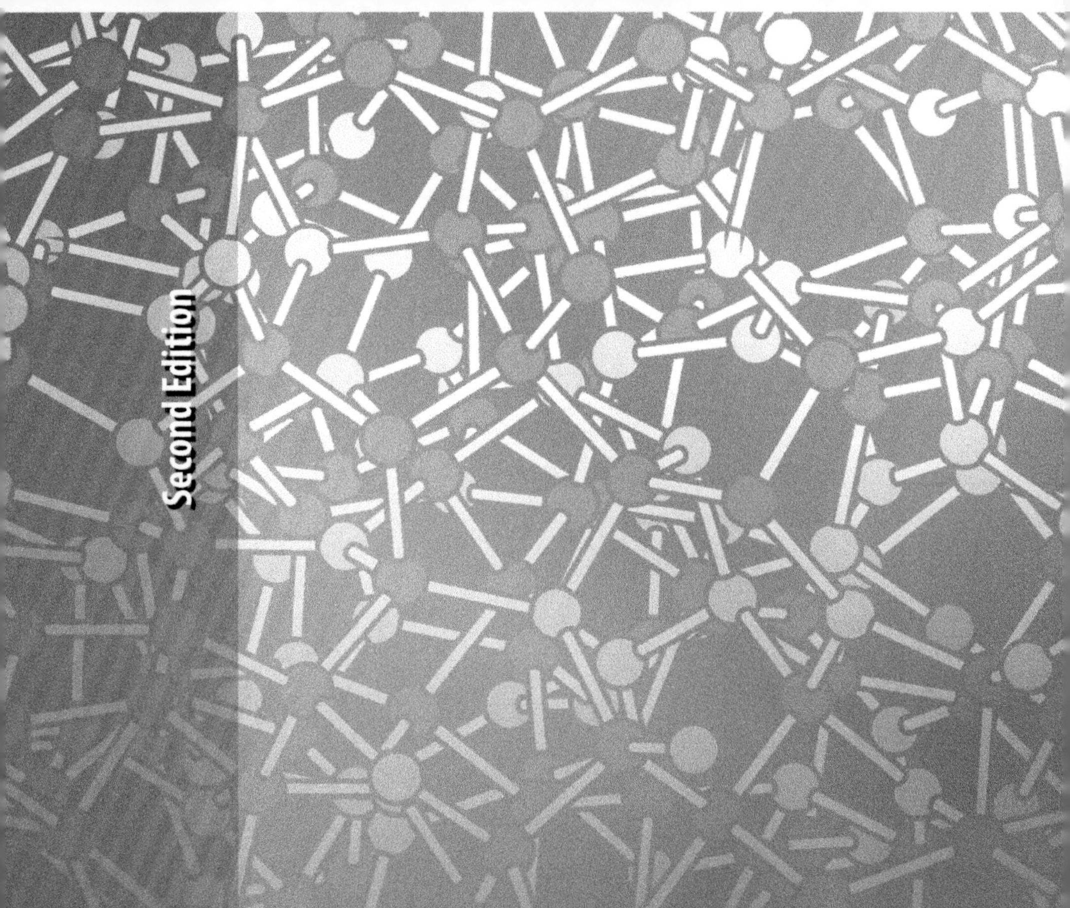

Second Edition

Disordered Semiconductors

Disordered Semiconductors

Physics and Applications

Second Edition

Anatoly Popov

Pan Stanford Publishing

Published by

Pan Stanford Publishing Pte. Ltd.
Penthouse Level, Suntec Tower 3
8 Temasek Boulevard
Singapore 038988

Email: editorial@panstanford.com
Web: www.panstanford.com

British Library Cataloguing-in-Publication Data
A catalogue record for this book is available from the British Library.

Disordered Semiconductors:
Physics and Applications (2nd Edition)

ISBN 978-981-4774-37-6 (Hardcover)
ISBN 978-1-315-10042-5 (eBook)

Contents

Preface

Six years have passed since the first edition of the book *Disordered Semiconductors: Physics and Applications* was published. In this second edition of the book, the developments from then on have been taken into account. The pace and extent of progress is evident from the new sections added to the book and additions made to the following sections of the book:

- 1.1 Definition of Disordered State,
- 1.3 Qualitative and Quantitative Characteristics of Glass Formation,
- 2.3 Methods of Investigation of Disordered System Structure,
- 4.4.3 Structural Modification at the Morphology/Heterogeneity Level,
- 5.1 Technological Distinctions of Chalcogenide Glassy Film Preparation,
- 5.2 Preparation of Hydrogenated Amorphous Silicon Films by Glow-Discharge Decomposition Method,
- 6.1.2 Semiconductor Materials for Photoreceptors,
- 6.1.5 Photo-thermoplastic Process,
- 6.2 Devices Based on Photo-induced Transformations in Chalcogenide Glasses,
- 8.1.3 The Second-Generation of Phase-change Memory Devices, and
- 8.2 Silicon Thin-Film Transistors.

New sections added in this edition are

- 2.5.4 Structure of Carbon-Based Amorphous and Nanocomposite Films,
- 2.5.5 Structure of Organic Semiconductors,
- 3.1.5 Electronic Structure of Organic Semiconductors,
- 3.2.5 Features of the Charge Carrier Transport in Organic Semiconductors,
- 4.5 Chemical Modification of the Atomic Structure of Disordered Semiconductors,

4.6 Structural, Chemical, and Phase Modification of Amorphous Diamond-Like Silicon–Carbon Films (ASCFs), which includes Section 4.6.1, Structural Modification of ASCFs; Section 4.6.2, Chemical and Phase Modification of ASCF; and Section 4.6.3, Nanostructuring of ASCF,

6.1.4 Electroradiographic Process,

7.1 Renewable Energy Resources,

7.2 Conversion of Solar Energy into Electricity,

7.4 Photovoltaic Cells Based on Silicon Heterojunction Technology, and

7.5 Organic and Hybrid Photovoltaic Cells.

Due to substantial updates, Section 3.2.2, Thermoelectric Power and Hall Effect, of the first edition of the book has been developed into Section 3.2.2, Thermoelectric Power, and Section 3.2.3, Hall Effect Anomaly, in this edition. Similarly, Chapter 7, Photoelectric and Electronic Devices Based on Disordered Semiconductors, of the previous edition has been remodeled as Chapter 7, Photoelectric Devices Based on Disordered Semiconductors, and Chapter 8, Electronic Devices Based on Disordered Semiconductors.

I am grateful to the Elsevier, Springer Science+Business Media, American Physical Society, Pleiades Publishing, Japan Society of Applied Physics, AIP Publishing, EDP Sciences, Nature Publishing Group, and IEEE, Inc., for granting permission to reproduce their copyright materials. I also thank Prof. E. Terukov and Dr. M. Kurashige for providing photographs that have not been published before and the Russian Foundation for Basic Research for providing financial support (Project No. 16-07-00147) to obtain the results outlined in Section 4.6.2. And I am particularly grateful to my son, Alex Popov, for his help in preparing the manuscript and thoughtful discussions.

Anatoly Popov

Summer 2018

Preface to the First Edition

The concepts of translational symmetry and long-range order of atomic arrangement present in single crystals are the pillars of modern solid-state physics. However, single crystals are a very special case of the solid state. Translational symmetry and long-range order of atomic arrangement are absent in most materials. Therefore, the future of solid-state physics is in developing a theory that would describe the different states of matter, including the theory of single crystals as a special case. Besides the academic interest, materials without long-range order in atomic arrangement are important from the application point of view. Devices based on disordered semiconductors have very wide applications. It is difficult to imagine modern life without electrophotographic printers and copiers, liquid-crystal displays and monitors, optical disks, cheap solar cells, and many other devices based on disordered semiconductors.

However, the peculiarities of disordered semiconductors call for new approaches to the device design as compared to traditional single-crystal materials. For example, the absence of long-range order in the structural network allows to apply new methods for control of properties without any doping (i.e., structural modification of properties). Moreover, with a correct approach, the so-called inherent disadvantages of disordered semiconductors can be used to the advantage of the device. For example, low charge carrier mobility allows to combine the high resistivity of the material with photosensitivity in the visible part of the spectrum and to create electrophotographic printers and copiers. Therefore, the goal of this textbook is to link the characteristic features of the atomic and electronic structures of disordered semiconductors to the device design process.

The textbook has eight chapters. In the first three chapters, we discuss the general concepts of disordered semiconductors, atomic structure of these materials, and the structure of energy bands, defects, as well as their electrical, optical, and photovoltaic properties. The success of the design of a semiconductor device is defined, to a large extent, by the possibility of controlling material properties. Since weak sensitivity to impurities is a distinguishing feature of disordered semiconductors, the methods of controlling property are discussed in Chapter 4. The use of disordered semiconductors in a device is usually limited to thin films. Such films must have predefined properties, and preparation methods must guarantee reproducibility of these properties. These questions are discussed in Chapter 5. Application of disordered semiconductors in various devices is considered in Chapters 6, 7, and 8. The book is based on lectures given by the author to students of electronics and nanoelectronics at the National Research University "Moscow Power Engineering Institute" and on the series of lectures given by the author at the Samsung Advanced Institute of Technology.

This book is intended for undergraduate, graduate, and postgraduate students in the field of electrical and electronic engineering, nanoelectronics, materials science, and solid-state physics. The book is also useful for researchers working in the field of disordered semiconductors.

I am grateful to Elsevier, Nature Publishing Group, Oxford University Press, Taylor & Francis Group, World Scientific Publishing Co., and Institute of Electrical and Electronics Engineering, Inc. for their permission to reproduce copyright materials. I also wish to thank Dr. Mitsuhiro Kurashige for his aerial photographs not published before. And I am particularly grateful to my sons Dr. Igor Popov and Mr. Alex Popov for their help in preparing the manuscript and helpful discussions.

Anatoly Popov

National Research University

"Moscow Power Engineering Institute"

Chapter 1

Introduction

An important event in the field of information technology occurred in 2009: the number of electronic devices connected to the network exceeded the earth's human population (Fig. 1.1). These electronic devices include not only various types of computers and mobile phones, but also various sensors and everyday objects. In 2015, the number of networked devices (25 billion) was 3.5 times larger than the earth's population, and by 2020, it is estimated to increase by more than twice (Doe, 2015). This event is important not only in terms of the information technology industry, but more importantly, from the point of view of the development of human civilization, because the role of electronic devices has radically changed. These devices have become indispensable to us and their role changed from merely making our lives easier to becoming a prerequisite for further development of human civilization. This new position of electronic smart objects requires new approaches to their design, manufacturing technologies, and expansion in the variety of materials on the basis of which they are made. One alternative relative to single-crystal semiconductors is disordered semiconductors. These materials and devices based on disordered semiconductors are discussed in this book.

Looking through the history of solid-state electronics, it is not hard to see that the main way of evolution (from the material science point of view) is the creation of highly ordered semiconductor materials. First, semiconductor devices (rectifiers) have been designed on the

Disordered Semiconductors: Physics and Applications (2nd Edition)
Anatoly Popov

ISBN 978-981-4774-37-6 (Hardcover), 978-1-315-10042-5 (eBook)
www.panstanford.com

base of polycrystalline copper oxides and polycrystalline selenium. Then an era of single crystals began. Since that time, one of the most important problems (of semiconductor material science) has been the following: how to bring to perfection the order in crystals. On the other hand, during the last 50 years, some attention has been focused on disordered semiconductors. But unlike "conventional" crystalline semiconductors, disordered (non-crystalline) materials have no long-range order of arrangement of atoms and translation symmetry. So the question is as follows. The attention to the disordered materials, is this step forward or step back from the point of view of solid-state physics and microelectronics?

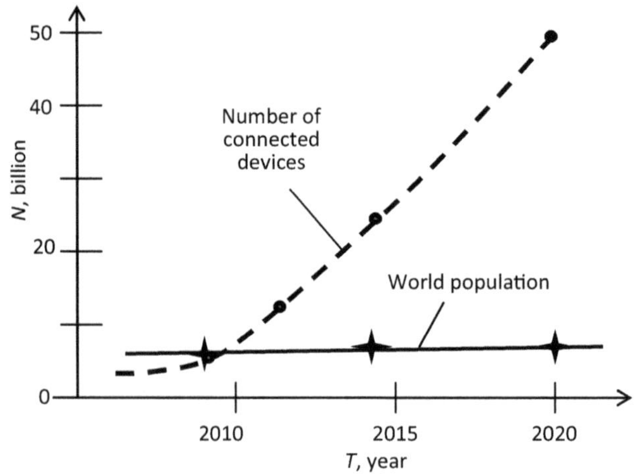

Figure 1.1 The number of connected devices and world population.

Let us start with the physics of semiconductors. It is common knowledge that the physics of condensed matter mostly succeeded in application to the greatly idealized object: single crystal. On the other hand, one can easily see that single crystal is not so common in nature as disordered systems: such as polycrystalline systems, disordered alloys, amorphous semiconductors, glasses, and liquids. And it becomes apparent that it is impossible to use even basic ideas of single-crystal physics to describe those materials because crystal physics is based on the crystalline latter or on the translation symmetry. As a result, we are facing a paradoxical situation when we cannot describe the simplest and obvious properties of non-crystalline materials on the basis of "classic" solid-state physics.

For example, it is known that optical transmission spectrum is determined by the forbidden gap. It means material should be transparent to photons with energy less than the width of the gap and opaque to the photons with energy greater than the width of the gap. The structure of energy bands and forbidden gap is derived from the Brillouin zone (forbidden gap is the outcome of Bragg reflection of electronic waves from the borders of Brillouin zone). In its turn, the concept of Brillouin zone is based on periodic crystal lattice. So far, since there is no periodic crystal lattice in disordered materials, it is impossible to use the notions of Bloch function and Brillouin zone to such an object. In this case, we should speak neither of the energy bands and forbidden gap nor of the optical transmission spectrum mentioned above. But then how could we explain why glass in the windows (non-crystalline material) is transparent to light?

Summing up all that has been mentioned, it is possible to say that the future of solid-state physics is in the development of the theory of condensed matter, which will include modern single-crystal physics as a particular case. The similar point of view has been expressed by Stanford Ovshinsky (1922–2012), former president of Energy Conversion Devices Company, USA, and prime mover in the development of amorphous solids and commercial application of these materials (Fig. 1.2). He said: "One does not understand amorphous materials because one understands crystallinity; rather, one can understand crystallinity as a special case when one understands the basic rules of amorphicity" (Ovshinsky, 1991).

Figure 1.2 Stanford Ovshinsky (1922–2012).

It is not an exaggeration to say that it was academic interest in the dielectric properties of a disordered solid in the last 1500 years that led to all our electronics and electrical engineering. That solid was amber and its Greek name "electron" has become almost a symbol of our age. So the answer to the first part of the question mentioned above is: disordered solids are to the same extent the past as they are the future of solid-state physics.

The second part of the question is linked to the application of disordered semiconductors in electronics. Disordered semiconductors such as chalcogenide glassy semiconductors, hydrogenated amorphous silicon and its alloys, and organic materials are used in various fields of electronics for the fabrication of:

- Thin-film transistor (TFT) matrices for liquid-crystal display application
- Solar cells on the base of hydrogenated amorphous and microcrystalline silicon and related materials
- Electrophotographic printers and copiers (Xerox process)
- Image sensors and optical recording devices (vidicons, optical disks, photothermoplastic optical recording media, photoresists, and electronic resists)
- Threshold and memory switches, including phase-change memory devices
- Sensors for medical diagnostics
- Infrared optical elements

One can say, all or almost all of these devices can be produced on the basis of crystalline semiconductors. So why should we consider use of alternative materials?

The advantages of disordered semiconductors stem from the following:

- A number of unique effects that are absent in crystals and that allow creation of devices on the new principles
- Possibility to make devices of large area and low cost
- Possibility to produce active electronics array on the non-crystalline substrate

1.1 Definition of Disordered State

The generally used rule for definition of something is the transition from the general to the particular case. But it is impossible to use this rule in this case because the starting point is physics of crystals. And a crystal is a particular case in comparison with disordered matter.

From the point of view of mutual arrangement of atoms, there are two cases. The first is crystals. In this case, atoms form a periodic lattice. It can be observed with high-resolution transmission electron microscopy (Fig. 1.3). In other words, crystals have long-range order of arrangement of atoms or translation symmetry. Diffraction pattern in this case consists of spots (single crystal) or sharp rings (polycrystal).

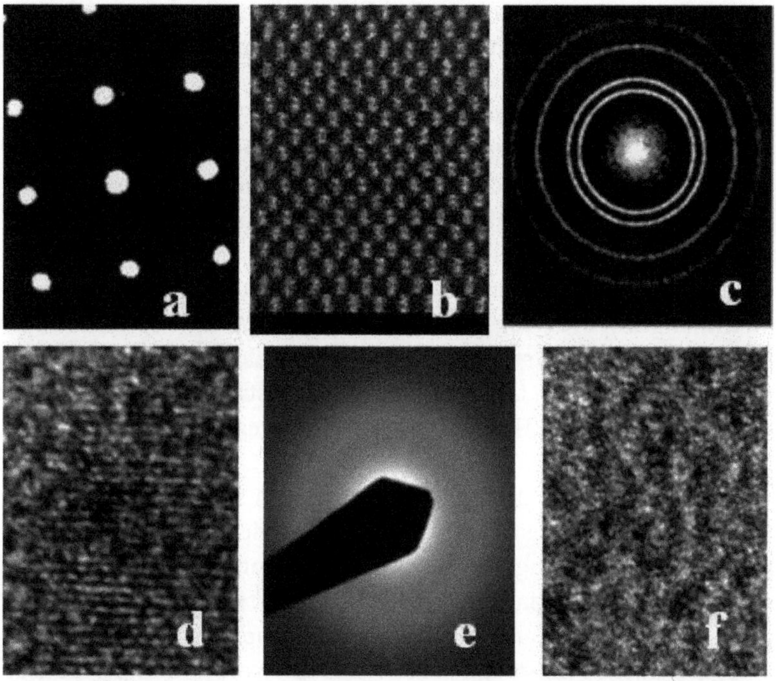

Figure 1.3 Diffraction patterns (a, c, and e) and high-resolution electron microscopic images (b, d, and f) of single crystal (a, b), polycrystal (b), polycrystal in amorphous matrix (d), and non-crystalline material (e, f).

The second case is non-crystalline (disordered) matter. In this case, long-range order of atomic arrangement is absent. Loss of long-range order leads to lack of periodic picture on high-resolution images in a transmission electron microscope, to disappearance of sharp reflex on diffraction pattern, and to appearance of diffused halos (Fig. 1.3).

So it is possible to give the following definition: **non-crystalline materials are characterized by the absence of long-range order or by the absence of translation symmetry.**

The negative character of this definition is not only contrary to the common rules (to define from the general to the particular) but also has very little useful data and sometimes does not give possibility to answer the question: is the sample a crystalline or non-crystalline material. Indeed, on one hand, only ideal single crystal with infinite dimensions has ideal long-range order. Any surface of real crystal is a breach of long-range order. On the other hand, the absence of long-range order in non-crystalline solids does not mean the total disorder of arrangement of atoms.

There is a short-range order (nearest neighbors) and a medium-range order of atomic arrangement in non-crystalline solids. Ordering in arrangement of atoms stretches for tens and hundreds Angstrom in non-crystalline solids. Hence, the answer to our question "is the given sample crystal or disordered solid" depends on the length of ordering in atomic arrangement, which one assumes to correspond to crystal. In other words, it depends on subjective estimation. Typically, the separation into crystalline and disordered systems is based on the size of crystalline regions in the material (Fig. 1.4).

There is another definition of disordered semiconductors. It is usually used in studies of electronic and electrical properties of these materials. Where long-range order is present in the arrangement of atoms, the potential energy of charge carriers in the total field of atoms is a periodic function of the coordinates (Fig. 1.5a). Violation of the long-range order leads to disruption of periodicity (Fig. 1.5b). On this basis, we can give the following definition: **disordered materials are materials with the potential energy of charge carriers expressed by a non-periodic function of the coordinates.**

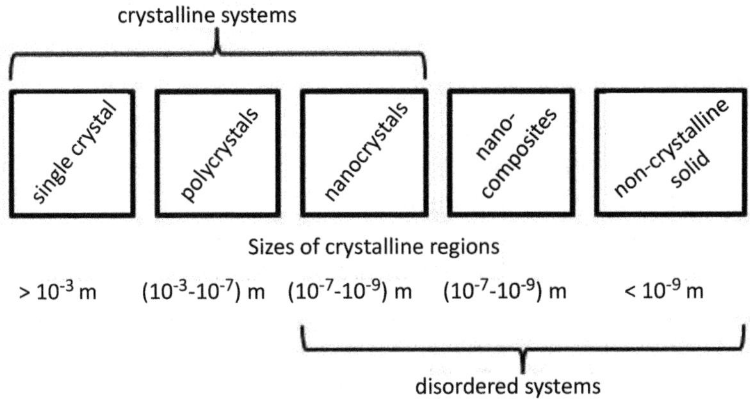

Figure 1.4 Crystalline and disordered systems.

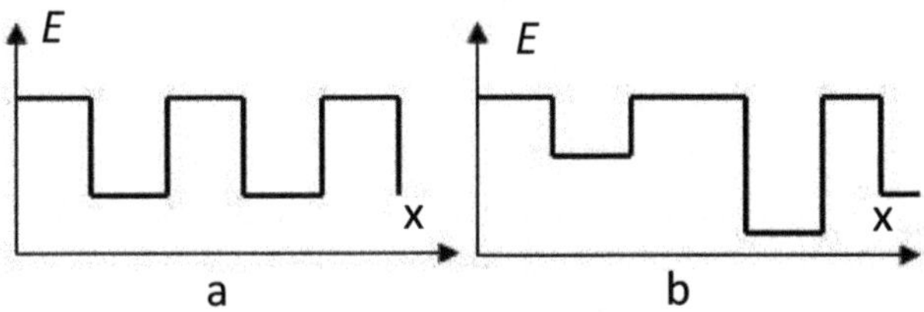

Figure 1.5 The dependence of the charge carriers' potential energy on coordinates in the case of crystal (a) and disordered material (b).

However, function $E(x)$ in a real single crystal always has a certain aperiodicity due to the presence of crystal lattice defects. Therefore, it is necessary to introduce a criterion to determine the value of the maximum permissible deviation of $E(x)$ from periodicity at which the material is still crystalline in order to use the above definition in practice.

The change in average energy of charge carrier $\boldsymbol{\Delta E}$ related to the violation of long-range order is used as such a criterion. The average energy of electrons in nondegenerate semiconductors equals kT (where k is Boltzmann's constant). Therefore, in the case of $\boldsymbol{\Delta E << kT}$, violation of the long-range order has little effect on the energy spectrum of charge carriers and the material is crystalline. But if $\boldsymbol{\Delta E \geq kT}$, then the material is disordered.

Using this criterion, we classify the following as disordered systems:

- Liquid semiconductors
- Non-crystalline semiconductors
- Heavily doped crystalline semiconductors (due to the random arrangement of the impurity atoms in the crystal)
- Surface of crystalline semiconductors
- Disordered crystalline semiconductor alloys (due to the unpredictability of presence in a given site of the crystal lattice atoms of a particular element).

This definition is very narrow; it is not devoid of subjectivity and, in some cases, is contrary to the previous (structural) definition.

1.2 Classification of Non-crystalline Systems

Classification is one of the natural stages of knowledge. It is necessary to systematize attributes of object prior to study. In other words, any classification is a certain ordering of knowledge about object. Therein is the paradox of this task: to order knowledge of the disordered systems. The problem consists in selection of general characteristics that could form the basis of classification. And it leads to a great variety of classifications. Let us consider the following basic characteristics: structural, mechanical, and thermodynamic.

First, the structural approach has been considered already; from the structural point of view, all materials can be divided into two groups (Fig. 1.6):

- Crystals, which have the long-range order of atomic arrangement (or translation symmetry)
- Non-crystalline materials, at which the long-range order of atomic arrangement is absent

Disordered or non-crystalline materials, from the mechanical point of view, can be divided into:

- Non-crystalline solids, where dynamic viscosity exceeds value $10^{14.6}$ poise
- Fluids (liquid, gases), which have dynamic viscosity less than this value

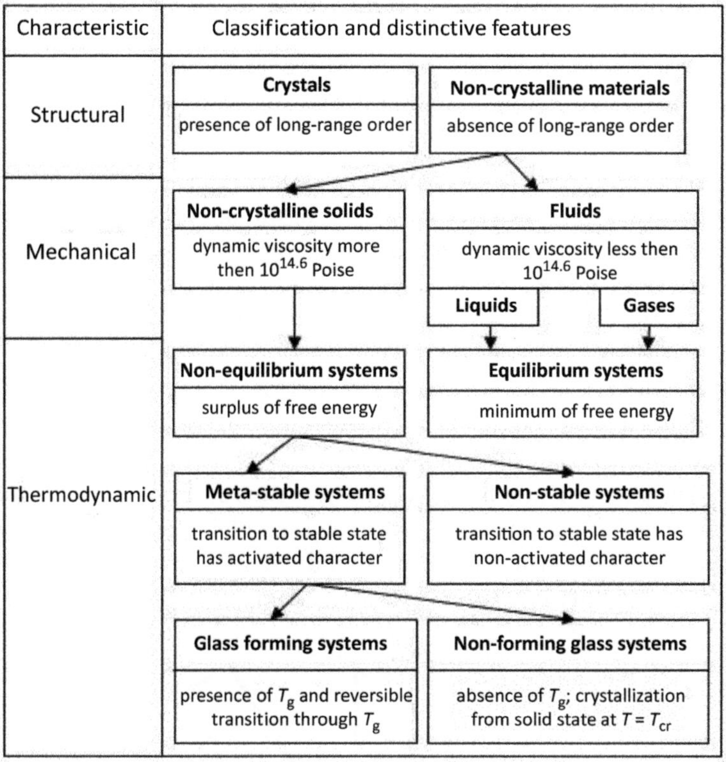

Figure 1.6 Classification of non-crystalline systems.

The boundary point (viscosity equal to $10^{14.6}$ poise) corresponds to the relaxation time of system equal to 24 h. Thus, the material is considered solid if, under the action of a small force within 24 h, residual deformation is not observed.

In the thermodynamic approach, the energy condition of a system is analyzed from the point of view of its stability. A stable (equilibrium) system has to have a minimum of free energy. Any changes of this system lead to energy increase and, therefore, to transference to non-equilibrium state.

Crystals, liquids, and gases may be in equilibrium state. Non-crystalline solids are non-equilibrium systems (Fig. 1.6). So there are two characteristic features of non-crystalline solids:

1. The absence of long-range order in atomic arrangement
2. Non-equilibrium state of system

Let us examine the non-equilibrium state at greater length. On one hand, there is the general rule that any system tends to attain a state of minimum free energy or equilibrium. On the other hand, it is well known that glasses remain in non-equilibrium glassy state for hundreds and thousands of years.

So it is necessary to consider the question of stability of non-equilibrium state. As a rule, two levels of stability of non-equilibrium materials are named (Fig. 1.6): meta-stable states and non-stable states.

In the meta-stable state, the system has to cross some intermediate non-stable state with higher energy before it reaches equilibrium state. In other words, there is an energy barrier between meta-stable and stable states. But there is no barrier in the transition from non-stable to stable state.

The situation may be illustrated by means of a simple mechanical analogue. Let us consider right-angled parallelepiped on the plane. Position on the wide side corresponds to stable state because the center of gravity is at the minimum height (Fig. 1.7a).

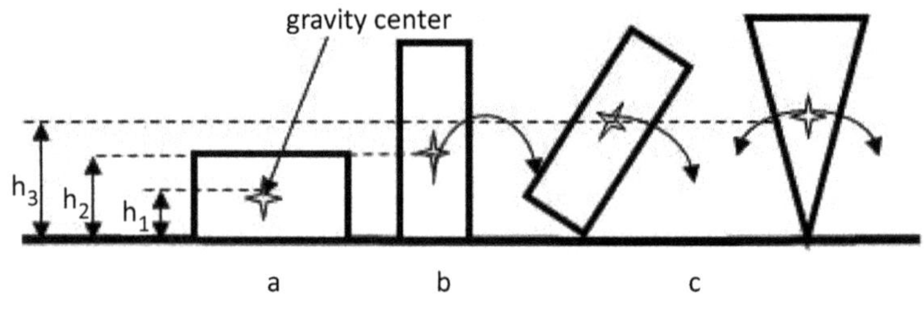

Figure 1.7 Illustration of stable (a), meta-stable (b), and non-stable (c) states (*h* is the gravity center height).

Position on the narrow side corresponds to the meta-stable state. In this case, it is necessary to get over the energy barrier connected with the center of gravity lifting to transfer parallelepiped to the stable state (Fig. 1.7b). The non-stable state may be presented as a pyramid set on its apex (Fig. 1.7c). There are no energy barriers in this case.

All transitions in solids are linked to the change in atom positions. That is why these transitions are illustrated by free energy versus configuration diagrams (Fig. 1.8). Point **C** gives the minimum energy level corresponding to the stable (equilibrium) crystalline state. On the other hand, points $\mathbf{M_1}$, $\mathbf{M_2}$, $\mathbf{M_3}$ are local minima corresponding to meta-stable states. Various points of local minima correspond to different meta-stable states, which are brought out by different conditions during the formation process or different treatment of material. The difference $E_M - E_C = \Delta E_1$ corresponds to the energy surplus of meta-stable state. This energy is exuded during phase transition (e.g., as crystallization heat). However, it is necessary to get over barrier ΔE_2 for this phase transition to take place (activation energy of crystallization). In the case of thermal excitation, the stability of meta-stable states depends on the ratio between ΔE_2 and kT.

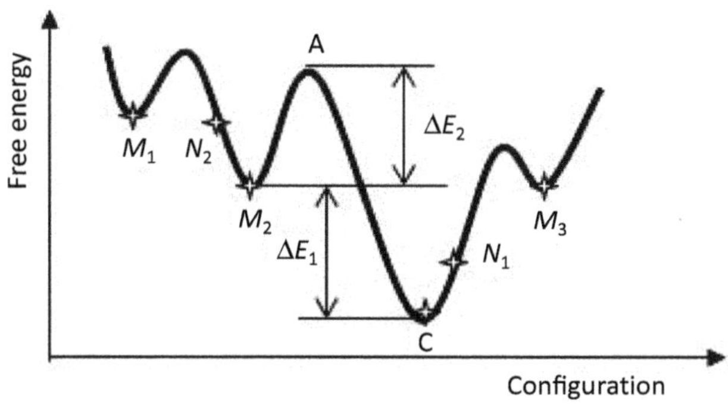

Figure 1.8 Free energy versus configuration diagram.

Let us define the probability of this transition. Transition from meta-stable to stable state means the displacement atoms from point **M** to point **C**. Before atoms reach point **C**, they have to get over the activated state (point **A**). The equilibrium constant of transition process from point **M** to point **A** is equal to atom concentration ratio in these states:

$$K = N_A/N_M \tag{1.1}$$

On the other hand, equilibrium constant is determined by the barrier height ΔE_2:

$$K = \exp(-\Delta E_2/kT) \tag{1.2}$$

In the first approximation, it is possible to assume that the rate of atom transition from point **M** to point **C** is proportional to the concentration of atoms in point **A** (in other words, all activated atoms pass into the stable state). In this case, the process rate may be written as:

$$\boldsymbol{V = \text{const}\bullet N_A = \text{const}\bullet K\bullet N_M = \text{const}\bullet N_M\bullet \exp(-\Delta E_2/kT)} \quad (1.3)$$

Taking into account that the barrier energy may be expressed by enthalpy and entropy changes:

$$\boldsymbol{\Delta E_2 = \Delta U_2 - T\bullet\Delta S_2} \quad (1.4)$$

we obtain

$$\boldsymbol{V = \text{const}\bullet N_M\bullet \exp\{(-\Delta U_2 + T\bullet\Delta S_2)/kT\}}$$
$$\boldsymbol{= \text{const}\bullet N_M\bullet \exp(\Delta S_2/k)\bullet \exp(-\Delta U_2/kT)} \quad (1.5)$$

or

$$\boldsymbol{V = D\bullet \exp(-\Delta U_2/kT)} \quad (1.6)$$

where $\boldsymbol{D = \text{const}\bullet N_M\bullet \exp(\Delta S_2/k)}$ is the temperature-independent pre-exponential factor.

Exponential dependence points to strong influence of temperature on the process rate. For usual solid values of $\boldsymbol{\Delta U_2}$ = 1.5–2.0 eV, the value of exponent is equal to 10^{-8}–10^{-10} at 1000 K and to 10^{-26}–10^{-33} at 300 K. It means that the duration of a process changes from 1 s at 1000 K to more than 10 billion years at 300 K. So meta-stable states may be rather stable in real conditions.

A typical example of a meta-stable system is glass. Alongside the absence of long-range order and thermodynamic non-equilibrium, **the distinctive sign of glass is the glass transition temperature T_g (or glass transition region) and the absence of phase changes at this point.**

Let us consider the dependence of some material property on temperature (Fig. 1.9). When the temperature decreases from their liquid phase, some materials do not crystallize at their melting temperature but they remain in the liquid (supercooled) state. As the temperature keeps decreasing, glass transition occurs, and below $\boldsymbol{T_g}$, these materials become solids (glass). So glassy materials assume the transition from liquid to solid state and back without appreciable changes in their properties. From the point of view of the free energy versus configuration diagram (Fig. 1.8), it means that the transition from solid to liquid state takes place at the temperature below one required to get over barrier $\boldsymbol{\Delta E_2}$. That is why the system remains in the local minimum **M** after cooling.

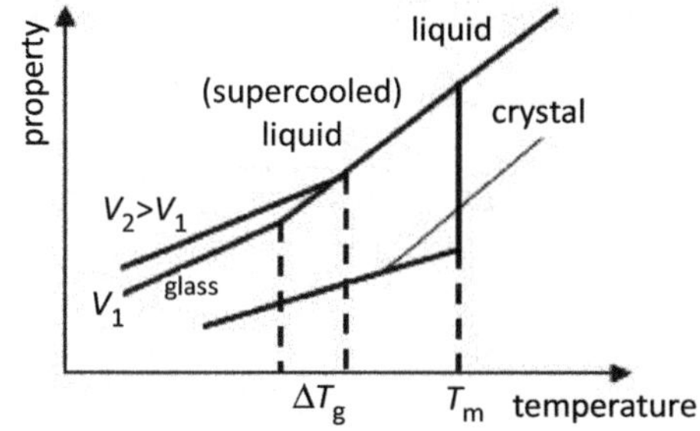

Figure 1.9 Schematic illustration of property versus temperature curve (V is cooling rate, T_m is melting temperature, ΔT_g is glass transition region).

Alongside glasses, there are non-crystalline solids (amorphous silicon, for example) that do not have glass transition region, but increasing the temperature to some critical value (**T_{cr}**) leads to rapid crystallization from the solid non-crystalline state. In other words, these materials are stable at temperatures less than the critical value, but at **$T = T_{cr}$**, irreversible phase transition takes place. The absence of glass transition temperature in meta-stable state is conditioned by features of the configuration diagram (Fig. 1.10): small value of barrier **ΔE_2** and the presence of a shelf in the region of point **M**. In this case, the system will be in a steady state as long as the atomic heat oscillation amplitude **Δ** is less than **Δ_{max}**. Increase in temperature and atomic heat oscillation amplitude up to the value **$\Delta = \Delta_{max}$** leads to system transition to crystalline state (point **C**).

Non-stable states correspond to points **N** in the free energy versus configuration diagram (Fig. 1.8). It should be mentioned that the non-stable state may be in the vicinity of the stable state (point **N_1**) or in the vicinity of the meta-stable state (point **N_2**).

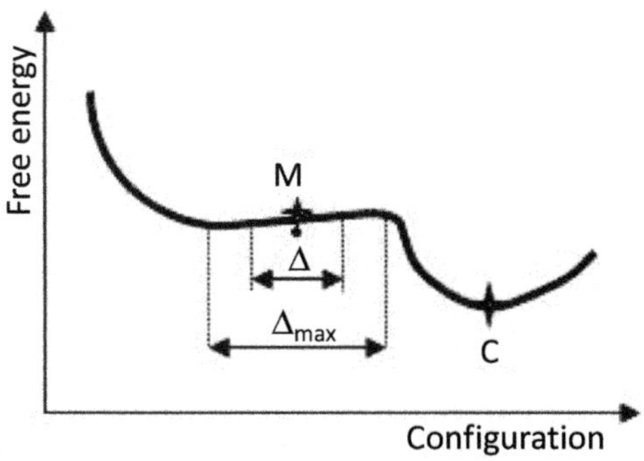

Figure 1.10 Free energy versus configuration diagram for amorphous silicon.

1.3 Qualitative and Quantitative Characteristics of Glass Formation

Glasses are widespread examples of disordered solids. The distinctive features of the glassy state are as follows: lack of long-range order in atomic arrangement; viscosity of the material more than $10^{14.6}$ poise; presence of the glass transition temperature region and reverse transition through this region. Typical mechanical properties of glass—fragility and conchoidal fracture—are usually added to the above criteria. A large group of glassy chalcogenide semiconductors can be distinguished from the general class of glasses based on electrical characteristics. These materials were discovered in 1954 by Prof. B. T. Kolomiets (Fig. 1.11) and have properties of semiconductor materials: electronic conduction mechanism; semiconductor character of the temperature dependence of conductivity; the presence of photovoltaic and thermoelectric effects, along with properties of a glassy material.

Figure 1.11 B. T. Kolomiets (1908–1989).

It is possible to prepare material in the form of glass only if crystallization is suppressed during the cooling of the melt. It means that the possibility of glass preparation depends on both the nature of material and the rate of cooling. The range of cooling rate is very wide: from 10^{-5} degree per second (change by 1° during 28 h—fritting of large optical glasses) to 10^{6} degrees per second and more (melt-spinning method). In the latter case, it is possible to prepare non-crystalline alloys of some metals. But these rates could not be provided for bulk samples. It is possible to prepare only thin films or powder in this case. Usually, **a material is classified as glass if the cooling rate in the tens of degrees per second is enough to prepare non-crystalline solid.**

Let us consider the following question: why some materials form glasses and others do not? There is no long-range order in the melt. As the temperature keeps decreasing, the rearrangement of atoms to crystal lattice occurs at the melt temperature $\boldsymbol{T}_{\mathbf{m}}$. The rate of rearrangement depends on the mobility of atoms or molecules. Hence, the condition of preparing glass is low mobility of atoms or molecules in the temperature range between melt temperature and glass transition temperature. On the other hand, the mobility of atoms or molecules is determined by the degree of polymerization. So the first necessary condition for glass formation may be formulated in the following way: a material has to consist of high polymer complexes. This condition is necessary but insufficient because any covalent bonded single crystal may be presented as high polymer complex.

The possibility of atom rearrangement into a crystal lattice during cooling of the melt depends on the amplitude of atom thermal vibrations as well. In turn, the amplitude of thermal vibrations is strongly dependent on the type of chemical bonds between the atoms. In the case of non-directional bonds between atoms (ionic, metallic, or van der Waals bonds), the amplitude of thermal vibrations is large enough to facilitate crystallization of the melt on cooling. On the contrary, directed covalent bond has a significantly higher force constant that limits the amplitude of thermal vibrations. Thus, at the same energy of covalent and ionic bonds (E_0) and the same level of heat exposure (E_T), the displacement of atoms in the case of covalent bond (Δx_c) is several times smaller than in the case of ionic bond (Δx_i) (Fig. 1.12) (Muller, 1960). As a result, the next condition for glass formation is the fact that glass-formation systems have to have a predominance of covalent bonds. According to Muller, the covalent component of the chemical bond in glass-forming materials must be at least 40–60%. But this condition is insufficient as well. For example, silicon has covalent bonds, but it does not form glass.

The third condition is connected with the rigidity of the structural network. A glass-formation system has to have a flexible network. It makes possibility for various atom positions in the matter below the melt temperature and glass transition temperature. So the third condition for glass formation is as follows: the connection of structural complexes has to be only through a single bridge bond. In other words, so-called swivel bonds must be present in the structure.

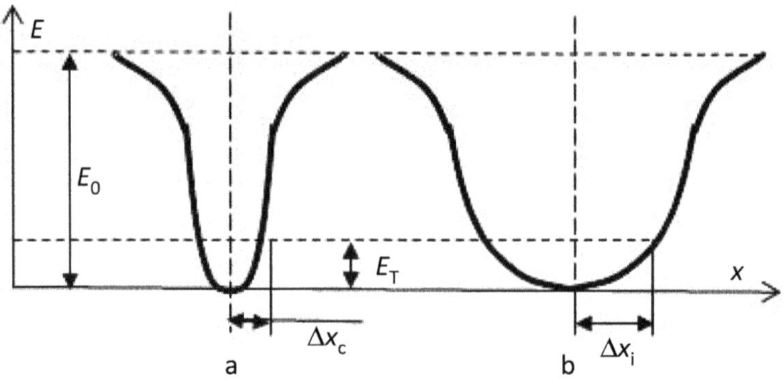

Figure 1.12 Energy versus configuration dependence in the case of covalent (a) and ionic (b) chemical bonds.

Summing up, we may formulate the necessary and sufficient qualitative conditions of glass formation:

- Material has to consist of high polymer complexes
- Predominant type of chemical bonds has to be covalent bonds
- Swivel (or bridge) bonds have to be present in the structure

Besides that, glass-forming ability increases if the number of components of a material increases and glass-forming ability decreases if the metallization or ionic degree of chemical bonds rises.

Let us now turn to quantitative characteristics. There are a lot of theories, concepts, criteria, and semi-empirical rules of glass formation (Minaev, 1991). All of them can be divided into three groups:

1. **Structural–chemical**, which addresses the impact on the glass formation of chemical bonds type, value of the first coordination number, chemical composition, and so on

2. **Kinetic**—based on the determination of the critical cooling rate of the melt, at which crystallization is absent
3. **Thermodynamic**—based on the relationship of glass-forming ability with such characteristics as melting entropy, liquidus temperature, and so on

It is necessary to note that the differences between these groups are rather indistinct. Very often concepts overlap from one group to another. Consider the structural–chemical criterion suggested by Dembovsky (1977) as an example. The author attempted to link the glass-formation ability to the outermost electron shell structure.

Let us consider the outermost electron shell structure in the case of elements of IV (silicon), V (arsenic), and VI (selenium) (Fig. 1.13).

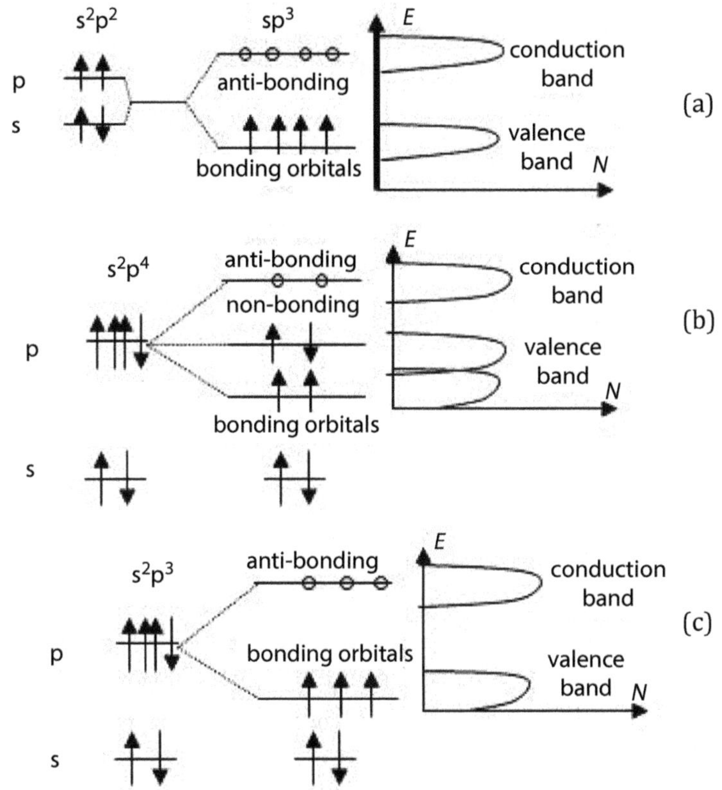

Figure 1.13 Schematic diagrams of the energy levels of atomic orbitals and energy bands for silicon (a), selenium (b), and arsenic (c).

The fourfold coordination of a silicon atom with its neighboring atoms formed by sp^3 hybridized orbitals gives four bonding orbitals and four anti-bonding orbitals (Fig. 1.13a). These orbitals form valence band and conduction band, respectively. In the case of selenium, the twofold coordination of atom, formed by p-electrons, gives two bonding orbitals, two anti-bonding orbitals, and two non-bonding orbitals (Fig. 1.13b). Non-bonding orbitals are occupied by lone-pair electrons. Anti-bonding orbitals form conduction band and non-bonding orbitals form upper part of valence band. Schematic diagrams of the energy levels of arsenic atomic orbitals are given in Fig. 1.13c.

Dembovsky found that the glass-formation ability of a material has good correlation with the difference between the average number of valence electrons (*NVE*) and the average coordination number (*N*), e.g., $NVE - N$. This difference is equal to the number of electrons on non-bonding orbitals or the number of lone-pair electrons. It was shown that in the case of glass-forming materials, the relative number of lone-pair electrons lies in a rather narrow range:

$$\psi = (NVE - N)/N = 0.50 \div 0.66 \tag{1.7}$$

For example, $As_{40}Se_{60}$, which is a good glass-forming material, has $NVE = 5 \times 0.4 + 6 \times 0.6 = 5.6$, $N = 3 \times 0.4 + 2 \times 0.6 = 2.4$, and $\psi = 0.57$. It corresponds to the middle of value spectrum in Eq. (1.7). The value of ψ for selenium equals 0.66. The preparation of glassy selenium is possible only at high quenching rate (cooling of the melt in the water). In the case of silicon, $\psi = 0$. This material does not form glass at all.

If ψ is less than 0.5, there are not enough swivel (or bridge) bonds in the structure of the material. Therefore, the material has a rigid structural network. If ψ is more than 0.66, the condition of high polymer complexes is broken. Thus, the parameter ψ is the quantitative criterion of glass-forming ability from the structural–chemical point of view.

In later works, Dembovsky made a number of additions to Eq. (1.7), which took into account the number of chemical components in the material, presence and number of different structural units, as well as the thermodynamic and kinetic aspects of glass formation (Dembovsky, 1981; Dembovsky and Chechetkina, 1981b).

As mentioned earlier, among the broad class of glass, systems that contain chalcogen atoms (sulfur, selenium, and tellurium) typically have semiconductor properties. And the number of such systems is large enough. Research (Vinogradova, 1984) published 30 years ago described more than 20 binary and more than 100 triple chalcogenide systems in which glass formation is observed. Since then their numbers increased substantially. The examples of glass-formation regions for binary ($A^{IV}B^{VI}$ and $A^{V}B^{VI}$) and triple chalcogenide systems according to Vinogradova (1984) are shown in Table 1.1 and Fig. 1.14.

It should be noted that the absence of glass-formation area in a binary system does not mean that glass-formation areas cannot exist in the triple systems based on it. So despite the lack of glass-formation region in the Sb–Se system (Table 1.1), in the triple system Sb–Se–Ag, glass formation is observed.

Table 1.1 Glass-forming regions in the systems $A^{IV}B^{VI}$ and $A^{V}B^{VI}$

System	Chalcogen content, atomic percentage
As–S	34–49, 55–95
As–Se	40–100
As–Te	42–54
Sb–S	66
Sb–Se	—
Sb–Te	—
Si–S	50–69
Si–Se	80–100
Si–Te	78–80
Ge–S	55–60, 66–90
Ge–Se	75–100
Ge–Te	78–88

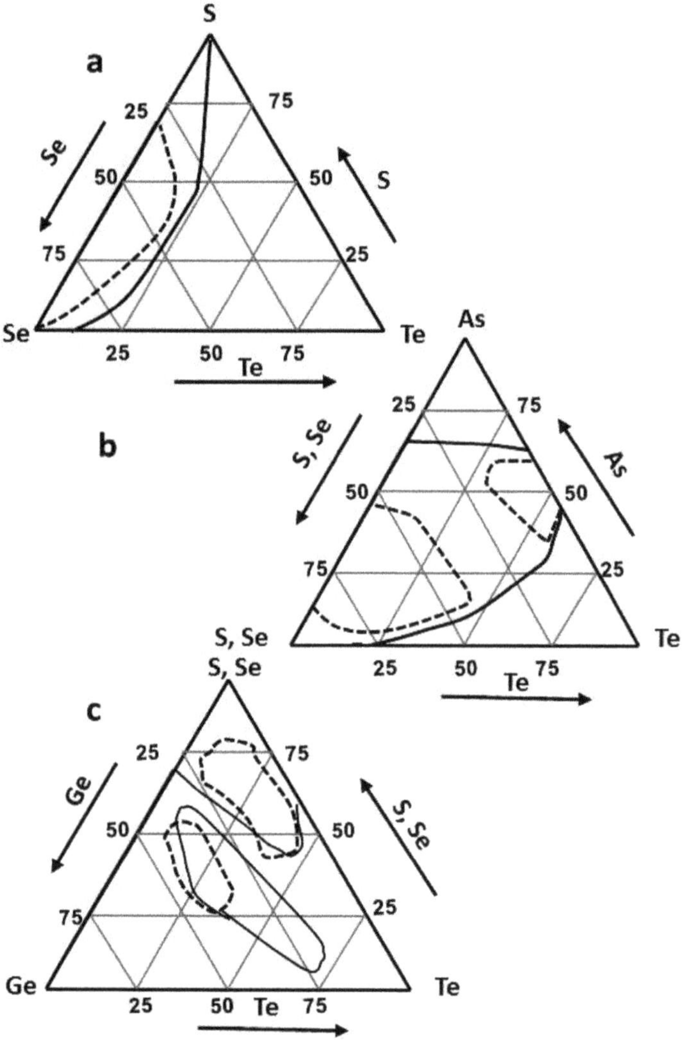

Figure 1.14 Glass-formation regions in the systems: (a) S–Se–Te (dotted curve—slow cooling; solid curve—quick cooling), (b, dotted curves) As–S–Te, (b, solid curves) As–Se–Te, (c, dotted curves) Ge–S–Te, (c, solid curves) Ge–Se–Te.

Chapter 2

Atomic Structure of Disordered Semiconductors

2.1 Structural Characteristics of Solids

The atomic structure defines all major properties of both crystalline and non-crystalline substances, so it is important to understand the structure. At the XIX International Glass Congress, it was pointed out that "commercial as well as research activity is taking place in the absence of anything like a complete understanding of the atomic structure of glass" (Cormack, Du, and Zeitler, 2001). Thus, above all, it is necessary to answer the following question: which characteristics of the structure are necessary and sufficient for defining non-crystalline state of solids?

Let us start with the simplest case. From the viewpoint of a structure, the simplest case is an ideal single crystal. In order to completely describe its structure, it is enough to know the structure of an elementary cell or a short-range order of the arrangement of atoms (Table 2.1) (Popov, 2004). It is necessary to add a defect subsystem for a whole definition of any real single crystal. To describe the structure of polycrystals, it is necessary to consider the morphology and heterogeneity of the material, that is, crystal size distribution, crystal texture, formation of spherulites, and so on in

Disordered Semiconductors: Physics and Applications (2nd Edition)
Anatoly Popov

ISBN 978-981-4774-37-6 (Hardcover), 978-1-315-10042-5 (eBook)
www.panstanford.com

addition to short-range order and defects. And for non-crystalline solids, four levels of structural characteristics for describing their structure should be taken into consideration (Table 2.1):

1. Short-range order of atomic arrangement
2. Medium-range order of atomic arrangement
3. Morphology and heterogeneity
4. Defect subsystem

One can see in the table that the number of characteristics necessary for describing the structure of a substance increases with growth in its complexity. Where one characteristic for describing the structure of an ideal single crystal is sufficient, it is necessary to use four characteristics to describe the structure of non-crystalline solids.

Table 2.1 Structural characteristics of solids

	Solid states			
Structure subsystems	**Ideal single crystal**	**Real single crystal**	**Polycrystal**	**Non-crystalline solid**
Short-range order	+	+	+	+
Defect subsystem	–	+	+	+
Morphology and heterogeneity	–	–	+	+
Medium-range order	–	–	–	+

It is evident from this that a more complete definition of non-crystalline solids in comparison with the definition based on the absence of long-range order of atomic arrangement may be formulated. **Non-crystalline solids are materials that require use of the parameters of short- and medium-range orders of atomic arrangement, morphology, and heterogeneity and defect subsystem to fully describe their structure.** At the same time, as regards non-crystalline substances, the very terms of short-range order and particularly medium-range order of atomic arrangement are presently under discussion.

2.2 Short-Range and Medium-Range Order

It is known that in the absence of long-range order of atomic arrangement in both non-crystalline solids and fluids, a certain "local order" remains. Unlike crystals, where the order of atomic arrangement at any level is predefined by translation symmetry, understanding of the local order in disordered systems requires specification: What is the dimension of fields of local order in atomic arrangement? Which parameters are required and sufficient in order to provide its full description?

When studying elements of order in non-crystalline materials, one can choose short-range order of atomic arrangement, determined by the chemical nature of atoms that forms the given substance (valency, bond length, bond angle). As a rule it is assumed that the field of short-range order includes the atoms that are the nearest to the atom chosen as a central one and that form the first coordination sphere (atoms 1 and 3 in respect to atom 2 in Fig. 2.1a, atoms 1, 3, 4, 5 in respect to atom 2 in Fig. 2.1b). The parameters of short-range order are as follows: the number of nearest-neighbor atoms (first coordination number), their type, the distance between them and the central atom (radius of first coordination sphere, r_1) (Fig. 2.1), their angle position in respect to the central atom defined by bond angles (valency angles φ).

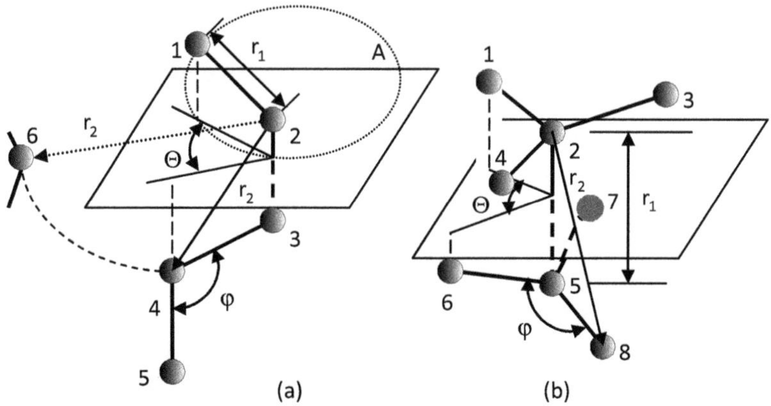

Figure 2.1 Characteristics of atoms' relative positions for linear (a) and tetrahedral structures (b) (r_1 and r_2 are the first and second coordination sphere radii, ϕ and θ are bond and dihedral angles).

This definition limits short-range order to the first coordination sphere. However, the above-mentioned parameters of short-range order define not only the first coordination sphere but at least in part the second one. Thus, the radius of the second coordination sphere, r_2 (Fig. 2.1), is determined by the radius of the first coordination sphere and valency angles:

$$r_2 = 2\, r_1 \sin(\phi/2) \tag{2.1}$$

On the other hand, atoms belonging to other molecules may be in the second coordination sphere as well (atom 6 in Fig. 2.1a, for example). Positions of these atoms do not connect with short-range order parameters at all.

This contradiction can be resolved if we depart from the geometric parameters of short-range order to the power parameters of interaction among atoms. We can include in the field of short-range order those atoms respective position of which is defined by the strongest interactions. For semiconductors with predominance of covalent type of chemical bonds, the strongest interactions are defined by the parameters of covalent relations (bond-stretching energy and bond-bending energy). Thus, the field of short-range order includes atoms of the first coordination sphere as well as those atoms of the second coordination sphere the position of which with respect to the chosen central atom is determined by covalent interaction.

The introduction of the notion of short-range order does not allow to describe in full the local order in atomic arrangement observed in disordered systems because it does not explain considerable length of the ordered fields in non-crystalline materials. Experimental proofs of quite long ordered fields triggered introduction of the notion of medium-range order in atomic arrangement in non-crystalline materials.

One can consider microcrystalline model of structure of disordered materials as the first attempt to explain the presence of medium-range order though the above-mentioned models arose historically before the introduction of the term medium-range order in atomic arrangement.

According to the microcrystalline model suggested by A. Lebedev in 1921, non-crystalline material consists of arbitrarily oriented microcrystalline particles. Microcrystalline particles are separated from one another by disordered layers. Initially, rather big dimensions of microcrystalline particles (up to 300 Å) were suggested. Later, diffraction analysis showed that there are no microcrystalline particles with dimensions larger than 15–20 Å in homogeneous non-crystalline materials. On the other hand, Laue patterns from the lower-dimension microcrystalline particles have insufficient contrast to be observed. In other words, experimental results indicate absence of microcrystalline particles with dimensions more than 15–20 Å in homogeneous non-crystalline materials but do not prove existence of smaller-dimension microcrystalline particles. However, transition to such small dimensions of microcrystalline particles leads to difficulties in interpreting structure on the basis of microcrystalline model. Decrease in dimensions of microcrystalline particles leads to increase in the number of surface atoms (Fig. 2.2). More than 50% atoms are on the surface of microcrystalline particles if their dimension is less than 50 Å.

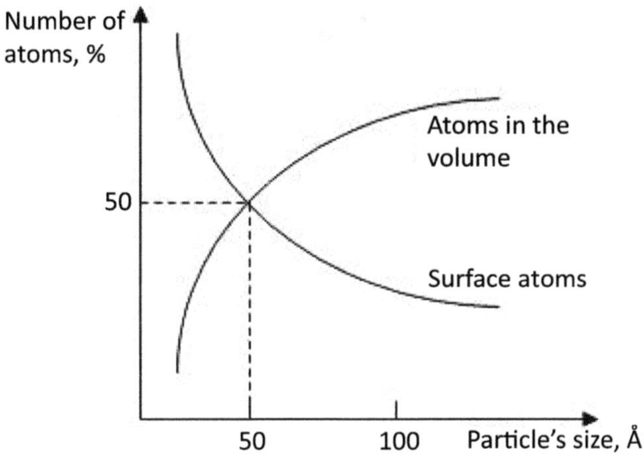

Figure 2.2 Number of atoms in the volume and on the surface of particles depend on particle's size.

In this case, disordered layers between microcrystalline particles become more significant than microcrystalline particles for describing structure.

In the case of tetrahedrally bonded semiconductors (amorphous silicon, amorphous germanium, and others), models of amorphous clusters (Sadoc and Mosseri, 1982) or amorphons (Grigorovici,

1974) have been proposed to explain medium-range order. According to these models, non-crystalline materials consist of clusters containing a few hundreds of atoms. Atoms are arranged in some regular (but different from crystalline) configuration in the cluster. It leads to a decrease in cluster free energy, but it does not give crystal-like diffraction pattern. These models were an advance over the microcrystalline model. However, the problem of defining clusters' junctions and their borders was not resolved.

Several papers developed these models further. Nakmanson et al. (2001) and Voyles et al. (2001) described paracrystalline model of amorphous silicon that consists of small grains, less than 30 Å in diameter, which have topologically crystalline bonding but within which the atoms are significantly displaced from their crystalline lattice positions by the strain exerted by the grain boundaries. This leads to the diffraction pattern of an amorphous solid. The grains are separated by a more disordered matrix. Popescu and Bradaczek (2001) proposed a micro-paracrystalline model for medium-range order in non-crystalline chalcogenides. The model is based on the assumption that paracrystallites in glassy chalcogenides consist of the backbones of the original stable crystal corresponding to the chemical composition of material (or crystal of the most similar composition).

A common shortcoming of these models is in the mechanical dissemination of short-range order notions to bigger but geometrically strictly limited regions. It does not allow to solve problems of mutual packing and connection of these regions. Lucovsky (1987) linked medium-range order with the distribution of dihedral angles (a dihedral angle is the angle between bonds of the nearest-neighbor atoms when projected onto a plane perpendicular to the common bond; Fig. 2.1, angle θ) for a distance of approximately 10 covalent-bonded atoms. Elliot (1987) divided the field of medium-range order into three levels: the field of local medium-range order (mutual disposition of neighboring structural units), medium field of medium-range order (mutual disposition of clusters), and long field of medium-range order linked to spatial order of different fields of a structural network. Voyles et al. (2001) offered paracrystalline model for explaining medium-range order. But the concrete definitions of this term so far remain open to discussion.

Another approach of defining medium- as well as short-range order of atomic arrangement is transition from geometrical characteristics to energy characteristics of mutual atom interaction (Popov, 2004). As mentioned earlier, short-range order is determined by the strongest interaction between atoms: bond-stretching energy (V_s) and bond-bending energy (V_b). For example, in the case of linear polymer (selenium, sulfur), short-range order includes first nearest neighbors (atoms 2 and 4, if atom 3 is considered the central one, Fig. 2.3) and those atoms of second coordination sphere that are of the same molecule as the central atom (atom 1 as its position with respect to atom 3 is defined by interaction of V_s and V_b). Interactions of atoms of the second order include relation of long-pair electrons of atoms that are in the same or different molecules and van der Waals interaction between atoms of neighboring molecules (V_3, V_4, $V_{v\text{-}v}$, Fig. 2.3). These interactions determine medium-range order of atomic arrangement. Therefore, medium-range order is formed by atoms that are partially positioned in the second coordination sphere (in case of linear polymer, atoms of neighboring molecules), and the atoms of coordination spheres of higher orders.

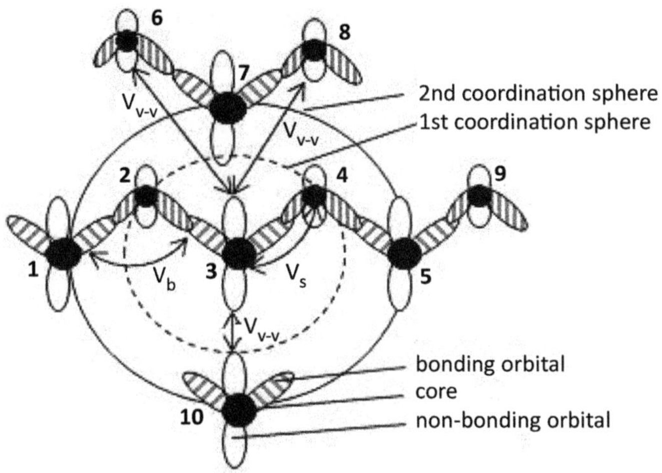

Figure 2.3 Interatomic interaction in the case of linear polymer (see text).

So there is no long-range order of atomic arrangement in non-crystalline materials, but there are short- and medium-range orders. In the case of semiconductor materials with predominance of covalent kind of chemical interaction, **short-range order is determined by the interaction of covalent-bonded atoms** and includes the first and partially second coordination sphere. **Medium-range order is determined by the interaction of long-pair electrons, van der Waals interaction**, and it is formed by atoms partially positioned in the second coordination sphere and atoms of coordination spheres of higher orders.

2.3 Methods of Investigation of Disordered System Structure

The absence of translation symmetry in disordered systems greatly complicates the task of investigation of their atomic structure as compared with crystals. The direct methods of investigating structure in disordered systems that are similar to crystals are **diffraction of short-wavelength radiation on the atoms of researched substance and X-ray spectroscopy**.

The core of diffraction methods is the registration of spatial picture of intensity of monochromatic radiation coherently dispersed by investigated object, transition from it to distribution of intensity in back space and calculation using Fourier transformation of microdistribution of density of the substance. However, if in the case of crystals, the data obtained in this way give total information about the spatial distribution of atoms in the object, for disordered systems it provides only **spherical symmetrical atomic radial distribution function (RDF)**, which is statistic in character and indicates probability of an atom being positioned at the given distance from the atom chosen as central one.

Three types of monochromatic radiation are used in diffraction techniques: X-ray, electron beam, and neutron beam. Interaction with substance is quite different in these cases. X-ray is scattered by electron shells of atoms and contains data about electron density distribution in the substance. Its peaks correspond to positions of atoms. Electron scattering is conditioned by Coulomb interaction and shows the potential distribution of structural network. Potential maximums correspond to atom cores. Neutrons are scattered by the delta potential of nuclear force. The interaction intensity of radiation with substance determines requirements to the samples for diffraction analysis. Sample thickness has to be 10^{-5} mm in the case of electron diffractometry, about 1 mm in the case of X-ray diffractometry, and up to 10 mm in the case of neutron diffractometry.

Let us consider intensity spatial distribution of coherent scattered monochromatic radiation. The radiation with incident direction $\vec{S}_0$ is scattered in a system of **N** atoms. Atom positions are described by a set of vectors $\vec{r}$. In this case, the scattered radiation intensity in the direction $\vec{S}$ is expressed by the following equation:

$$I(s)=\sum_{i=1}^{N}\sum_{j=1}^{N} f_i \cdot f_j \cdot \exp\left[is\left(\vec{r}_i-\vec{r}_j\right)\right] \qquad (2.2)$$

where $s=\left|\vec{S}-\vec{S}_0\right|=4\pi\cdot\dfrac{\sin\Theta}{\lambda}$, $\boldsymbol{\lambda}$ is the wavelength, $\mathbf{2\Theta}$ is radiation deflection angle, $\boldsymbol{f}$ is the atomic scattering factor, and $\boldsymbol{i}$ is a complex unity. Double sum is taken over all atoms. In the case of $\boldsymbol{i}=\boldsymbol{j}$ vectors of atoms equal as well ($\vec{r}_{\mathrm{i}}=\vec{r}_{\mathrm{j}}$) and exponent equals unity.

In the case of disordered systems, vectors of every interatomic distance $\vec{r}_{\mathrm{ij}}$ may have arbitrary orientation relative to the direction $\vec{S}_0$. Therefore, spatial averaging of Eq. (2.2) is necessary. Furthermore, the summation in Eq. (2.2) changes to integration because of a considerable amount of interaction with radiation atoms. Then we obtain

$$\bar{I}(s)=Nf^2+Nf^2\int_0^\infty 4\pi r^2\left[\rho(r)-\rho_0\right]\cdot\frac{\sin(sr)}{sr}dr \qquad (2.3)$$

where ρ_0 is the average density of atoms and $\rho(r)$ is the radial distribution of atom density. Some manipulation yields

$$s\cdot i(s)=\int_0^\infty 4\pi r\left[\rho(r)-\rho_0\right]\cdot\sin(sr)dr \qquad (2.4)$$

where $i(s)$ is called structure factor

$$i(s)=\frac{\bar{I}(s)}{Nf^2}-1$$

Fourier transformation of Eq. (2.4) provides possibility to calculate RDF from the experimental intensity distribution of scattered radiation:

$$4\pi r^2\rho(r)=4\pi r^2\rho_0+\frac{2r}{\pi}\int_0^\infty s\cdot i(s)\cdot\sin(sr)ds \qquad (2.5)$$

In the case of multi-component material, Eq. (2.5) is given by

$$4\pi r^2 \rho(r)\sum_m K_m = 4\pi r^2 \rho_0 \left(\sum_m K_m\right)^2 + \frac{2r}{\pi}\int_0^\infty s\cdot i(s)\cdot \sin(sr)ds; \quad (2.6)$$

where m is the number of components,

$$i(s) = \frac{I(s) - \sum_m f_m^2}{f_e^2};$$

$$f_{e\partial}^2 = \frac{\sum_m f_m^2}{\sum_m K_m^2};$$

$K_m = (Z_m/Z_l)^{0.75}$, f_m and Z_m are atomic scattering factor and atomic number of component m, Z_l is the atomic number of the lightest component of material, and f_e is the value of atomic scattering factor falling at one electron.

A typical RDF of a glassy semiconductor (bulk and film samples of As_2S_3) is given in Fig. 2.4. Disposition and area of the first peak of RDF allow determination of the number of nearest neighbors of an atom (first coordination number) and distances to them (radius of first coordination sphere, r_1). In the case of elementary materials and double compounds of stoichiometric composition in which only different types atoms are connected chemically, disposition of the second peak (r_2) and its halfwidth (δ_2), as a rule, give information about the average meaning of the bond angle ϕ [Eq. (2.1)] and about the range of its changes, $\Delta\phi$:

$$\delta_2 = r_1 \, \Delta\phi \cos(\phi/2) \quad (2.7)$$

The values of coordination numbers and radii of coordination spheres of the third and higher order do not give enough direct information about the spatial distribution of atoms beyond the first coordination sphere.

Usually, the diffraction patterns carry averaged information about mutual arrangement of the atoms in a sufficiently large area of material. The reduction in the electron beam diameter allows to reduce the analyzed area. Hirata et al. (2011) showed that by using the nanobeam electron diffraction method with diameter of electron beam equal to 0.36 nm, it is possible to obtain diffraction patterns from individual atomic clusters and, therefore, receive direct information on the local atomic order in amorphous material.

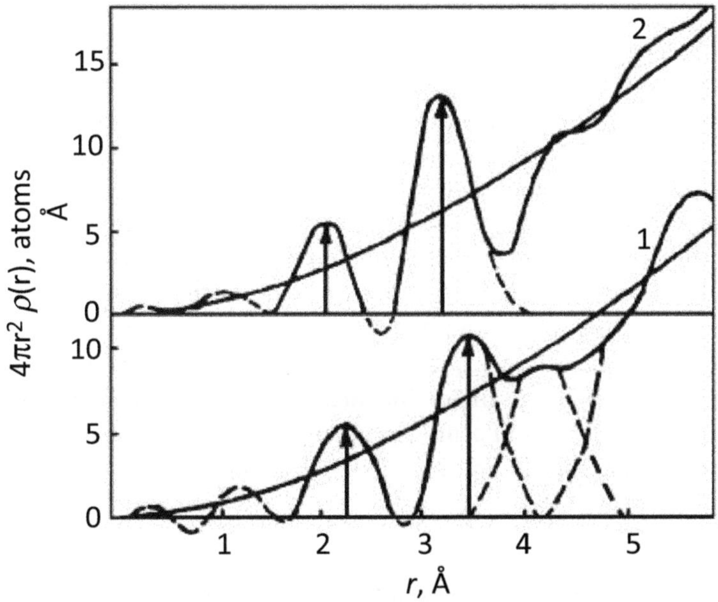

Figure 2.4 Radial distribution function of bulk (**1**) and thin film (**2**) a–As_2S_3 (Smorgonskaya and Tsendin, 1996).

As for binary materials of nonstoichiometric composition and multi-component non-crystalline materials, RDF received from the results of diffraction measuring does not give enough information even for interpretation of short-range order parameters. In this case, the method of X-ray absorption fine structure (XAFS) spectroscopy is more useful. The fine structure of X-ray absorption spectrum is determined as oscillations of the absorption coefficient (μ) relative to a smooth curve, which should be observed behind the absorption edge (as a rule, k-absorption) of isolated atom (Fig. 2.5). The sharp peak at the absorption edge is called the white line. EXAFS spectrum is usually divided into two parts: X-ray absorption near edge structure (XANES) and extended X-ray absorption fine structure (EXAFS). XANES includes X-ray absorption spectrum of 10 eV before the absorption edge, the absorption edge area, and the area up to 50 eV after the absorption edge. EXAFS is typically observed from

50 to 1000 eV above the absorption edge. It should be noted that at present the quantitative analysis of EXAFS spectra is well developed and is a standard procedure. However, quantitative analysis of the XANES spectra is not a trivial task (Bunker, 2010). Analysis of the extended X-ray absorption factor allows to calculate short-range order parameters around an atom absorbing X-ray emission quantum. In the case of multi-component material investigation, one can get EXAFS separately for atoms of each element by changing the energy of X-ray radiation. This allows to calculate RDF around atoms of each element. But in this case too, the information is limited by parameters of short-range order in atomic arrangement.

Thus, **diffraction and X-ray spectral analysis methods allow to define parameters of short-range order in atomic arrangement and give some, mainly qualitative, information about structure beyond short-range order but are not sufficient to reproduce spatial disposition of atoms in non-crystalline materials on the base of experimental data.**

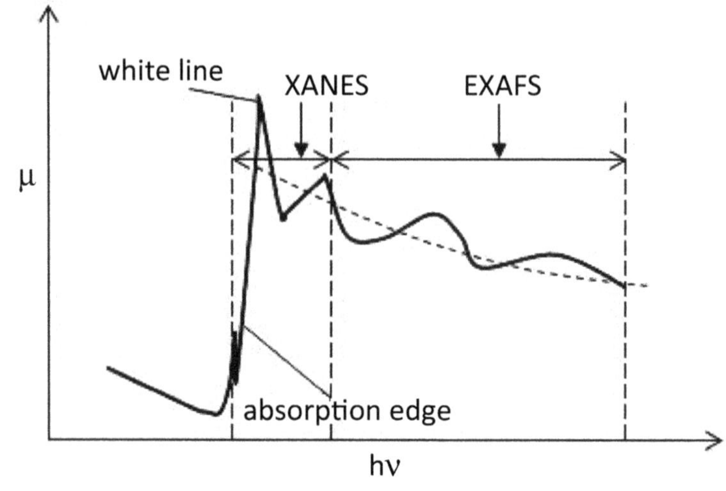

Figure 2.5 Illustration of the X-ray absorption fine structure (XAFS) spectrum.

Another group of structural investigation methods of disordered systems is **vibration spectroscopy, including spectroscopy of infrared absorption and Raman scattering**, as a rule in the frequency range of 400 to 44 cm^{-1}. In both cases, details of received spectra are connected with the vibration of atoms and chemical bonds in structural network of materials and, therefore, give information about forces, operating within structural units. However, infrared and Raman spectra do not duplicate each other, since their rules of selection for transition between oscillatory levels are different.

In the case of crystals, harmonic approach is used for decoding spectra of vibration spectroscopy. In this case, definite bands of absorption (also called group frequencies or characteristic lines) in vibration spectrum correspond with single bonds and groups of atoms inside structural units of different chemical compounds.

Transition from crystal to disordered systems furthers and significantly complicates vibration spectrum calculation due to the loss of long-range order. This is why the method of comparative analysis of the obtained spectra with vibration spectra of crystals analogous in chemical composition is usually used for interpreting experimental spectra of non-crystalline materials. The comparison of frequencies of characteristic lines and their intensity in both spectra allows to draw definite conclusions about the changes of atomic interaction during transition from crystalline state to disordered.

Thus, **methods of vibration spectroscopy give information about the presence and character of certain bonds and groups of atoms in structural units of investigated substance. At the same time, there is no theoretically substantiated analytical correlation that would allow to reliably calculate short interatomic distances (not to mention parameters of medium-range order) based on the results of vibration spectroscopy**. This is due to the fact that interatomic distances are only one of many parameters that are the part of kinematic factors of interaction in this complicated vibration spectrum.

In addition to diffraction methods and vibration spectroscopy, so-called **indirect methods** can give certain information about the structure of disordered systems and its change under the influence of different factors. Indirect methods are based on structural dependencies of physical and chemical properties of a substance. As such, dependencies exist practically for all material properties and success in using indirect methods is defined by the choice of the most structurally sensitive property of a substance. In the case of glassy substances, structurally sensitive characteristics include, in the first place, density, viscosity, solubility, thermal conductivity, heat capacity, sound velocity, refracting index, and their temperature dependencies. Since electrical conductivity changes, as a rule, by several orders of magnitude during transition from glassy to crystalline state, its measurement allows to investigate kinetics of such phase transitions.

The analysis of experimental investigation methods of non-crystalline material structure shows that **in the best case, they give information about short-range order in atomic arrangement but none of them separately nor all of them combined provide comprehensive data about medium-range order and, therefore, they do not allow to reproduce spatial disposition of atoms in disordered systems.** At present, a simulation of disordered material structure is used for solving this problem.

2.4 Simulation of Disordered Material Structure

To determine the spatial distribution of atoms in disordered systems, one can use structure simulation. The results obtained through such structure simulation can then be compared with the same characteristics of the modelled object defined experimentally. There are two methods of creating structural models of non-crystalline materials: physical simulation, that is, creation of model from physical objects (wire, tubes, balls, and so on) with further check of model adequacy and correction of atomic coordinates; computer simulation based on some basic atomic distribution and its further transformation until characteristics of model and simulated object are matched. Both approaches are not free from a number of defects. In the physical simulation for model construction, it is necessary to use the rules of construction based on common conception about the structure of substance (possible mutual disposition of structural units, the meaning of dihedral angles, and so on) in addition to the data received from direct experiment (first coordination number, radius of first coordination sphere). Thus, the model is based on both the objective data and subjective perceptions of the author. Some methods of computer simulation are free from such drawbacks. However, insufficient amount of border conditions received from direct experiments causes final distribution of atoms in a model to become only one of multiple possible distributions that match the simulated object.

A physical continuous random network model of amorphous silicon was constructed by Polk (1971). The model contained 440 atoms represented by balls and stems. All atoms (with the exception of atoms on the model surface) have tetrahedral bonds. The author allowed for small bond length fluctuations (±1%) comparative to crystal bond length. Mean bond angle was assumed to be equal to bond angle value in crystal (109° 28′). A random network arose from the absence of limitations on the value of dihedral angle.

The model contained two types of cells (Fig. 2.6):

- "Staggered" cells with dihedral angle of 60° and six-membered rings typical for crystalline silicon
- "Eclipsed" cells with dihedral angle of 0° and five-membered rings

The formation of "eclipsed" cells distorted the bond angle. The mean value of bond angle fluctuation was ±10%. The calculated density of the model was 97 ± 2% in comparison to the actual crystal density. The RDF of the model adequately reflected the experimental RDF of amorphous silicon.

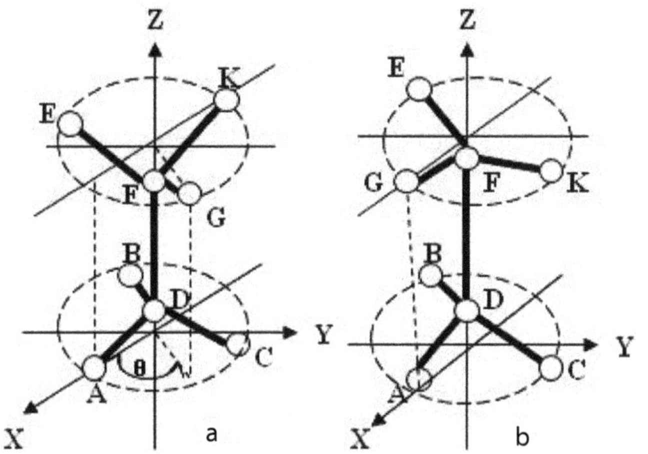

Figure 2.6 "Staggered" (a) and "eclipsed" (b) cells of amorphous silicon.

The structural random network model of threefold coordinated amorphous arsenic containing 553 atoms was constructed by Davis and Greaves (1976). The authors tolerated ±20° distortion of the bond angle and placed no restriction on the dihedral angle. One of the objectives during the construction was to leave no unsatisfied bonds in the interior of the model. In contrast to the model of tetrahedrally coordinated material, it was necessary to keep apart atoms not directly bonded. Therefore, the interlayer space in the model was not allowed to be less than 3 Å. An important feature of the model was that over short distances, it was layer-like but on a larger scale, it was isotropic; that is, the layers were so contorted and interconnected that it was impossible to follow them over distances exceeding 15 Å. This feature may be important for reconciling molecular-like, layer-like, and isotropic-like features inferred from certain experimental observations.

A structural model of non-crystalline selenium that has 539 atoms was built by Long and associates (1976). They did not introduce any limitations on the dihedral angle value in their model. Tetrahedral cells that had two hard covalent bonds and two flexible bonds representing interaction between chain molecules were used for modelling. When designing the model, the following limitations were used: presence of large cavities was not permitted; presence of regions with parallel packing of chain molecules similar to microcrystals was not allowed; large deviation of covalent bond lengths and angles as well as break up or rolling of chain molecules within model were excluded. Atomic coordinates of the built model were put into a computer and model strain resulting from deformation of length and angles of covalent bonds as well as intermolecular interaction was minimized. After minimization of strain in the model, the mean square deviation of length of covalent bonds was 0.89% and bond angle was 3.6% (in respect to 105°). The density of the model was different from the density of non-crystalline selenium by less than 3%. The RDF of the model coincides with the experimental RDF of non-crystalline selenium within the limits of the first and second peaks.

The main advantage of physical models is their illustrational value. However, construction of models with a large number of atoms requires considerable effort. It is for this reason that physical models are used for representing mainly elementary cells of non-crystalline substance when interpreting first peak of radial distribution function. As mentioned in Section 2.3 for multi-component and even nonstoichiometric binary non-crystalline materials, RDF received from diffraction measurements does not give enough information for definitive interpretation of the elements of short-range order. Constructing physical models of elementary cells, comparing areas and coordinates of first peaks calculated on the basis of the models with similar parameters of experimental RDF, allows to determine mutual atomic configuration at least within the first coordination sphere. Coincidence of calculated and experimental parameters of RDF proves correctness of the chosen model to a certain extent.

In the computer simulation of atomic structure of glassy materials, molecular dynamics and Monte Carlo methods are widely used. The method of molecular dynamics is used for studying the kinetic properties of the matter (e.g., for simulating processes of phase transition: crystallization, melting, glass transition) and for constructing models of atomic structure of glassy materials (Cormack, Du, and Zeitler, 2001). The method is based on the assumption that movement of atoms can be described by Newton's equations. The force acting on an atom is the sum of vectors $\sum_{j}^{N-1} \nabla\Phi(ij)$, where N is the amount of atoms in system, Φ(ij) is the pair-wise interaction potential of atoms (assumed that type of potential is defined), and ∇ is Hamiltonian. The starting coordinates of atoms are defined pseudo-randomly (that is, with the additional condition of prohibition to place two or more atoms at one point of space) or by the periodic crystalline network configuration. Initial speeds of atoms have random directions and equal absolute meanings chosen so that full kinetic energy of the system is true for the given temperature. When the initial coordinates and speeds are set, the atoms are set free one by one and the system begins to approach equilibrium state. The result of simulation is a series of atomic configurations matching different points in time.

Unlike purely deterministic equation of molecular dynamics, the **Monte Carlo method** is a method of numeral calculation in which probability elements are included. The characteristic feature of the method is construction of statically random process where separate states represent various configurations of the examined system that are obtained by random removal of its particles. Every new configuration is accepted or rejected. The criterion for a decision is probability of existence of a new configuration estimated by Boltzmann factor $\exp(-\Phi_{Nj}/kT)$ (Φ_{Nj} is the potential energy of the given configuration) or by the similarity of RDF calculated for the given configuration with experimental RDF.

The main shortcoming of the Monte Carlo method is that the computer experiment is very time consuming, which leads to limitation of the size of created models. The **gradient method of atomic structure simulation** was developed to overcome this shortcoming. Unlike the Monte Carlo method based on random search for optimum position of atoms in a model, the gradient method undertakes targeted search for the final disposition of atoms. This allows to reduce the time needed to obtain adequate models of structure significantly, especially in the case of materials where various types of chemical bonds coexist.

Let us examine the gradient method in simulating the structure of glassy selenium (Vasil'eva et al., 1982) as an example. The simulation of structure by the gradient method includes several stages. First, the number of atoms in the model is chosen, which allows us to calculate the size of the model while taking into consideration the atomic density of the simulated matter. The average density is defined by

$$U = \rho/M\, m_{\mathrm{H}}, \tag{2.8}$$

where ρ is the experimental density of simulated substance, M is atomic weight, and $m_{\mathrm{H}} = 1.65\times10^{-24}$ g is the mass of hydrogen atom. In addition to the size of the model, it is necessary to define its form. Cubic or spherical forms are widely used, and the choice of a particular form is mainly determined by the method of taking into consideration of the model's finite size. In the case of spherical form, various correction factors in calculating characteristics of the model are used to take into account finite size. So, for example, in calculating RDF, errors caused by the finite size of a model are compensated by the division of RDF by the correction factor defined by

$$D(r) = 1 - 1.5\frac{r}{d} + 0.5\left(\frac{r}{d}\right)^3, \tag{2.9}$$

where d is the model's diameter and r is the radial distance from the center of the model. At the same time, not every characteristic of a model can be corrected satisfactorily with the help of correction factors. This is why only the internal part of the model and not its whole volume is used for calculation, so that any atom in it has a normal environment that corresponds to an atom in the volume of material. For all of the above reasons, the examined model of non-crystalline selenium took form of a sphere with a diameter of 2.9 nm and included 420 atoms.

After defining the size and form of the model, the next stage is the creation of initial disposition of atoms. Pseudo-random distribution of atoms in a model is the most acceptable as a starting disposition because it excludes the influence of subjective initial assumptions on the results of simulation. During the formation of initial disposition of atoms, two limitations should be observed: The number of atoms must equal the predefined number of atoms in the model, and the distance between any atoms must not be less than a minimum distance defined by the position of the first peak in experimental RDF.

In the third stage of simulation, the initial atomic disposition is rearranged in order to match RDF of the model with the experimental RDF. The degree to which they are different is estimated by the mean square deviation of these functions:

$$\Delta = \sum_{i=1}^{N} [\mathrm{RDF_E}(I) - \mathrm{RDF_M}(I)]^2, \qquad (2.10)$$

where N is the number of points, where RDF is compared and $\mathrm{RDF_E}(I)$, $\mathrm{RDF_M}(I)$ are the experimental and calculated values for the model RDF at point I. The purpose of rearranging atoms in the model is to minimize the value of Δ. When it is necessary to obtain the best possible fit between the RDF of the model and experimental RDF at a certain range (e.g., in the region of RDF first peak), different weight factors for different ranges of values (I) can be included in Eq. (2.10):

$$\Delta = K_1 \sum_{I=1}^{N_1} [\mathrm{RDF_E}(I) - \mathrm{RDF_M}(I)]^2 + K_2 \sum_{I=N+1}^{N_1} [\mathrm{RDF_E}(I) - \mathrm{RDF_M}(I)]^2, \qquad (2.11)$$

where K_1 and K_2 are weight factors.

When rearranging the initial disposition of atoms using the gradient method (as opposed to the Monte Carlo method where new atomic coordinates are defined with the help of generation of random numbers), the new atomic disposition corresponding to the minimum value of mean square deviation (Δ) is found as follows. Gradient Δ in every direction of coordinate axes is determined for each atom. Based on the obtained values of gradient, the direction in which the mean square deviation decreases with the largest speed is chosen. The new disposition of the atom in the model that corresponds to minimum Δ in the chosen direction is defined as follows (Fig. 2.7): A cylinder with a radius that equals minimum distance between atoms is built in the direction of the maximum value of mean square deviation gradient. The numbers of atoms that have their centers covered by the cylinder are defined. Then spheres with a radius that equal minimum distance among atoms are built around these atoms. The areas of direct line of maximum gradient Δ covered by these spheres are prohibited for placement of the rearranged atom. After that the values of mean square deviation at the points of overlap between the spheres and the line of maximum gradient Δ are calculated sequentially and an allowed area with the disposition of atom equaling minimum mean square deviation is chosen. In Fig. 2.7, it is the part between points 2 and 3.

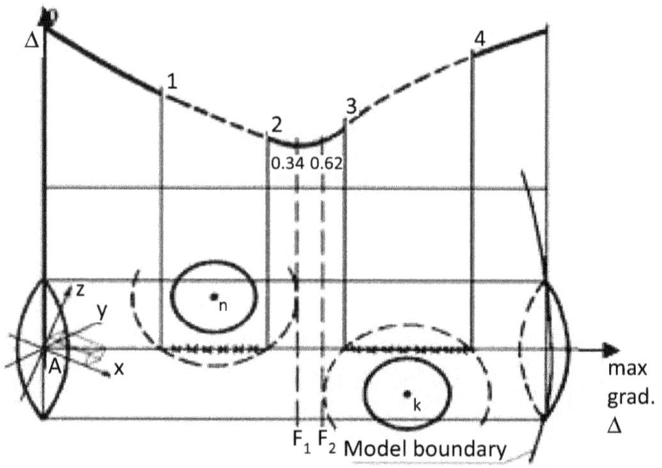

Figure 2.7 Procedure of atom positioning corresponding to mean square deviation (Δ) minimum value finding.

The minimum Δ between points 2 and 3 is defined by the method of "golden section." If moving an atom in any direction does not lead to a decrease in the mean square deviation, then this atom is left in its initial position. The procedure is repeated sequentially for each atom until a desirable similarity between the RDF of the model and the simulated object can be achieved.

The fourth stage is the finding of covalent-bonded atoms or uniting atoms in molecules. Undertaking this stage after minimizing the value of the mean square deviation of RDFs is justified by the fact that the rearrangement of free atoms (i.e., atoms that are not linked in molecules) at the previous stage allows to achieve better similarity of experimental and model RDF faster.

Thus, after the fourth stage, the model reflects the spatial disposition of atoms united into molecules with RDF coinciding with experimental RDF to a given degree of accuracy. However, the obtained spatial atomic disposition is not stable, because the energy of such a system does not correspond with local or basic minimum. This is why at the fifth stage of simulation, it is necessary to carry out relaxation of the model in order to minimize the total energy of the system, which allows to achieve a realistic atomic disposition.

In the case of tetrahedral bonded systems (silicon, germanium), bond-stretching and bond-bending energy is determined by the Keating equation (Keating, 1966)

$$E = \frac{3\alpha}{16d_0^2}\sum_{ij}\left[\left(\vec{R}_i-\vec{R}_j\right)^2-d_0^2\right]^2+\frac{3\beta}{8d_0^2}\sum_{ijl}\left[\left(\vec{R}_j-\vec{R}_i\right)\left(\vec{R}_l-\vec{R}_i\right)-\frac{d_0^2}{3}\right]^2 \quad (2.12)$$

where d is the equilibrium bond length, determined by the position of the first peak of the experimental RDF; i, j, and l are atom numbers; $\vec{R}_{i(j,l)}$ is the vector of atom i (j,l); α and β are bond-stretching and bond-bending constants.

But in the case of twofold coordinated systems (selenium), the total energy of atoms consists of four components:

1. Bond-stretching energy between the given atom and its covalently bonded nearest neighbors

$$E_s = 0.5\alpha\sum_{ij}(R_{i,j}^2-d_0^2)^2 \quad (2.13)$$

where $R_{i,j}=\left|\vec{R}_i-\vec{R}_j\right|$ is the bond length between atoms i and j.

2. Bond-bending energy

$$E_b = 0.5\beta\sum_{ijl}\left[\left(\vec{R}_{l,i}\vec{R}_{l,j}\right)-c\right]^2 \quad (2.14)$$

where $\vec{R}_{l,i}=\vec{R}_l-\vec{R}_i$ and $\vec{R}_{l,j}=\vec{R}_l-\vec{R}_j$ are vectors binding atom l with atoms i and j; c is a constant selected in such a way as to make E_b equal to zero at the given bond angle φ_0. The meaning of φ_0 is selected as a rule to be equal to a bond angle in the appropriate crystalline material (for selenium, $\varphi_0 = 105°$).

3. Energy of the van der Waals interaction between the given atom and the atoms that have no covalent bond with the given atom

$$E_V = \sum_i E_{l,i}, \tag{2.15}$$

$$E_{l,i} = \begin{cases} -\dfrac{A}{R_{l,i}^6} + \dfrac{B}{R_{l,i}^{12}}, & \text{if } R_{l,i} \le R_c, \\ 0 & \text{if } R_{l,i} > R_c, \end{cases}$$

where A and B are the constants defining the repulsion and attraction of atoms and selected in such a way that E_{li} equals the minimum value possible with the given distance $R_{li}{}^0$ between the atoms not directly connected by covalent bonds (in the case of selenium, the meaning of $R_{li}{}^0$ is determined in accordance with the position of the second peak of the experimental RDF and is 3.7 Å). R_c is the maximum distance where the van der Waals interaction is taken into consideration (as a rule 4–5 Å).

4. Bond-twisting energy

$$E_d = \gamma \sum_i [(\vec{R}_{ij} \cdot \vec{R}_{jk}) \cdot (\vec{R}_{jk} \cdot \vec{R}_{kl}) - K]^2 \tag{2.16}$$

where γ is a constant; i, j, k, and l are the indexes of sequentially bonded atoms in the molecule; and K is the constant, the value of which should provide minimum meaning E_d at the given meaning of dihedral angle $\theta = \theta_0$. Value θ_0 is selected, as a rule, as being equal to the dihedral angle in a certain crystalline modification of the matter (for selenium, $\theta_0 = 102°$).

When designing models like these, substantial uncertainty exists in the selection of values for constants α, β, and γ. This is why these values are usually selected to provide the required level of similarity between the experimental and model RDF.

Energy is minimized by the sequential movement of each of the atoms in the direction of the quickest decrease in its total energy. This direction is determined by calculating the energy gradient of atom, that is, by calculating the force that influences the atom. The search for the position of an atom that would correspond to its minimal total energy in the selected direction is carried out with the help of the algorithm used at the third stage of simulation.

It should be noted that, as a rule, the similarity between the model and experimental RDF degrades after the system's energy is minimized. In order to enhance the similarity between the experimental and model RDF and to further decrease the total energy of the model, the third, fourth, and fifth stages of simulation process are repeated to achieve the desired value of MSD at the minimum of total energy of the model (Fig. 2.8).

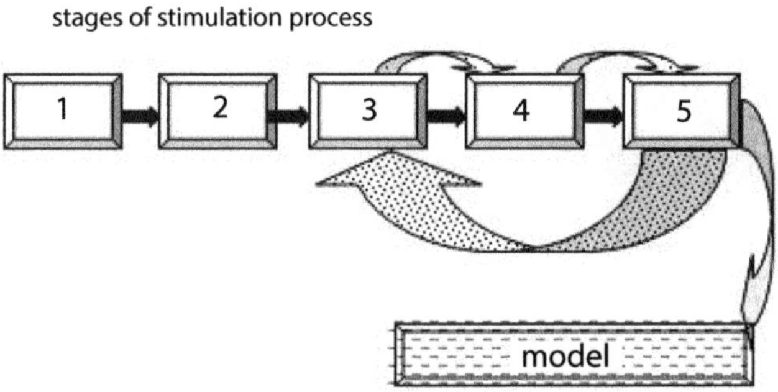

Figure 2.8 Algorithm of atomic structure simulation process.

2.5 Results of Structural Research of Disordered Semiconductors

2.5.1 Atomic Structure of Non-crystalline Selenium

Selenium is a group VI member of the periodic system. The structure of the outermost electron shell is $4s^2p^4$. Hybridization of electron orbitals in selenium is small. Therefore, as a rule, only *p*-electrons form chemical bonds. In elementary selenium, two *p*-electrons of each atom form covalent bonds, creating molecules in the shape of rings or high polymer chains, and the other two *p*-electrons stay in a non-bonding state as lone-pair electrons. Intermolecular interactions in selenium are caused by van der Waals forces.

Selenium exists in several allotropic crystalline and non-crystalline forms. Crystalline forms of selenium have been (Table

2.2, Baratov and Popov, 1990) studied quite well. The elementary cells of trigonal and monoclinic crystalline selenium are shown in Fig. 2.9. Trigonal selenium, formed by spiral chain molecules Se_n (Fig. 2.9a), is the thermodynamically stable form. All other forms of selenium turn into trigonal modification when exposed to thermal treatment.

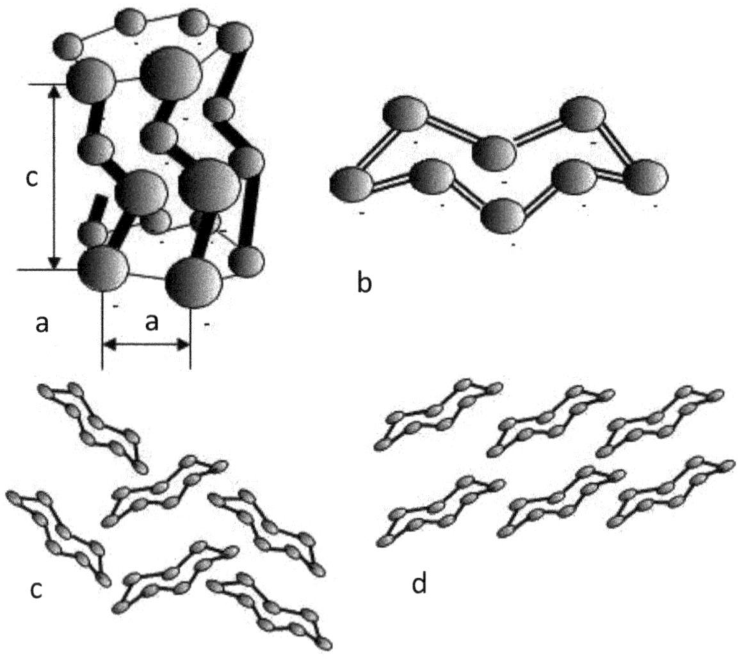

Figure 2.9 The elementary cells of the selenium crystal modifications: (a) trigonal elementary cell; (b) ring molecule; (c, d) packing ring molecules in the crystal lattice of the α and β monoclinic selenium (*a* and *c* are constants of the crystal lattice).

At the same time, a strict classification of allotropic non-crystalline forms is absent. There are at least three allotropic forms of solid non-crystalline selenium: red amorphous, black amorphous, and glassy. Red amorphous selenium is produced by chemical restoration, for example, H_2SeO_3, or by sharp quenching of superheated vapor of selenium. It is unstable even at a temperature of about 300 K. The structural models of red amorphous selenium are rather contradictory. However, analysis of the obtained data allows to suggest that red amorphous selenium consists of ring-shaped molecules Se_6. At 30–40°C, red amorphous selenium is turned into black amorphous modification. This transition has an irreversible endothermic effect, which is probably connected with the splitting of ring molecules. Information about black amorphous form of selenium is limited by the fact that there is no long-range order in atomic arrangement in this material.

Table 2.2 Allotropic forms of crystalline selenium

N	Form	Type of molecules	Bond length (nm)	Bond angle (degrees)	Lattice constants (nm)		
					a	*b*	*c*
1	Trigonal	Spiral chains Se_n	0.233	103.1	0.436	—	0.495
2	α-monoclinic	Rings Se_8	0.232	105.9	0.905	0.908	1.160
3	β-monoclinic	Rings Se_8	0.234	105.5	1.285	0.807	0.931
4	α-cubic	—	0.297	—	0.297	—	—
5	β-cubic	—	0.248	—	0.575	—	—
6	Rombo-hedral	Rings Se_6	0.235	101.1	1.136	—	0.442
7	Ortho-rombic	—	—	—	2.632	0.688	0.434

Glassy selenium is the widest spread non-crystalline form of selenium. At the same time, information about the structure and properties of glassy selenium is greatly different and sometimes even contradictory. It is caused by the fact that the glassy state includes various forms of selenium that differ from each other in ratio, sizes, forms, and mutual packing of structural units and, therefore, have different properties.

While examining the molecular shapes of crystalline forms of selenium (Table 2.2), one can suggest that the same molecules (rings Se_8, rings Se_6, spiral polymer chains Se_n) are present in the glassy matter too. The dependence of glassy selenium properties on the conditions of its preparation is explained by changes in the ratio of different molecules and changes in the level of polymerization of chain molecules. To illustrate this, films of glassy selenium that were prepared in research by Cherkasov and Kreiter (1974) consisted of only ring molecules Se_8 (amorphous analogue of monoclinic selenium) or of only chain molecules Se_n (amorphous analogue of trigonal selenium) and had, as a result, widely different properties.

At the same time, the absence of long-range order in glassy selenium opens possibility for wider changes in the molecular structure rather than a simple mixing of molecular forms of different crystalline modifications. This is primarily illustrated by the possibility of changing the value or the sign of dihedral angle in molecules and by the possibility of forming defects so characteristic of glassy semiconductors.

The absolute values of dihedral angles for trigonal and monoclinic modifications are close to each other (102° and 101°). The difference is that the sign of dihedral angle in the ring molecules of monoclinic form is changed under transition to each next atom, so atoms are placed in *cis*-coupling configuration and are closed into rings (Fig. 2.10a). The sign of dihedral angle in chain molecules of trigonal form is constant for the whole molecule. In this case, atoms are in *trans*-coupling configuration and form an endless spiral chain (Fig. 2.10b).

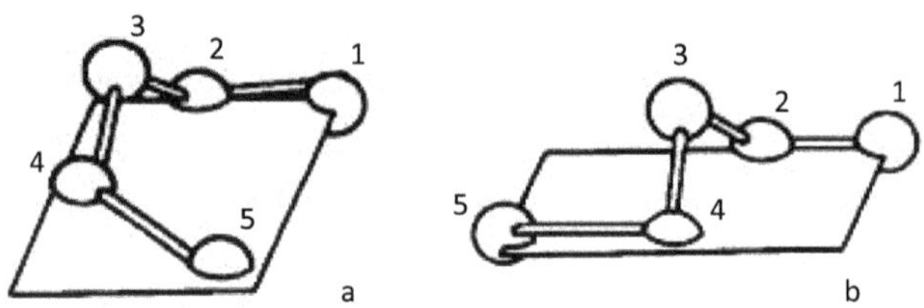

Figure 2.10 (a) *Cis*- and (b) *trans*-coupling configuration for molecular bonding in monoclinic and trigonal selenium.

When long-range order is lost, the requirement to have a strictly defined absolute value and sign of dihedral angle is no longer necessary. This is why models of flat zigzag chains (dihedral angle equals zero) (Richter and Breiting, 1971), free rotation chain model in which dihedral angle can have any value (Malaurent and Dixmier, 1977), disordered chains, where changes only to sign but not to value of dihedral angle are permitted (Lukovsky and Galeener, 1980), were suggested for describing the structure of selenium (Fig. 2.11). In the latter case, a molecule can have elements of both monoclinic form (rings) and elements of trigonal form (spiral chains).

Other factors that influence glassy selenium structure are quasi-molecular defects and valence-alternation pairs (VAP). VAP concentration in selenium can achieve the value comparable with the concentration of molecules (5×10^{18} to 5×10^{19} cm^{-3}) that must have an effect on the structure of matter due to the formation of intermolecular bonds by Se_3^+ centers. The introduction of quasi-molecular defects also leads to the formation of intermolecular bonds and to increasing selenium atom coordination (see Chapter 3).

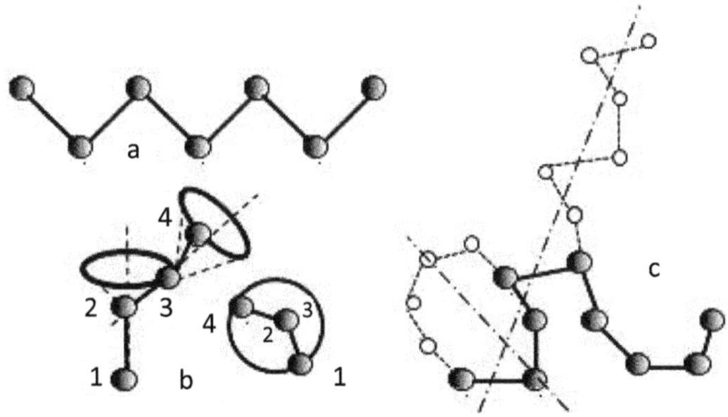

Figure 2.11 "Flat zigzag chain" (a), "free rotation chain" (b), and "disordered chain" (c) models of non-crystalline selenium molecules.

The structure of glassy selenium was widely investigated with the help of diffraction methods and the analysis of EXAFS. The first peak of RDF (which is determined by the nearest neighbors in the same molecule) is well isolated from the rest of this curve. The first coordination number in non-crystalline selenium is usually more than two (which is the value of the first coordination number in its crystalline modification) and equals 2.1–2.2 (Poltavzev, 1984; Kolobov et al., 1997).

The second peak of RDF has a shoulder on the side of large "*r*." It is caused by the fact that contribution to the second peak is made both by the atoms situated in the molecule (partial coordination number equals two) and the atoms of neighboring molecules. Decomposition of the second peak into two sub-peaks of Gaussian form under the condition that the area of the first sub-peak corresponds to the coordination number two allows to estimate the average value of bonds angle (100°–105°) and the value of deflection from medium meaning (~8°). A number of trigonal form coordination radii

(Table 2.2) is absent in the RDF of glassy selenium. An example could be radius 4.36 Å that fits the network constant "*a*" in the trigonal form. The absence of appropriate coordination sphere is due to the loss of periodical disposition of molecules. Disposition of the third peak of RDF of glassy selenium (4.7–4.9 Å) is close to the value of the network constant "*c*" of the trigonal form (4.95 Å). This coordination radius can prove that there exist at least some fragments of spiral chain molecule in non-crystalline substances.

As the diffraction method alone does not allow to determine the atomic structure of non-crystalline matter unambiguously, it is necessary to use other methods alongside the diffraction method. Basic details of the infrared absorption and Raman scattering spectroscopy of glassy, α-monoclinic, and trigonal selenium are given in Table 2.3 (Baratov and Popov, 1990).

Table 2.3 Features of selenium vibration spectroscopy spectra

	Selenium		
Spectroscopy	**Glassy**	**α-monoclinic**	**Trigonal**
Infrared absorption, Location (cm^{-1}) / Type	95 / peak	92–97 / doublet	—
	120 / shoulder	120 / peak	—
	135 / peak	—	144 / peak
	230 / shoulder	—	230 / peak
	254 / peak	254 / peak	—
Raman scattering, Shift (cm^{-1}) / Type	110–115 / peak	113 / peak	—
	140 / shoulder	—	143 / peak
	235 / shoulder	—	237 / peak
	250 / peak	250 / peak	—

Comparison of the infrared absorption spectra shows that the peaks of absorption in glassy selenium at 95 cm^{-1} and 254 cm^{-1} as well as shoulder at 120 cm^{-1} are in good agreement with the bands of fundamental absorption in α-monoclinic form of selenium. This served as the basis for an assumption that a considerable amount of Se_8 molecules is present in glassy selenium. On the other hand, the deep band of absorption at 135 cm^{-1} in glassy selenium is close to peak 144 cm^{-1} in trigonal selenium, and the shoulder at 230 cm^{-1} directly corresponds to the peak in trigonal selenium, which is interpreted as proof of presence of spiral chain molecules. Analysis of Raman scattering (Table 2.3) leads to similar conclusions. The glassy selenium spectrum is characterized by a wide peak of complicated form with the maximum at 250 cm^{-1} and a shoulder at 235 cm^{-1}, the position of which covers the peaks at 237 and 250 cm^{-1} on the spectra of trigonal and monoclinic forms. Other peaks of glassy selenium spectrum are close to fundamental modes of crystalline forms as well. Thus, when the results of IR- and RS-spectroscopy are interpreted, the structure of glassy selenium is viewed as a mixture of molecular forms of monoclinic and trigonal modifications with considerable amount of ring molecules Se_8. But such interpretation is not in agreement with other experiments. So, for example, the high viscosity of selenium at the temperatures of glass transition range proves predomination of polymeric molecules rather than monomer ring molecules in the substance.

Lucovsky and Galeener (1980) interpreted the Raman scattering spectra of selenium from the position of the disordered chain model. They showed that in the case of disordered chain selenium molecule (Fig. 2.11c), all atoms take part in vibrations with the frequency of 256 cm^{-1} (displacement in the bonding plane) and only part of atoms in the fragments of rings (cis-coupling configuration, Fig. 2.10a) contributes to the mode of 113 cm^{-1} (displacement perpendicular to the band plane). This explains the considerably higher intensity of peak 250 cm^{-1} in Raman scattering spectra of glassy selenium.

Differential solubility of glassy selenium in CS_2 or CH_2J_2 is used as an alternative method of estimating the ratio of ring Se_8 and polymer Se_n molecules. As these solvents dissolve monoclinic form well and do not dissolve trigonal form of crystalline selenium at all, it is assumed that ring monomers Se_8 are turned into solution. In the early works using this method, the amount of matter that turned into solution was up to 20–40%. But later on, it was defined more exactly that the solution of glassy selenium in CS_2 depends on the degree of illumination of a sample in the process of dissolving, which is probably caused by the structural changes in a substance under the influence of radiation. When radiation has no influence on the dissolving of glassy selenium in the mentioned solvents, the amount of substance that turns into solution is, as a rule, less than 5–10%. This testifies to a low concentration of monomer molecules. Predominance of polymer molecules in glassy selenium is also confirmed by the high viscosity of the material.

The results of viscosimetry are often used to obtain information about the degree of selenium chain molecule polymerization. Generally, the viscosity of selenium should be determined by the following factors: strength of intermolecular and intramolecular interactions, ratio of the amounts of ring and chain molecules, and degree of polymerization of the latter, types and concentration of defects. The level of influence that these factors have on the viscosity of selenium will be different under various temperatures. Analysis of temperature dependence of non-crystalline selenium viscosity (Popov, 1980) showed that at 60–80°C, the viscosity of selenium is determined mainly by two factors: intermolecular interaction and molecular structure of material. So in this temperature range, the degree of chain molecule polymerization (the amount of eight-atom monomers in a molecule) may be determined by the expression:

$$\lg \eta = A(T)P^{1/2}, \tag{2.17}$$

where $A(T)$ is factor dependent on the temperature of measuring, P is the degree of polymerization molecules, and η is the viscosity. As a rule, the degree of polymerization value is within 10^3–10^4.

The examined results of research on glassy selenium structure allow to determine accurately enough the short-range order in atomic arrangement, as well as to confirm low content of monomer ring molecules; it also allows to estimate the degree of polymerization of chain molecules and finally to select the most probable models of chain molecular structure—the models of free rotating and disordered chains. At the same time, there are not enough data to describe the spatial disposition of atoms in the material. As mentioned in Section 2.4, this task is completed with the help of simulation of non-crystalline material structure.

Vasil'eva et al. (1982) used the gradient method for simulating the structure of glassy selenium layers obtained by the vacuum evaporation method at different substrate temperatures. The typical distributions of bond lengths, bond angles, and dihedral angles for one of the models are shown in Fig. 2.12. The distribution of the bond lengths in the model has a clear maximum that corresponds to both form and disposition of the first maximum of the experimental RDF. The spread of bond length values increases with the decrease in substrate temperature within forming layers. The distribution of bond angles is represented by curves with the range of values between 60° and 180° with the maximum in the interval of 100–110°. Although the difference between bond angle average values in the model and the value of 105° characteristic for crystalline form increases slightly with the decrease in the substrate temperature, the distribution of bond angles as a whole in the models produced at different substrate temperatures differs only slightly. The values of dihedral angles are in the range of 10° to 180°, and their distribution does not have explicit maxima, which confirms the presence of free rotation chain model with arbitrary value of dihedral angles in the researched samples.

Analysis of the data proves that the structure of investigated layers of glassy selenium consists of disordered chain molecules with a small predomination of *trans*-coupling configuration in comparison with *cis*-coupling configuration. The level of structural ordering increases with the increase in the substrate temperature. The latter is proved not only by the statistic characteristics of models but also by analysis of their energy characteristics: the total energy of system and its four components (energy of distortion of bond lengths, bond angles, dihedral angles, and energy of van der Waals interaction). It is necessary to point out that the main difference in the meaning of total energy is determined by the difference of the van der Waals component, that is, the difference of the energy of intermolecular interaction. It corroborates that the changes in the degree of structural ordering with the changes in the preparation conditions are determined mainly by the changes in the mutual packing of molecules, that is, in medium but not in the short-range order in atomic arrangement.

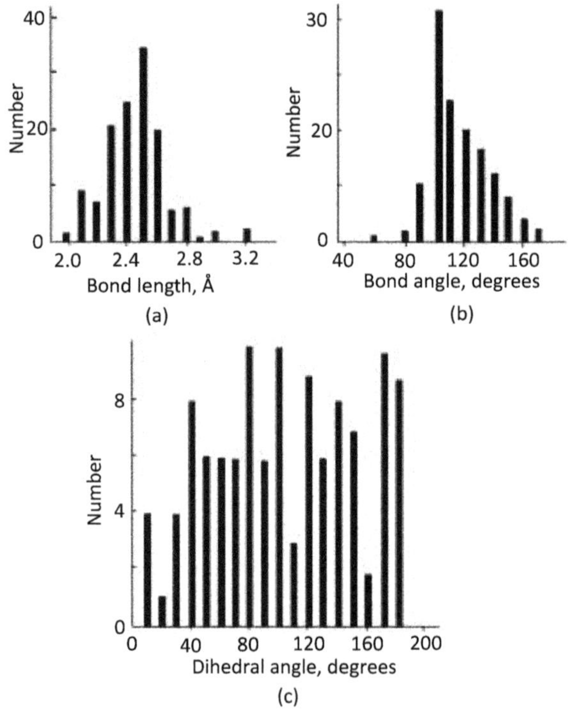

Figure 2.12 Distribution of bond lengths (a), bond angles (b), and dihedral angles (c) in the model of a–Se.

2.5.2 Atomic Structure of Chalcogenide Glasses

The structure of **non-crystalline arsenic chalcogenides** is explained on the basis of continuous network concepts. In terms of chemical order, that is, correlation between heteropolar (arsenic–chalcogen) and homopolar (arsenic–arsenic, chalcogen–chalcogen) chemical bonds, two extreme cases are possible: completely random network, in which chemical bond distribution is purely statistic, and chemically ordered network where heteropolar bonds realize anywhere if permitted by the chemical composition and by the requirement of continuity of the network. In real materials, there is some intermediate position determined by the difference of energy value of various bonds, the proportions of components, and to a large extent by the conditions of material preparation. For example, according to the data of different researchers, the concentration of homopolar bonds in films As_2Se_3 differs from 10–15 to 35%.

Most investigations of glassy arsenic chalcogenide structure were carried out on materials of stoichiometric composition $\mathbf{As_2X_3}$ (where X is a chalcogen: sulfur, selenium, tellurium). A particular feature of such compounds is a laminated atomic distribution and the predominance of covalent atomic bonds in layers, which is confirmed by isolation of the first peaks of RDF (Fig. 2.4). The position of the first peak in Fig. 2.4 corresponds to the sum of the covalent radii of arsenic and sulfur atoms. This indicates that heteropolar bonds between atoms of different types are most common, and the basic structural unit is pyramidal block $AsX_{3/2}$.

There are fewer investigations devoted to the structure of nonstoichiometric compositions of non-crystalline arsenic chalcogenides. The pyramidal blocks in glasses rich in selenium are assumed to be connected to each other with additional atoms of selenium that increase their free mutual orientation. Free orientation of pyramidal blocks decreases, and the horizontal length of layers increases in line with the growth of arsenic content.

Direct interpretation of the area of RDF peaks to determine coordination numbers within binary compositions is possible only for samples of stoichiometric compositions (if we suppose that there are only heteropolar bonds) for the first peak. In order to correctly interpret the results of diffraction investigation of nonstoichiometric binary compounds, one needs additional information about the interaction of atoms in the substance within an elementary cell due to the presence of homopolar bonds in nonstoichiometric compounds. Such information is obtained from the spectra of Raman scattering that allow to identify the structural units of the investigated substance.

The Raman scattering spectra of glassy arsenic selenides are given in Fig. 2.13 (Popov, 2004). The Raman scattering spectrum of glassy arsenic consists of peaks at 200, 240, 285 cm^{-1} and a system of peaks in the region of 110–160 cm^{-1}. Well-prepared volume specimens of glassy As_2Se_3, as a rule, have only one peak at 227 cm^{-1}, which is evidence to the presence of one kind of structural unit $AsSe_{3/2}$. This, in turn, confirms correct interpretation of RDF of this compound based on the assumption about the existence of heteropolar bonds only.

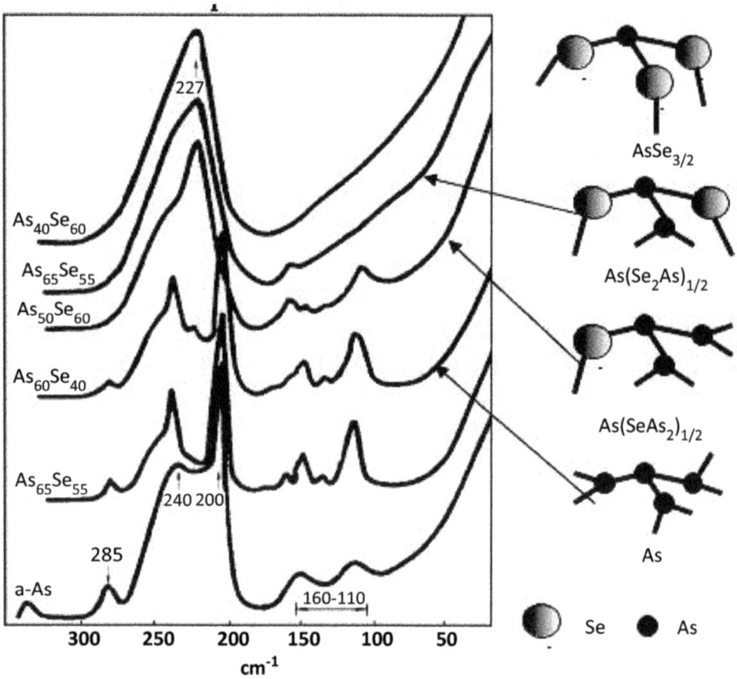

Figure 2.13 Raman spectra of glassy arsenic chalcogenides (a–As after Graves et al., 1979) (left panel) and structural units in the system As–Se (right panel). Arrows indicate compositions where these structural units first appear with increasing content of arsenic.

Comparison of the Raman scattering spectra of specimens of the As–Se system with different component ratios shows that the peak of the stoichiometric compound As_2Se_3 at 227 cm^{-1} is in the spectra of all specimens to $As_{65}Se_{35}$. However, the intensity of this peak decreases with the increase in arsenic content and only the shoulder is observed in the spectra of specimen $As_{65}Se_{35}$ under the given frequency shift. A small excess of arsenic content over stoichiometric composition (sample $As_{45}Se_{55}$) leads to the appearance of a peak at 160 cm^{-1} and a shoulder in the region 240 cm^{-1}. Emergence of these details in the spectrum is linked to the origin of As–As bonds in the glass network. However, the absence of a considerable number of peaks characteristic to the Raman scattering spectrum of non-crystalline arsenic in the spectrum of $As_{45}Se_{55}$ testifies to the absence of structural arsenic units (atom As linked with three atoms of As) in this sample. Persistence of a large intensity peak at 227 cm^{-1} along with the emergence of the above-mentioned details allows to assume that pyramidal blocks $AsSe_{3/2}$ remain predominant structural units in a sample of composition $As_{45}Se_{55}$ and that new structural units appear alongside in which one atom of selenium is replaced by one atom of arsenic—$As(Se_2As)_{1/2}$.

The rise in arsenic content to 50% makes a shoulder at 240 cm^{-1} in Raman scattering spectrum more expressed and leads to a series of peaks in the zone of 100–160 cm^{-1}, which could be a reflection of the fact that the amount of structural units $As(Se_2As)_{1/2}$ and the possibility of presence of structural units consisting of only one atom of selenium $As(As_2Se)_{1/2}$ increase. At the same time, the absence of a peak at 200 cm^{-1} in the Raman scattering spectra typical to non-crystalline arsenic testifies to the absence of arsenic structural units in this matter. The Raman scattering spectra of samples with 60 and 65% arsenic are characterized by a great reduction in the intensity of the peak at 227 cm^{-1} typical to As_2Se_3 (which confirms a small concentration of structural units $AsSe_{3/2}$), rise in the intensity of peaks at 240 cm^{-1} and at a range of 110–160 cm^{-1}. Besides, peaks at 200 and 285 cm^{-1} become apparent and thus all peaks typical to non-crystalline arsenic are present in these spectra. The obtained data testifies to the predominance of structural units $As(Se_2As)_{1/2}$ and $As(As_2Se)_{1/2}$ in these substances as well as to the existence of arsenic structural units in them.

Analysis of Raman scattering results in substances of systems As–S and Sb–S at a range of compositions from $Sb_{36}S_{64}$ to $Sb_{43}S_{57}$ shows that changes in spectra together with change in component ratio are analogous to the ones considered for the system As–Se.

Additional information about the structure of arsenic chalcogenide glasses can be obtained by indirect methods by measuring various physical and chemical properties of the matter. Qualitative results of alteration of heteropolar and homopolar bond ratio in the materials of systems As–Se, As–S can be obtained using the method of differential dissolution. It is known that alkaline solvents have more impact on heteropolar bonds (As–Se, As–S) than on homopolar ones (As–As, Se–Se, S–S) in the process of dissolving of arsenic sulfides and selenides.

Investigation of As–Se materials' solubility in 10% KOH showed that the speed of dissolving falls significantly and the quantity of insoluble residue rises with the increase in arsenic content over stoichiometric composition ratio. It conforms to the conclusion made on the basis of analysis of Raman scattering spectra that the amount of As–As bonds increases in this case.

The chalcogenides of group IV elements have smaller regions of glass formation and demand harder conditions of melt quenching in comparison with arsenic chalcogenides. In their glassy state, silicon and germanium have valency 4. The main features of structural changes with different composition are similar to those for arsenic chalcogenides observed above. Tetrahedrons $GeX_{4/2}$, $SiX_{4/2}$, which consist of a central atom germanium (silicon) bonding with four atoms of chalcogens, are the structural units of the composition GeX_2, SiX_2 (Fig. 2.14a). In compositions enriched by chalcogens, the number of tetrahedrons is proportional to the atomic concentration of the group IV elements and surplus atoms of chalcogen are united in chains linking tetrahedrons among themselves. Under higher concentration of chalcogen, separate molecules of the latter appear and form a solid solution with structural network of material (Fig. 2.14b). In the materials enriched with germanium, the peak associated with the structural units $Ge(X_3Ge)_{1/2}$ are observed in the Raman spectra (Fig. 2.14c).

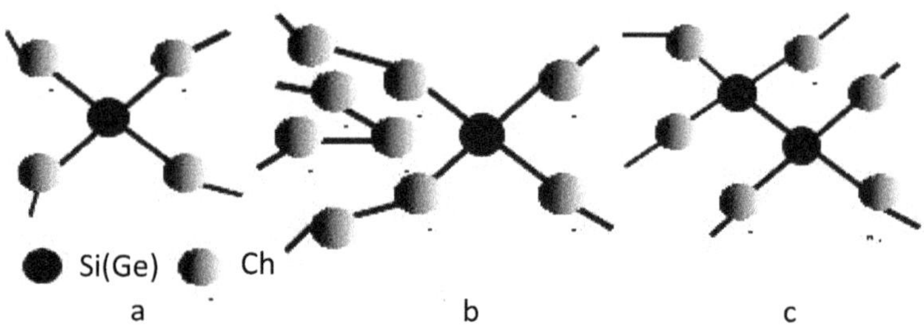

Figure 2.14 The structural units of glassy silicon (germanium) chalcogenides of various compositions: stoichiometric compound $Si(Ge)X_2$ (a), excess chalcogen content (b), and excess silicon (germanium) content (c).

2.5.3 Atomic Structure of Amorphous Silicon

Tetrahedral bonded semiconductors (Si, Ge, SiC, and others) in non-crystalline state may be prepared in the form of thin films only.

The first coordination number in these non-crystalline films equals 4±0.05 and the first coordination sphere radius corresponds with bond length in crystalline form (with deviation of no more than ±0.06 Å). The same applies for the mean value of bond angle. This indicates that tetrahedral structure and short-range order in non-crystalline states remain the same as in crystals. However, the mutual packing of tetrahedrons differs in the case of non-crystalline and crystalline states. It leads to the loss of long-range order in non-crystalline state and is an essentially different form of RDFs beginning on the third maximum. Taking these facts into consideration, a number of structural models have been created (Sections 2.2 and 2.4).

It is necessary to note that all these models represent idealized atomic structure of the material because neither dangling bonds inside model nor inside clusters were permitted. In reality, rigid covalent network of tetrahedral bonded disordered semiconductors contains a considerable amount of dangling bonds that strongly influence properties of the material. Addition of hydrogen atoms in amorphous silicon films leads to a sharp decrease in dangling bond concentration and to changes in the material's properties (see Chapter 3). The content of hydrogen in amorphous silicon films may amount to 30 at.%. In fact, it is a new material: silicon–hydrogen alloy or hydrogenated amorphous silicon (a–Si:H). In comparing a–Si:H structure to a–Si structure, it is necessary to answer the following questions: (1) What is the influence of hydrogen atoms on silicon's structural network? (2) What are hydrogen atom surroundings in the structural network of hydrogenated amorphous silicon? Hydrogen atomic scattering factor is negligible in comparison with the scattering factor of silicon atoms in the case of diffraction experiment. Therefore, hydrogen atoms do not give visible contribution in the diffraction pattern. That is why diffraction methods make it possible to calculate the RDF of silicon atoms in a–Si:H. Vibration spectroscopy methods are usually used to answer the second question.

The dependencies of the first coordination number, the first coordination sphere radius, and the density of hydrogenated amorphous silicon on hydrogen concentration are shown in Fig. 2.15 (Aivasov et al., 1995).

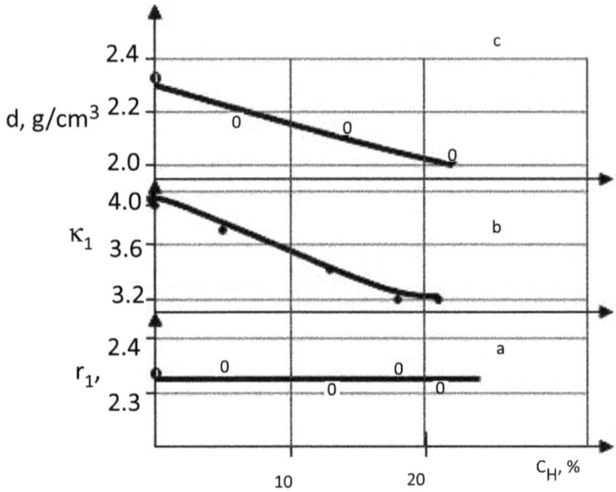

Figure 2.15 Dependencies of the first coordination sphere radius (a), the first coordination number (b), and the density (c) of hydrogenated amorphous silicon on hydrogen concentration.

One can see that the first coordination sphere radius remains constant with increasing hydrogen concentration. At the same time, values of the first coordination number and density of material decrease. The reduction in the material's density as a result of hydrogen concentration increase is obvious. The decrease in the first coordination number is conditioned by the formation of bonds Si–H instead of some of the bonds Si–Si. The following configurations of atoms are possible in this case: silicon atom bonded with three atoms of silicon and one atom of hydrogen (Si–H, Fig. 2.16a), silicon atom bonded with two atoms of silicon and two atoms of hydrogen ($Si{-}H_2$, Fig. 2.16b), silicon atom bonded with one atom of silicon and three atoms of hydrogen ($Si{-}H_3$, Fig. 2.16c), and polysilane chain ($(Si{-}H_2)_n$, Fig. 2.16d). These configurations have different group frequencies (vibration modes) in the infrared absorption (Table 2.4) and Raman scattering spectra (Lukovsky and Pollard, 1984). Thus, the location and intensity of appropriate peaks in the vibration spectra make it possible to identify and estimate the concentration of the structural units mentioned above.

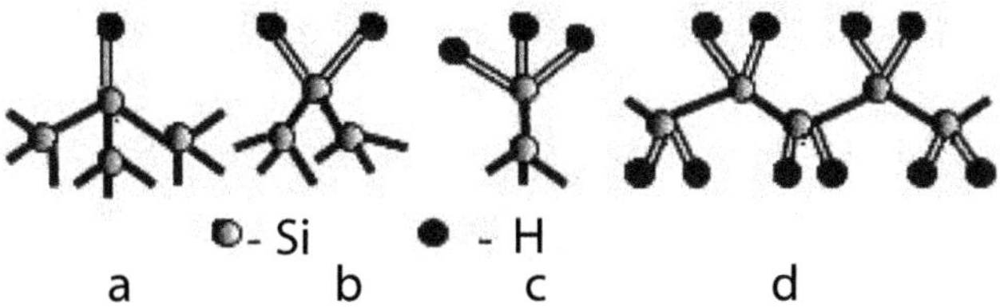

Figure 2.16 Configurations of atoms in a–Si:H:Si–H (a), Si–H_2 (b), Si–H_3 (c), and polysilane chain (Si–H_2)$_n$.

Table 2.4 Absorption peaks of a–Si:H in the infrared spectra

Configuration of atoms	Si-H	Si-H_2	(Si-H_2)$_n$
Wave number, cm^{-1}	2000	2090	2100
	630	875	890
		630	845
			630

For example, hydrogen atom concentration in various structural units may be determined from infrared absorption data as follows:

$$C_{\mathrm{H}} = \frac{1}{\Gamma}\int \frac{\alpha(\nu)}{\nu} d\nu \qquad (2.18)$$

where C_{H} is hydrogen atom concentration, ν is wave number, α is absorption coefficient, and Γ is trapping section. The ratio of various structural units is determined primarily by methods and conditions of material preparation and exerts determinant influence on the material's properties.

2.5.4 Structure of Carbon-Based Amorphous and Nanocomposite Films

It can be concluded from the previous sections that the basic features of the atomic structure are determined by the configuration of the outermost electron shell of atoms. Therefore, it suffices to know the material's group of the periodic system to make conclusions about the main features of the atomic structure:

- Group IV atoms are four coordinated and, as a result, a rigid three-dimensional covalently linked structural network occurs.

- In group V, atoms have the first coordination number of three, and this leads to the formation of layered structure.
- Group VI atoms, as a rule, have two covalent bonds and form a chain or ring molecules bound together by weak van der Waals interactions.

A stark exception to this pattern is carbon, which belongs to group IV of the periodic system and has s^2p^2 structure of the outer electron shell. Due to the varying degrees of hybridization of the electron orbitals, carbon in its crystalline state develops allotropic forms with different atomic structures and diametrically opposite properties:

- Diamond, in which due to sp^3 hybridization of electron orbitals, all four outermost shell electrons form covalent bonds with neighboring atoms (like silicon structure). This leads to the formation of one of the hardest materials and the best of solid dielectrics.
- In graphite, sp^2 hybridization of electron orbitals is realized, and as a result, three covalent bonds of each atom are formed, leading to the creation of a layered structure similar to group V elements. As a result, graphite is one of the softest materials and the best conductors.
- In carbines, due to sp hybridization of electron orbitals, only two electrons form covalent bonds, creating a chain structure similar to group VI elements. As a result, these materials have properties typical for polymers.

It is reasonable to assume that the structural units corresponding to various crystalline modifications can be realized in the amorphous state. This assumption is confirmed by the analysis of the atom RDFs in the amorphous carbon films. A model of amorphous carbon atomic structure constructed on the basis of this analysis is shown in Fig. 2.17. As seen from the figure, structural units of all crystalline modifications are present in the material. The ratio of different structural units is determined by the methods and regimes of amorphous carbon films fabrication. It means that the material of the same chemical composition can be used to create very hard dielectric films [amorphous diamond-like carbon (DLC) film] and the highly conductive (graphite-like) film with a band gap close to zero. Furthermore, it is possible to produce films with any properties that lie between these two edge cases. The implementation of this approach is discussed in Section 4.4.

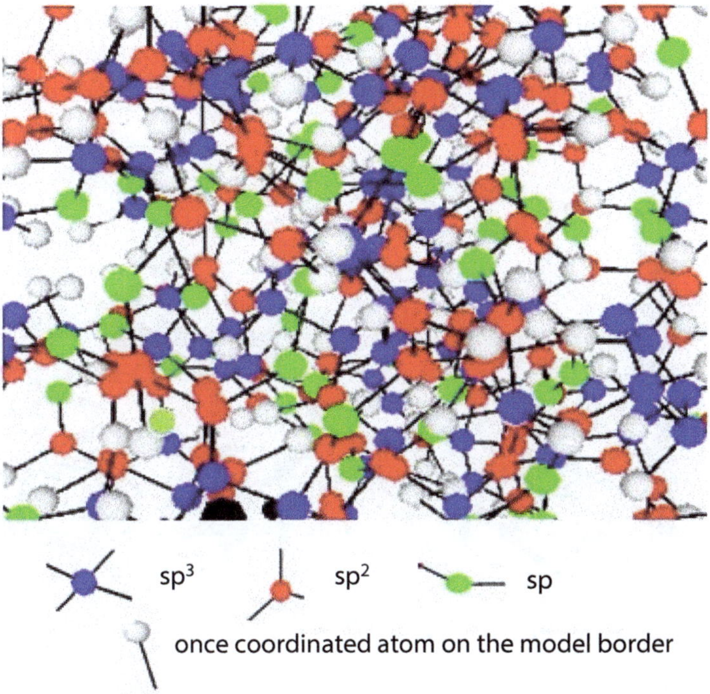

Figure 2.17 Model of atomic structure of amorphous carbon films, created on the basis of the radial distribution function.

The most widespread are amorphous DLC films (Robertson, 2002; Popov and Shupegin, 2012). These films consist of carbon network, mainly in the form of sp^3 hybridization of electron orbitals, and are widely used as protective coatings for optoelectronic, magnetic recording and other devices due to their good optical transparency, high hardness, and low friction. However, these materials have some limitations, particularly poor adhesion of the coating to the substrate and a limited range of operating temperatures. It has been shown (Won et al., 2003) that thermal treatments at temperatures above 200°C modify of both the structure and properties of amorphous DLC films.

Materials that retain the advantages of amorphous DLC films, on the one hand, and are largely free from the drawbacks of DLC, on the other hand, are amorphous silicon–carbon films (SCFs) (Dofman, 1992; Meškinis and Tamulevičienė, 2011). SCF is obtained by plasmo-chemical decomposition of silicon-containing organic matter: polyphenylmethylsiloxane (PPMS) $(CH_3)_3SiO(CH_3C_3H_5SiO)_3Si(CH_3)_3$ (Wan et al., 1997), mixture of methane (CH_4) and tetraethylorthosilicate $(C_2H_5O)_4Si$ (Won et al., 2003), and others.

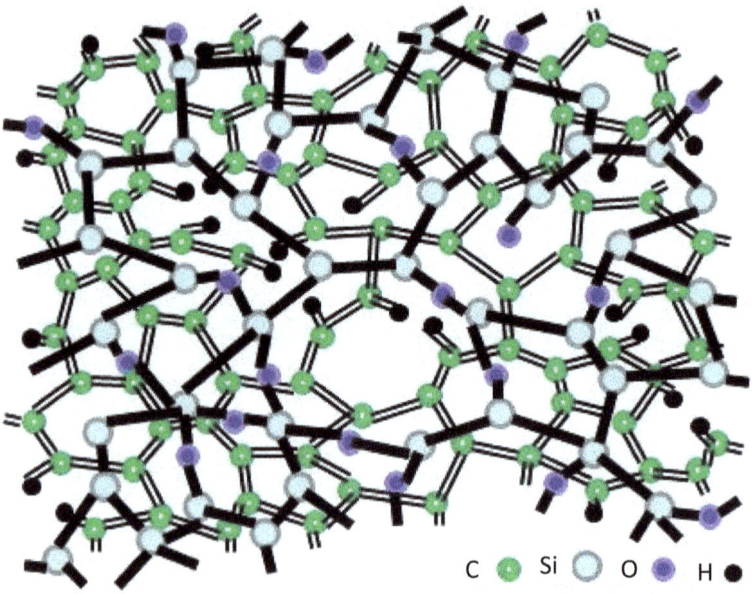

Figure 2.18 Atomic structure of amorphous silicon–carbon film.

These materials consist of an amorphous carbon (mainly diamond like) network (*a*–C:H) and an amorphous silica network (*a*–Si:O) that mutually stabilize each other (Fig. 2.18). This leads to a significant reduction in the intrinsic stress between films and substrates, improving adhesion of the films. It also leads to a substantial increase in the amorphous state stability of SCFs. Comparison of some properties of amorphous DLC films and SCFs is presented in Table 2.5 (Popov and Shupegin, 2012).

Table 2.5 Properties of DLC and SCF

Properties	DLC	SCF
Film thickness (micron)	≤2	≤100
Hardness (GPa)	20–30	6–20
Friction coefficient	<0.15	0.01–0.1
Roughness (nm)	>0.4	0.2–0.4
Resistivity (Ω cm)	10^6–10^8	10^{10}–10^{14}
Adhesion	Limited	High to various substrates
Chemical resistance	High	High

Stability of the amorphous state of SCFs makes it possible to introduce a large amount of additives in the amorphous matrix in order to change the material's properties (e.g., transition metals up to 70 at. %) (Venkatraman et al., 1997). Consider, as an example, the formation of nanocomposites by introducing tantalum into the silicon–carbon matrix (Presnyakov et al., 2013). Films of silicon–carbon matrix nanocomposites containing tantalum were obtained by the simultaneous decomposition of polyphenylmethylsiloxane and magnetron sputtering of tantalum. Experimental setup made it possible to obtain samples with different content of tantalum in a single technological cycle, depending on the position of the substrate relative to plasmatron and magnetron (Shupegin, 2013). Furthermore, it was possible to obtain both samples with a constant concentration of tantalum across the film thickness and samples with a predefined distribution of tantalum over the film's thickness (nanostructured films) (Popov et al., 2014a).

Electron diffraction and high-resolution electron microscopy helped demonstrate (Presniakov, Popov, and Vasiliev, 2013; Presnyakov et al., 2014) that the silicon–carbon matrix remains in the amorphous state in a wide range of tantalum concentrations. Bright-field high-resolution transition electron microscopy image of the silicon–carbon matrix is shown in Fig. 2.19a. The inset in the upper part of the figure shows a two-dimensional Fourier spectrum, indicating that the material is in the amorphous state.

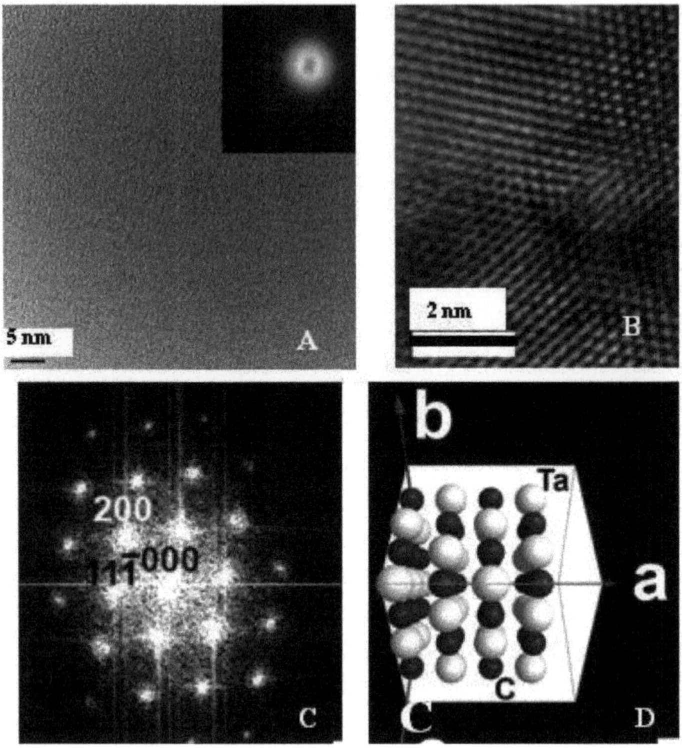

Figure 2.19 Bright-field high-resolution transmission electron microscope (HRTEM) image of the silicon–carbon matrix (two-dimensional Fourier spectrum of this image in the inset) (A); bright-field HRTEM image of TaC nanocrystal (B); two-dimensional Fourier transformation, acquired from the nanocrystal region, observed in projection [011] (C); and the model of the crystal TaC lattice (D). Reprinted from Presniakov et al. (2014), with permission from Springer.

However, the addition of tantalum results in the formation of nanocrystals in the matrix, and their concentration and size depend on the metal concentration in the film. Nanocrystal sizes are within 2 to 10 nm. Bright-field high-resolution transition electron microscopy image of the nanocrystal is shown in Fig. 2.19b and two-dimensional Fourier spectra, acquired from the nanocrystal region, observed in projection [011] is shown in Fig. 2.19c. The study of Fourier spectra has shown that their crystalline lattice is cubic and lattice parameter a = 0.446 nm, which corresponds to that of TaC crystals. The structural model of this composition along the [011] projection is shown in Fig. 2.19d.

Similar results were obtained when tungsten, molybdenum, hafnium, and other carbide-forming metals were introduced into the silicon–carbon matrix. Thus, during the formation of the nanocomposite films, metal atoms extract carbon from the silicon–carbon matrix to form the metal carbide nanocrystals. However, if the metal does not form a carbide (such as platinum), pure metal nanocrystals are formed in the film. But in all cases, the silicon–carbon matrix remains in the amorphous state.

2.5.5 Structure of Organic Semiconductors

Combinations of carbon with hydrogen alone or with hydrogen and some other elements (e.g., oxygen or nitrogen) are called organic materials. Organic materials are divided into two classes:

1. Materials consisting of relatively small, fixed-sized molecules: benzene molecule has six carbon atoms (benzene ring) and six hydrogen atoms; phthalocyanine molecule has eight benzene rings, eight nitrogen atoms, and two hydrogen atoms; and others.
2. Polymeric materials consisting of "infinitely" long molecules, such as polyacetylene.

In both cases, the materials are characterized by strong covalent interaction and ordered arrangement of atoms within the molecules and weak van der Waals interaction between molecules. Due to this, electrons in organic semiconductors are localized on the molecules. This represents an intermediate situation between the crystals (in which the electrons are delocalized throughout the volume) and non-crystalline semiconductors with the localization of electrons on the point states. A distinctive feature of organic materials is the presence of conjugated chemical bonds. In Chapter 3, it is shown that the presence of conjugated bonds determines the possibility of the existence of semiconducting properties in organic materials.

Certain articles on the semiconductor properties of organic materials have been published since the middle of the last century. However, a significant development of organic semiconductors occurred due to the successful synthesis and controlled doping of conjugated polymers in the late 1970s. It has been shown (Chiang et al., 1977) that doped polyacetylene forms a new class of conducting polymers in which the electrical conductivity can be systematically and continuously varied over a range of 11 orders of magnitude. These works were honored with the Nobel Prize in chemistry in 2000.

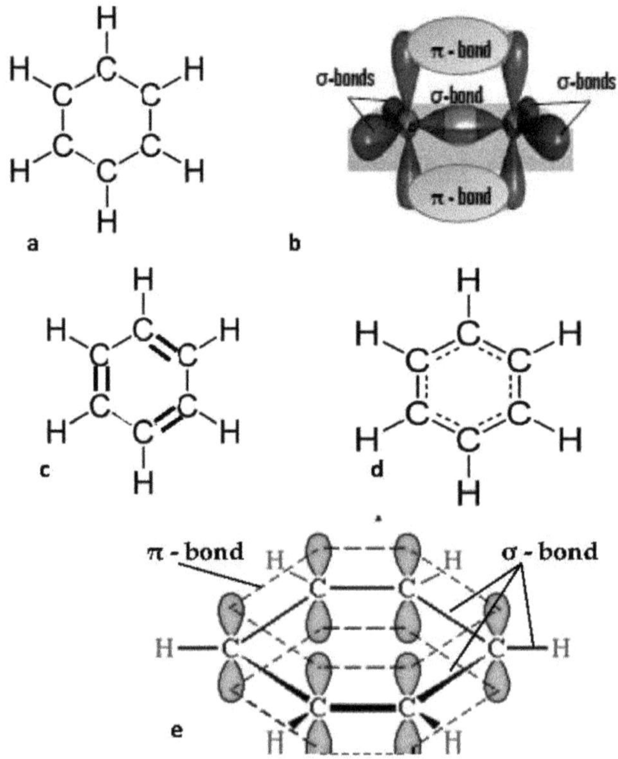

Figure 2.20 The structure of the benzene molecule: σ-bonds in the benzene molecule (a); σ-bonds and π-bonds between carbon atoms in the benzene molecule (b); model of benzene molecule by Kekule (c); corrected model of benzene molecule (d); σ-bonds and π-bonds in the benzene ring (e).

Let us consider the structure of conjugated organic materials on the example of the benzene molecule. The chemical formula of benzene is C_6H_6. The carbon atoms in benzene have sp^2 hybridization of electron orbitals to form three covalent σ-bonds. The fourth electron remains on the *p*-orbital placed perpendicular the plane of σ-bonds. Taking into account only σ-bonds, benzene molecule can be represented as a regular hexagon, where all bonds are coplanar

(Fig. 2.20a). Additionally, electrons on the p-orbitals of neighboring atoms interact with each other and form π-bond (Fig. 2.20b). Based on the existence of σ-bonds and π-bonds in the benzene ring, German chemist F. A. Kekule assumed in 1865 that single and double chemical bonds alternate with each other in this molecule (Fig. 2.20c). In this case, the length of a single bond between the carbon atoms should be 0.154 nm and the double-bond length should be 0.134 nm. However, the length of all σ-bonds between carbon atoms in a molecule of benzene is the same and equal 0.140 nm. This corresponds to an intermediate value between the lengths of single and double bonds and indicates that the electrons on the p-orbitals form a common bond between all the carbon atoms of the benzene ring. Therefore, the Kekule model of the benzene molecule was corrected later (Fig. 2.20d). Thus, the benzene ring consists of carbon atoms linked together by strong σ-bonds that are in the plane of sp^2-orbitals and π-bonds forming a single cyclic π-cloud located above and below the plane of σ-bonds (Fig. 2.20e).

Chapter 3

Electronic Structure and Properties of Disordered Semiconductors

3.1 Electronic Structure

3.1.1 Localized States in Disordered Semiconductors

Solid-state physics provides us with a simple understanding of the electronic structure of crystalline semiconductors. However, solid-state physics is based on the periodic crystal lattice (see Chapter 1). This means it has nothing to say about the electronic structure of non-crystalline semiconductors that do not have the periodic lattice. On the other hand, the main features of the electronic structure of atoms in the solid are determined by interaction between nearest neighbors or, in other words, by the short-range order of atomic arrangement. One could deduce that if the short-range order remains the same in the crystalline and non-crystalline states of the material, then the main features of the electronic structure of the material remain similar as well. Of course, similar does not mean the same. Absence of long-range order leaves traces on the spectrum of the electronic states of non-crystalline semiconductors. The main distinguishing feature of the electronic spectrum of non-crystalline semiconductors is localized states.

There are two definitions of localized states suggested by P. Anderson and by N. Mott (Fig. 3.1). The first one states: the state

Disordered Semiconductors: Physics and Applications (2nd Edition)
Anatoly Popov

ISBN 978-981-4774-37-6 (Hardcover), 978-1-315-10042-5 (eBook)
www.panstanford.com

is localized if electron with energy $E \pm \Delta$, located in volume large enough to satisfy the uncertainty principle, does not diffuse from this volume. Mott mentioned the other feature of localized states: the contribution of electrons in localized states to direct current conductivity reduces to zero at low temperatures.

(a) (b)

Figure 3.1 Winners of the Nobel Prize in physics in 1977: P. Anderson (a) and N. Mott (b).

From the point of view of physics, this can be explained as follows. The density of electronic states at the conduction and valence bands is high. It means that the states at the same energy are separated by very thin barriers. Electrons can tunnel through these barriers under the influence of concentration gradient or weak electrical field (Fig. 3.2a). If the density of states decreases, the states of the same energy become well separated in the lattice site, so that the tunneling probability between these sites is zero (Fig. 3.1b). The states become localized. Electrons in these states cannot diffuse (Anderson's definition) or take part in electrical conductivity (Mott's definition). Only over-barrier electron transitions are possible in this case.

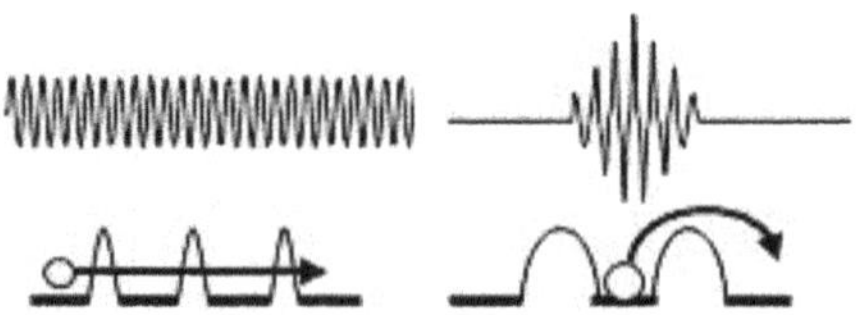

(a) (b)

Figure 3.2 The wave functions and transitions of electrons in the delocalized (extended) states (a) and localized states (b).

3.1.2 Models of Energy Bands

The localized states mentioned above are present not only in disordered semiconductors but also in crystals as well. The donor and acceptor states in crystals are localized states. For example, an electron at donor level cannot transfer in crystal until its excitation to conduction band. Another example of localized states in crystals is deep levels in the gap caused by crystalline lattice defects. The presence of a point defect (vacancy or interstitial defect) leads to the change in the interaction of neighboring atoms and to the shift in the energy of atomic orbitals.

The absence of long-range order in non-crystalline semiconductors leads to shifts in the electronic states in conduction and valence bands as well. Some of these states may shift to the gap. As a result, the tails of conduction and valence bands in the gap are formed (Fig. 3.3) (Gubanov, 1963). The tail states are localized and are separated at E_c and E_v from the extended (delocalized) states.

This model is rather clear from the physics point of view, but it cannot explain some experimental facts. For example, the absence of localized states in the middle of the gap does not explain the effect of the so-called fixation of Fermi level in the middle of the gap, which is experimentally observed in chalcogenide glasses.

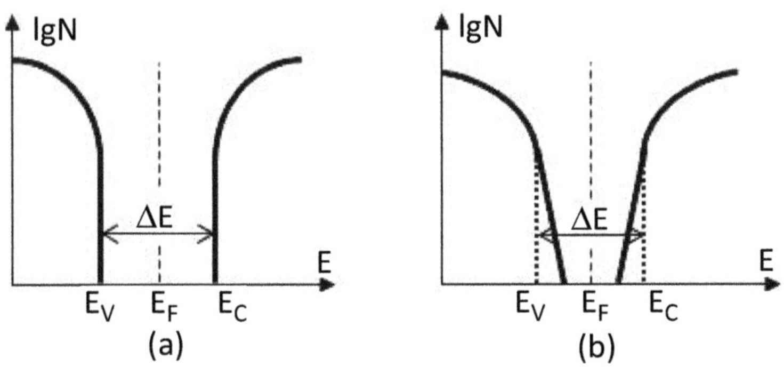

Figure 3.3 Schematic illustration of energy bands in the case of crystalline (a) and non-crystalline (Gubanov model) (b) semiconductors.

It is known that the Fermi-level position depends on temperature. In turn, the value of the Fermi-level temperature shift depends on the density of electronic states. The shift has bigger value in the case of lower density of electronic states in the region of the Fermi-level position. However, the Fermi level in non-crystalline semiconductors is situated near the middle of the gap and its position only changes slightly even with significant temperature variation. This can only happen when a small shift in the Fermi-level position leads to a considerable change in the ratio between the occupied (lower E_F) and the empty (higher E_F) states or when the density of states in the region of the Fermi-level position is high. But the density of electronic states in the middle of the gap equals zero in the Gubanov model (Fig. 3.3b).

That is why Cohen, Fritzscher, and Ovshinsky (CFO) (1969) proposed another model of energy bands for multicomponent chalcogenide glasses. They suggested that due to the additional chemical disorder of these materials, the tails of localized states have larger length and overlap in the middle of the forbidden gap (Fig. 3.4a). But in this case, the concept of forbidden gap makes no sense at all because electronic states are present at any energy. That is why the authors introduced the forbidden gap of mobility instead of the forbidden gap of electronic states (Fig. 3.4b). The forbidden gap of mobility is limited to the values of energy E_c and E_v, which separate the extended (high mobility) and localized (low mobility) states. Mobility in localized states tends to zero at low temperatures.

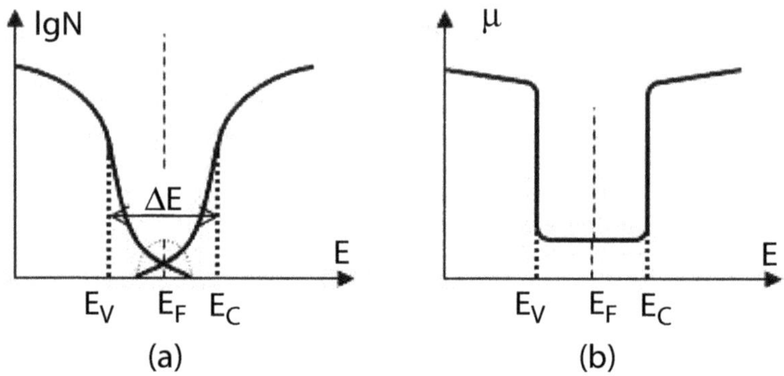

Figure 3.4 Schematic illustration of density of states (a) and carrier mobility (b) in the CFO model.

The overlapping of valence and conduction band tails leads to the formation of charged centers (positively charged higher Fermi-level position and negatively charged lower Fermi-level position). It explains the effect of fixation of Fermi level in the middle of the gap. But this model is in contradiction with some other experimental data because the density of states in the forbidden gap is too large (e.g., some optical properties and temperature dependence of AC conductivity).

Davis and Mott (1970) made the next step. Their model (Fig. 3.5) has short tails of localized states and narrow region with high density of localized states in the middle of the gap, which fixes the Fermi level. This region cannot be explained by the fluctuation shift in electronic states from conduction and valence bands. The authors suggest that it consists of electronic states of structural defects. This model is generally used now. Let us consider the possible types of intrinsic defects in non-crystalline semiconductors.

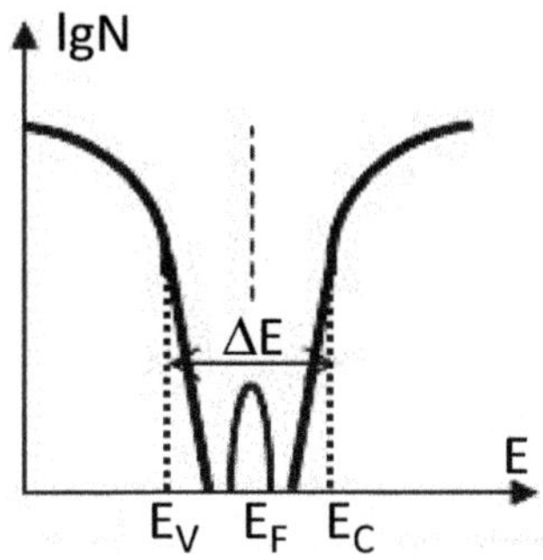

Figure 3.5 Schematic illustration of density of states after the Mott–Davis model.

3.1.3 Defect States in Disordered Semiconductors

The absence of long-range order causes much more variety of defect types in non-crystalline solids in comparison with crystals. Point defects in crystals (vacancy or interstitial defects) represent the variation of coordination number of several atoms simultaneously. For example, vacancy in tetrahedral crystalline lattice leads to the formation of four atoms with coordination number of three (Fig. 3.6a). By contrast, the variation of coordination number of three, two, or one atom is possible in non-crystalline tetrahedral material (Fig. 3.6b).

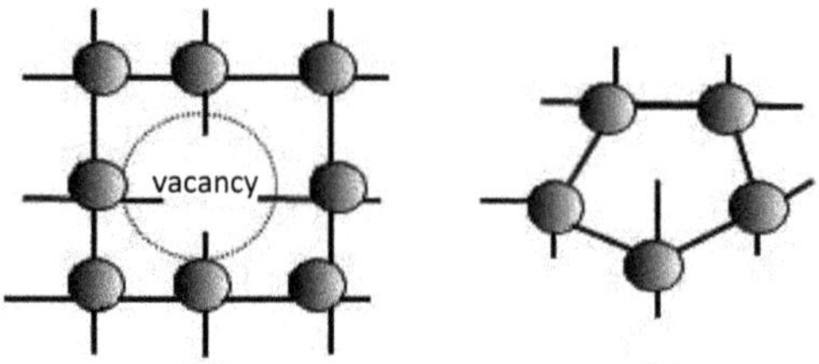

Figure 3.6 Variation of atom coordination number in the point defect region in the case of tetrahedral crystal (a) and tetrahedral non-crystalline material (b).

The equilibrium concentration of defects both in crystals and non-crystalline materials is determined by free energy minimum condition:

$$G = H - TS \tag{3.1}$$

where H is enthalpy and S is entropy. Defect formation leads to increase in both enthalpy (on the value of defect formation work) and entropy because entropy may be written as statistical weight of system:

$$S = k \cdot \ln \frac{N!}{(N-n)!} \tag{3.2}$$

where n is the concentration of defects, N is the number of possible places of defect distribution, and k is the Boltzmann's constant.

Therefore, equilibrium defect concentration differs from zero at any temperatures except 0 K and increases exponentially with rising temperature. In the case of disordered solids, the concentration of defects has to be more than the equilibrium concentration because of nonequilibrium preparation conditions. Experimental results of drift carrier mobility, photoconductivity, value of Stokes shift, effect of Fermi-level fixation, and others confirm the presence of high defect concentration and appropriate localized states in the gap of disordered semiconductors.

Let us consider the electronic structure of the common type of defects in non-crystalline semiconductors—abnormal coordinated atom. The example of this type of defect is the atom with a dangling bond at the end of the chalcogen chain molecule. The atom inside the chain molecule forms two covalent bonds. As a result, two p-electrons pass on bonding orbitals and two p-electrons remain on non-bonding orbitals (Fig. 1.13b). The atom at the end of the chain molecule forms only one covalent bond. Therefore, only one *p*-electron passes on the bonding orbital and three *p*-electrons remain on the non-bonding orbitals (Fig. 3.7). Because there is an odd number of electrons on the non-bonding orbitals, this defect has to have non-paired spin. Taking into consideration the high concentration of defects and non-paired spins, it is reasonable to expect the presence of an electron spin resonance (ESR) signal in these materials. However, ESR signal is absent in chalcogenide glasses and glassy selenium in normal conditions.

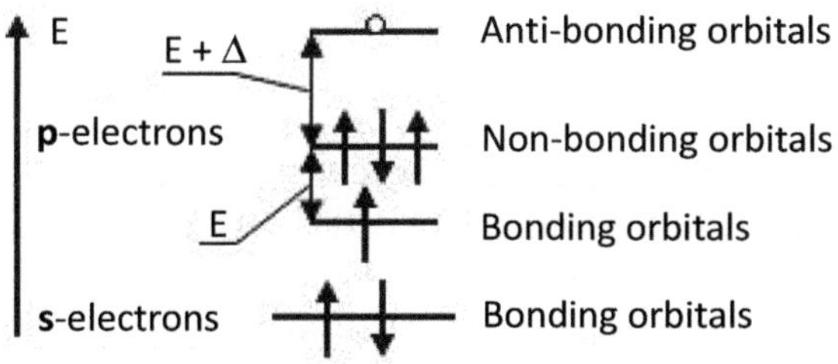

Figure 3.7 Schematic diagram of orbitals in the case of atom at the end of chalcogen chain molecule.

The absence of ESR signal means that localized states in the gap of chalcogenide glasses are either empty or occupied twice. But it means that these centers have a positive (non-paired electron removed) or a negative (extra electron added) electric charge. It is rather difficult to explain such situation from the point of view of crystal physics. Charge state in crystals has greater energy in comparison with neutral state because of the repulsive Coulomb interaction. The energy difference between charge and neutral states is named correlation energy U and always has positive value in crystals (Fig. 3.8).

Anderson (1975) has suggested the concept of negative correlation energy. He pointed out that in the case of flexible structural network of chalcogenide glasses, electron–phonon interaction gives rise to an attractive interaction between electrons overcoming the repulsive Coulomb interaction. As a result, charged defects have to have less energy as compared with neutral defects and correlation energy becomes negative.

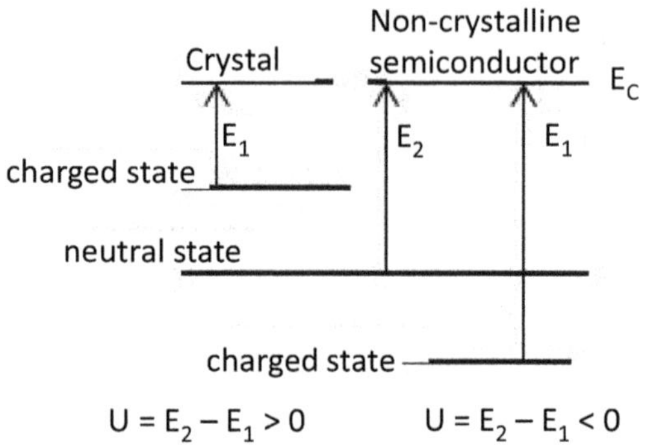

Figure 3.8 Schematic illustration of correlation energy determination (see text).

Concerning the nature of center with negative correlation energy, Street and Mott (1975) proposed that dangling bonds are negative correlation energy defects. Neutral dangling bonds (D^0) become positively (D^+) and negatively (D^-) charged as a result of network distortion, which brings about negative correlation energy. That is why

$$2D^0 \rightarrow D^+ + D^- \tag{3.3}$$

is an exothermal reaction. In other words, the first ionization of negatively charged center ($D^- \rightarrow D^0 + e$) requires energy E_1 greater than the energy E_2 ($D^0 \rightarrow D^+ + e$) of second ionization (Fig. 3.8). As a result, the correlation energy $U = E_2 - E_1$ has negative sign, and defects in chalcogenide glasses have to be positively or negatively charged. The neutral state D^0 can be achieved only by excitation. Electron transitions between valence band and defect states are shown in Fig. 3.9a.

The lower parabola represents D^+ center and the filled valence band. The upper parabola represents D^0 center with a hole at the top of the valence band. Transition ***A*** corresponds to the excitation of electron from the valence band into D^+ center, leaving a hole in the valence band. After the excitation of electron into D^+, the charge of state is changed ($D^+ + e = D^0$). This leads to a rearrangement of the structural network and to the decrease in the energy of the polaron value w^+. Transition ***C*** corresponds to the recombination of D^0 with a hole in the valence band and the appropriate rearrangement of the structural network and energy decrease of w^+. Both transitions are vertical (optically allowed) transitions, following the Frank–Condon principle.

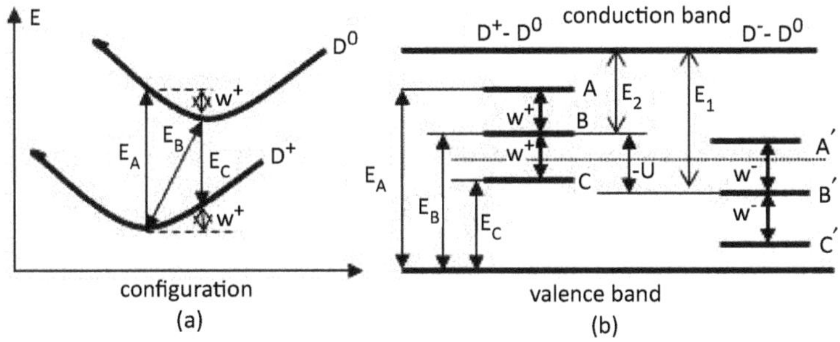

Figure 3.9 Configuration (a) and level (b) diagrams for defect levels with negative correlation energy. Reprinted from Street and Mott (1975), with permission from American Physical Society.

Transition energies E_A, E_B, and E_C are shown in the level diagram (Fig. 3.9b). Similar levels for the case $D^- - D^0$ transitions are shown on the right side of the figure. The energy difference between ***B*** and ***B'*** levels ($E_2 - E_1$) corresponds to the negative correlation energy. Thus, Stokes shift is equal to the doubled polaron energy ($E_A - E_C = 2w^+$). Transition ***B*** corresponds to the thermal excitation of electron in the valence band into D^+ center.

Kastner, Adler, and Fritzscher (1976) have analyzed defects D^+ and D^- from the point of view of chemical bonding theory. They proposed that the positively charged defect D^+ has minimum energy in the case of threefold coordinated (over-coordinated) chalcogen atom C_3^+. On the contrary, the negatively charged defect D^- has minimum energy in the case of one-fold coordinated (under-coordinated) chalcogen atom C_1^- (superscripts and subscripts denote charge states and coordination number, respectively). In this case, expression (3.3) may be rewritten as follows:

$$2C_2^0 \rightarrow C_3^+ + C_1^- \tag{3.4}$$

It is necessary to note that the total number of atom covalent bonds in reaction (3.6) does not change (and equals four). However, the coordination number of each atom changes. That is why this type of defects was called **valence-alternation pairs** (VAPs). Let us consider the VAP formation process in glassy selenium.

Two fragments of selenium chain molecules and the electron configuration of selenium atoms are shown in Fig. 3.10. As mentioned above, atoms inside the chain have two *p*-electrons on bonding orbitals and two *p*-electrons on non-bonding orbitals. Atoms at the end of the chain have only one *p*-electron on the bonding orbital and three *p*-electrons on the non-bonding orbitals (including one electron with unpaired spin) (Fig. 3.10a). The electron with unpaired spin can interact and form covalent bond with the lone-pair electron of the neighboring atom and (Fig. 3.10b):

$$C_1^0 + C_2^0 \rightarrow C_3^0 + C_2^0 \qquad (3.5)$$

as a result, the threefold coordinated atom C_3^0 arises. This atom has three electrons on the bonding orbitals and one electron on the anti-bonding orbital due to the transition of two electrons from non-bonding to bonding orbitals (energy reduction on the value $2E$) and one electron from the non-bonding to the anti-bonding orbital (energy increase on the value $E + \Delta$). The total energy change is equal to $-E + \Delta$. Thus, under the assumption of $\Delta << E$, reaction (3.5) is accompanied by a considerable energy gain.

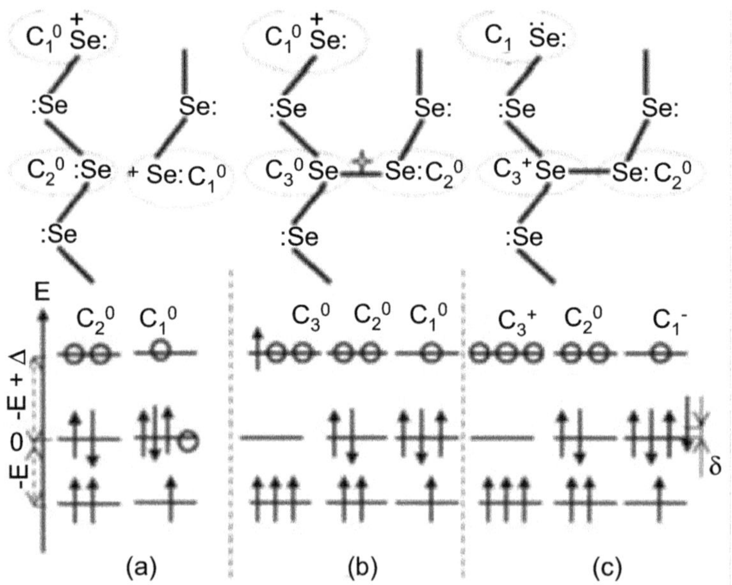

Figure 3.10 Formation of valence-alternation pair in glassy selenium. See the text for elucidation (: lone-pair of electrons, + non-paired electron on non-bonding orbital, and ✧ electron on anti-bonding orbital).

The energy of an electron in the anti-bonding state is close to conduction band. So there is the probability of electron transfer from the anti-bonding state of C_3^0 atom to the non-bonding state of C_1^0 atom (Fig. 3.10c):

$$C_3^0 + C_1^0 \rightarrow C_3^+ + C_1^- \quad (3.6)$$

The formation of C_1^- requires additional energy δ associated with the Coulomb interaction of electrons on non-bonding orbitals. Thus, if $E > \Delta + \delta$, the formation of VAPs C_3^+ and C_1^- is an exothermic process and these defects are stable configurations.

The presence of dangling bonds is not an indispensable condition for VAP formation. Figure 3.11 shows the example of **close** C_3^+–C_1^- pair formation by exciton self-trapping (Street, 1977). The rearrangement of only one bond and the associated distortions of the atoms are necessary for this transformation. The Coulomb interaction makes the energy of the close C_3^+–C_1^- pair even smaller than the energy of the distant C_3^+–C_1^- pair. The close C_3^+–C_1^- pairs are meta-stable states and may help to understand the nature of photo-structural change effect in chalcogenide glasses (see Section 6.2.1).

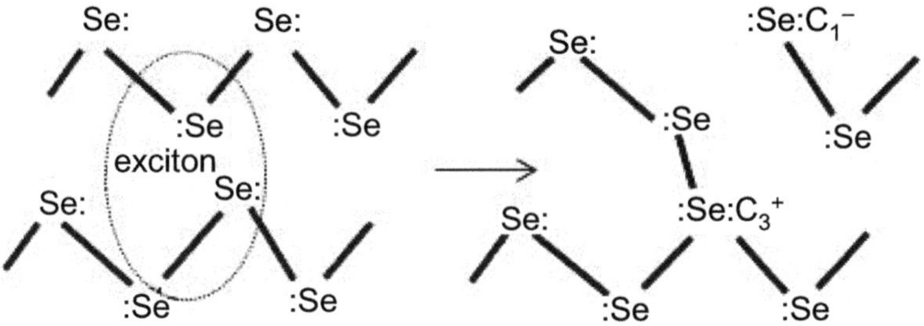

Figure 3.11 Schematic diagram of the transformation of an exciton into the close C_3^+–C_1^- pair.

A more stable configuration of close VAP states, named **intimate valence-alternation pairs** (IVAPs), has been suggested by Kastner (1977) and Kastner and Hudgens (1978). The C_3^+–C_1^- pair forms diamagnetic complex with the total charge equaling zero in this model (Fig. 3.12).

An alternative model of defect states in chalcogenide glasses is the model of **quasi-molecular defects** suggested by Popov (1980, 1981). The model is based on the electron-surplus bond concept (Pimentel and Spratley, 1970) and assumes the presence of multi-center orbital-deficiency bonds in chalcogenide glassy semiconductors. The formation of the simplest three-center quasi-molecular defect in selenium is shown in Fig. 3.13.

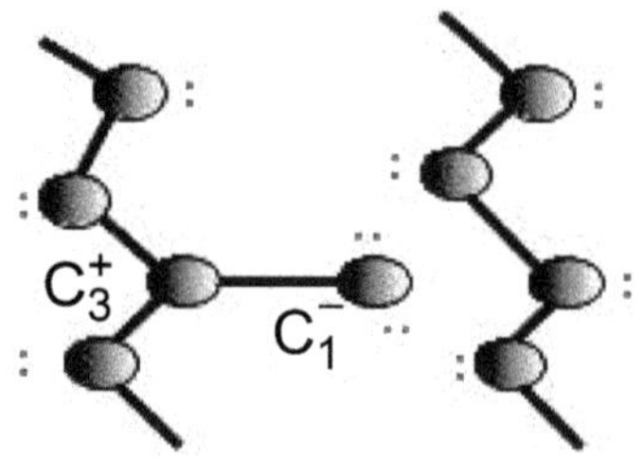

Figure 3.12 Intimate valence-alternation pair.

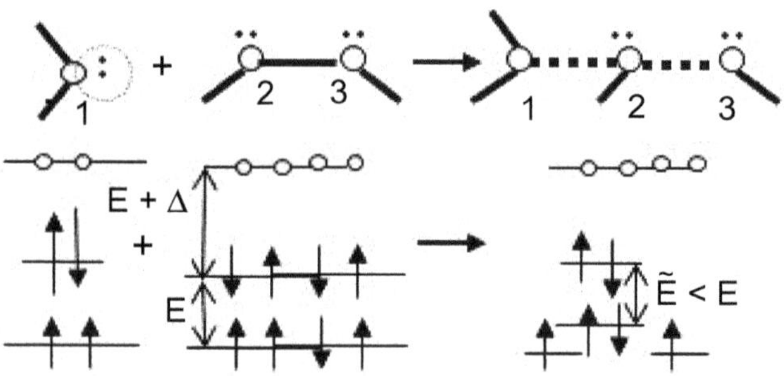

Figure 3.13 Schematic diagram of the quasi-molecular defect formation.

Two electrons from the non-bonding orbitals of atom number 1 (points in the circle in the figure) and two electrons from the bonding orbitals of atoms 2 and 3 (solid line in the figure) form an orbital-deficiency bond of three selenium atoms (dotted line in the figure). At the same time, these electrons occupy two non-bonding orbitals and two bonding orbitals (energy $\tilde{E} < E$) as before bond formation. Thus, two atoms with coordination number three and one atom with coordination number two appear instead of three twofold coordinated atoms. Quasi-molecular defect has no electrical charge but has dipole moment because of redistribution of electron density. The constant value of electrons on bonding and non-bonding orbitals as well as the absence of electrons on anti-bonding orbitals leads, in the author's opinion, to less defect formation energy in comparison to the VAP model.

Based on quantum-chemical examination, Dembovsky (2007) concluded that the minimum energy is observed in a neutral four-coordinated quasi-molecular defect Se_4^0 (Fig. 3.14). It should be noted that the four bonds of Se_4^0 are not actually equivalent. There are two "long" bonds in one line (i.e., 180° inter-bond angle) and two "short" bonds (with 100° angle). The presence of hypervalent defects may explain the higher value of the first coordination number of non-crystalline selenium (2.1–2.2) compared to the crystal modifications of the material (2.0) (see Section 2.5.1).

The VAP, IVAP, and quasi-molecular defect models do not negate but supplement one another. It is not unreasonable to suppose that all types of defect and transformation of one type of defect into another are possible in chalcogenide glassy semiconductors. Moreover, Dembowski (2007) suggested that a neutral quasi-molecular defect Se_4^0 may be formed from a pair of charged VAP defects Se_3^+ and Se_1^- (Fig. 3.14).

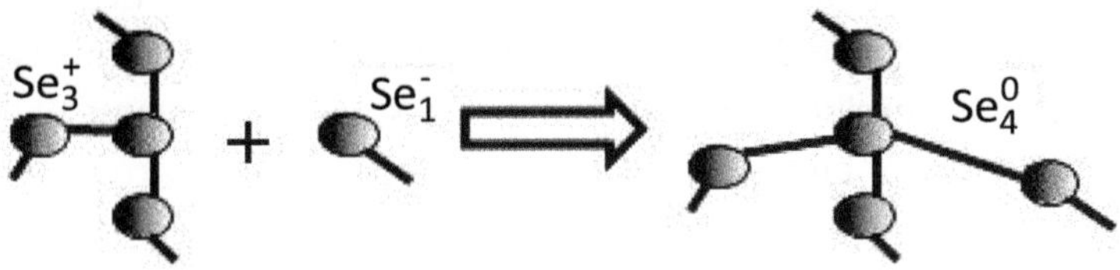

Figure 3.14 Formation of hypervalence Se_4^0 state when approaching the VAP defects Se_3^+ and Se_1^-.

The creation of defects such as VAPs is possible in the atoms of the fifth group of periodic table (arsenic, antimony) as well. But in this case, the transformation is more complicated because over-coordinated defects can only arise in the case of orbital hybridization.

The normal state of V group atom (pnictide) is ${}_pP_3^0$, where p-electrons form three covalent bonds (Fig. 3.15a). It has been shown (Elliott and Davis, 1979) that the lowest energy defects are under-coordinated defect ${}_pP_2^-$ (Fig. 3.15b) and over-coordinated defect ${}_{sp}P_4^+$. Over-coordinated defect forms due to *sp*-hybridization (Fig. 3.15c).

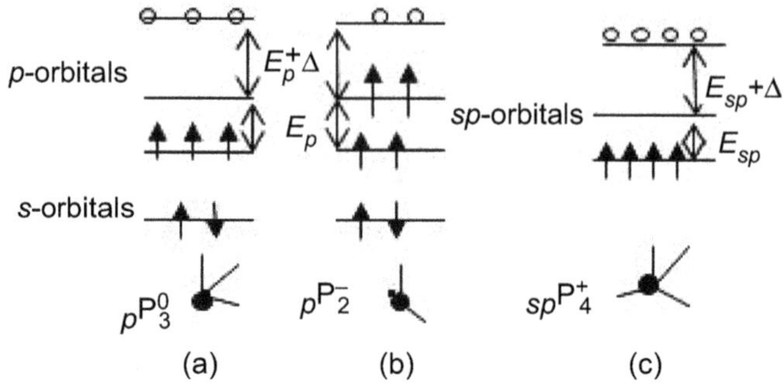

Figure 3.15 Energy levels of pnictide atom in normal (a), under- (b), and over-coordinations (c).

In the case of chalcogenide glasses that contain atoms of both group elements, all defects (C_3^+, C_1^-, P_4^+, P_2^-) may be present. But the concentration of different types of defects depends on defect formation energy. It has been shown that the main under-coordinated defect is C_1^-. Over-coordinated defects C_3^+ and P_4^+ have approximately equal values of formation energy.

The presence of lone-pair electrons on non-bonding orbitals is an *indispensable* condition for VAP formation. This is why this type of defects is possible in the case of group V and VI elements of the periodic system, but it cannot be formed on atoms of group IV in the absence of lone-pair electrons. Consequently, only under-coordinated defects may be present in this case. The electronic structure of feasible defects in amorphous silicon has been analyzed by Adler (1978). Silicon atom in normal configuration has a coordination equaling four as a result of sp^3-hybridization of electron orbitals $_{sr}T_4^0$ (Fig. 3.16a). In the author's estimate, the lowest energy defect in amorphous silicon is the neutral twofold coordinated atom $_pT_2^0$ (Fig. 3.16b). The second level of energy has charged defects $_{sp}T_2^+$ and $_pT_3^-$ (Fig. 3.16c,d). The $_{sp}T_2^+$ center contains an unpaired spin, which gives an ESR signal. It is important to note that the neutrality of the lowest energy defect signifies the positive correlation energy of its formation. It may be explained by significantly more rigid structural network of amorphous silicon in comparison with chalcogenide glassy semiconductors.

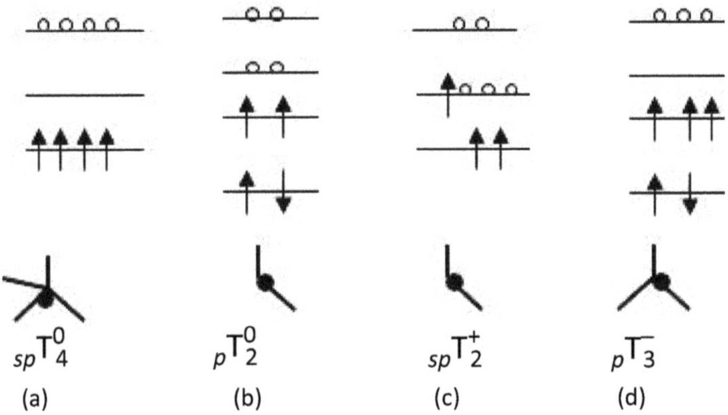

Figure 3.16 Energy levels of silicon atom in normal (a) and defect (b–d) configurations.

3.1.4 Electronic Structure of Hydrogenated Amorphous Silicon

The distinctive feature of disordered semiconductors is the weak influence of impurity on the electrical properties of these materials (see Chapter 4 for details). In the case of amorphous silicon, this fact is explained by the high density of intrinsic defects due to the presence of a large number of dangling bonds in the fourfold coordinated rigid disordered structural network. As a result, the density of localized states in the gap of amorphous silicon reaches 10^{20} cm^{-3}eV^{-1}. The concentrations of donor or acceptor levels after doping proved to be less than the intrinsic defect concentration (Fig. 3.17a) and impurities do not influence the Fermi-level position and the electrical properties of semiconductor.

Hydrogen introduction in amorphous silicon leads to the formation of silicon–hydrogen (Si–H) bonds instead of dangling silicon bonds. Bond creation is accompanied by electron transition from the non-bonding orbitals of the dangling silicon bond and hydrogen atom to the bonding orbitals. Because the energy of the Si–H bond exceeds the energy of the Si–Si bond, these bonding orbitals shift to the valence band and anti-bonding orbitals are disposed in the conduction band (Fig. 3.17b). It leads to a sharp decrease in the localized state density in the gap of hydrogenated amorphous silicon (Fig. 3.17a) and to the possibility of impurity level density exceeding that of intrinsic defect levels.

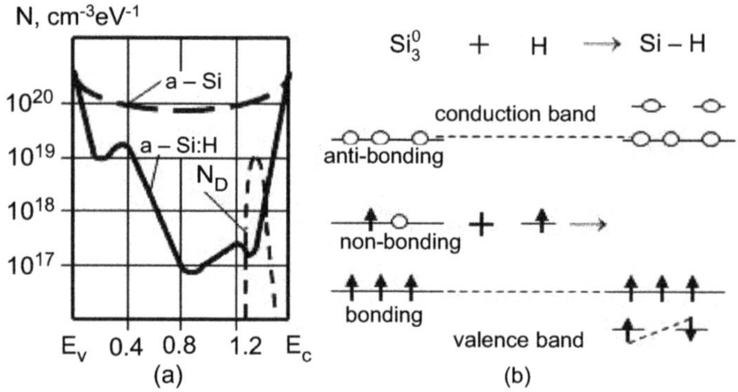

Figure 3.17 Density of localized states in the gap of a–Si and a–Si:H (a) and illustration of Si–H bond formation.

Tiedje et al. (1981) analyzed the effect of hydrogen concentration on the density of gap states in amorphous silicon films prepared at different hydrogen partial pressures. It has been shown that the density of localized states near midgap decreases almost exponentially to less than 10^{15} states cm^{-3} eV^{-1} (Fig. 3.18).

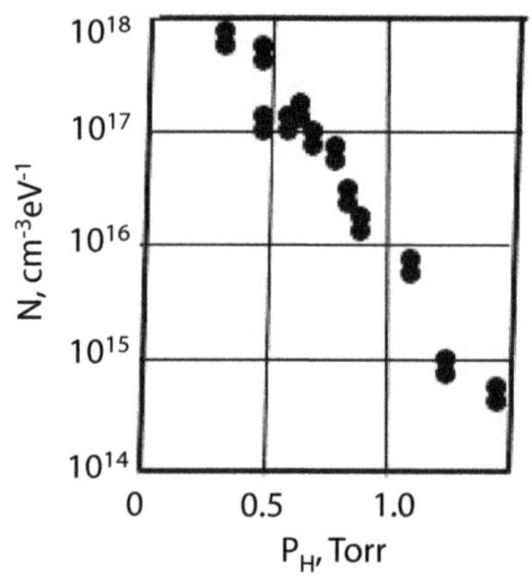

Figure 3.18 Dependence of the density of localized states near the midgap of a–Si:H on hydrogen partial pressures. Reprinted from Tiedje et al. (1981), with permission from American Physical Society.

3.1.5 Electronic Structure of Organic Semiconductors

It was noted in Section 2.5.5 that carbon atoms in organic materials usually have sp^2-hybridization of electronic orbitals. As a result, three of the four electrons of the outermost electron shell are on the sp^2-orbitals and one electron is on the *p*-orbital (Fig. 2.20b and Fig. 3.19). Electrons located on sp^2-orbitals form σ-bonds that create the backbone of molecules and occupy bonding σ-orbitals (HOMO or highest occupied molecular orbital), forming the σ-valence band. Anti-bonding σ-orbitals (LUMO or lowest unoccupied molecular orbitals) form the conduction band (Fig 3.19). Due to the significant energy of σ-bonds, the difference between the HOMO and LOMO σ-orbital energies (forbidden gap) is large. The material with such a value of forbidden gap would not have semiconductor properties. However, the electrons on the *p*-orbitals of the carbon atoms form significantly weaker π-bonds, creating bonding and anti-bonding π-orbitals (Fig. 3.19).

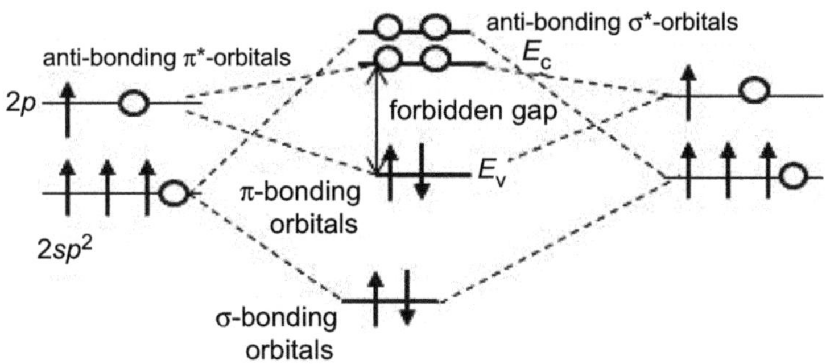

Figure 3.19 Schematic diagram of orbitals in the case of conjugated chemical bonds between two carbon atoms.

Therefore, the lowest electronic excitations of conjugated organic molecules are the π–π^* transitions with energy gap typically between 1.5 and 3.0 eV. These values correspond to semiconductor materials and lead to light absorption or emission in the visible part of the radiation spectrum (Brutting, 2005).

3.2 Electrical Properties of Disordered Semiconductors

Three physical characteristics—electrical conductivity, thermoelectric power, and Hall effect—are usually used to investigate electrical charge transport in semiconductors. Knowledge of these characteristics and their temperature, frequency, and field dependencies allows to determine practically all parameters of electrical charge transport in crystalline semiconductors: type of conductivity, charge carrier concentration, carrier mobility, activation energy of conductivity, and others.

The absence of long-range order of atom arrangement in non-crystalline semiconductors exerts strong influence on electrical transport. This is why different theoretical and experimental approaches to electrical transport analyses have been created. There is a number of electrical transport models in non-crystalline semiconductors at present, but only the most common models are considered in this section.

3.2.1 Electrical Conductivity

We will discuss the electrical conductivity on the basis of the Mott–Davis model of energy bands (see Section 3.1.2). The characteristic features of the model are as follows (Fig. 3.20):

- Presence of mobility edges at the energies E_C and E_V separating localized and extended states
- Presence of localized state tails of conduction band (between energies E_C and E_A) and valence band (between energies E_V and E_B)
- Presence of narrow band of localized states in the middle of gap (between energies E_1 and E_2)

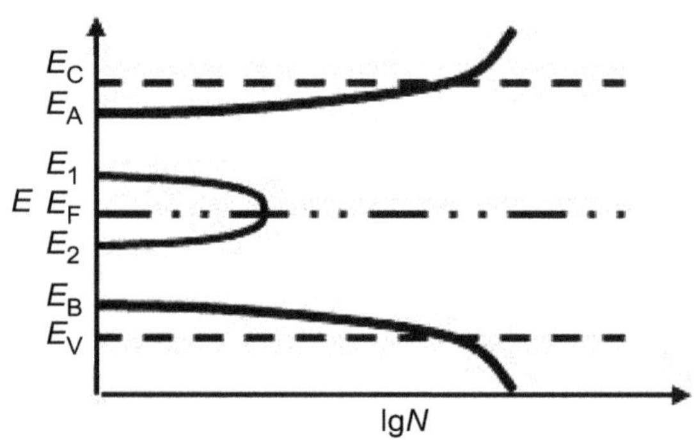

Figure 3.20 The Mott–Davis model of energy bands.

In accordance with this model, three types of electron transport are possible:

1. Carrier transport in extended (delocalized) states (carrier energy more than E_C and less than E_V)
2. Hopping conduction in the tails of conduction band and valence band (carrier energy between E_C and E_A and between E_V and E_B)
3. Hopping conduction in localized states near the Fermi level (carrier energy between E_1 and E_2)

It is clear that the contribution of each type of transport to material conductivity depends on temperature, frequency of electrical field, and so on. Let us begin from temperature dependence of direct current conductivity.

3.2.1.1 Conductivity in extended states

Conductivity in extended states is observed at a sufficiently high temperature because carriers have to excite to delocalized states. In this case, conductivity is similar to the conductivity of crystalline semiconductors. The temperature dependence of p-type conductivity is given by

$$\sigma = \sigma_0 \exp\,[-(E_F - E_V)/kT] \tag{3.7}$$

and in the case of n-type conductivity is given by

$$\sigma = \sigma_0 \exp\,[-(E_C - E_F)/kT] \tag{3.8}$$

where pre-exponential factor σ_0 = 10 – 10^3 Ω^{-1} cm^{-1} in the case of non-crystalline semiconductors. Because the value of gap has linear dependence on temperature, the temperature dependence of $(E_F - E_V)$ is given by

$$E_F - E_V = (E_F - E_V)_{T=0\,K} - \gamma T \tag{3.9}$$

where $\gamma = dE/dT$ is the band temperature coefficient. So Eq. (3.7) may be written as follows:

$$\sigma = \sigma_0 \exp(\gamma/k) \exp[-(E_F - E_V)_{T=0\,K}/kT] \tag{3.10}$$

and plot ln $\sigma(1/T)$ is the straight line with slope equal to

$$(E_F - E_V)_{T=0\,K}/k$$

and with point of ordinate axis intersection at $\sigma_0\exp(\gamma/k)$. Carrier mobility on the analogy of crystalline semiconductors slightly decreases with the increase in temperature

$$\mu \sim 1/kT \tag{3.11}$$

and lies in the range from 1 to 10 cm^2/V·s at 300 K.

3.2.1.2 Conductivity in the tails of bands

Conductivity in the tails of bands is realized by carriers excited to the states in band tails. Because these states are localized, electrical transport may be realized only by electron hopping from site to site (Fig. 3.2b) with the assistance of phonons (so-called hopping conductivity). In this situation, the mobility of carriers is much smaller in comparison to the conductivity in extended states (approximately 10^{-2} cm^2/V·s at 300 K) and depends on the value of activation energy of the hop between two sites:

$$\mu \sim (1/kT)\cdot\exp(-w_1/kT) \tag{3.12}$$

where w_1 is the activation energy of hops. Thus, there is exponential dependence of mobility on temperature in this case. As a result, the activation energy of conductivity is changed as well. The conductivity of p-type is given by

$$\sigma = \sigma_1 \exp\,[-(E_F - E_V + w_1)/kT] \tag{3.13}$$

and n-type conductivity is given by

$$\sigma = \sigma_1 \exp\,[-(E_A - E_F + w_1)/kT] \tag{3.14}$$

where the pre-exponential factor is $\sigma_1 = (10^{-2} - 10^{-4})\sigma_0$ due to low mobility value. So in this case, the activation energy of conductivity is the sum of the activation energy of carrier concentration ($E_F - E_V$) and the activation energy of mobility (w_1).

3.2.1.3 Conductivity in localized states at the Fermi level

Conductivity in localized states at the Fermi level is a hopping conductivity as well. There are two types of carrier transport in this case: nearest-neighbor hopping and variable-range hopping.

If temperature is high enough, then hopping is most probable between nearest neighbors with activation energy w_2 (Fig. 3.21) because the transfer integral between these sites is rather large. This hopping is called nearest-neighbor hopping. This type of hopping is similar to the conduction through impurity levels in heavily doped compensated crystalline semiconductors. The temperature dependence of conductivity due to nearest-neighbor hopping has activation character and may be written as follows:

$$\sigma = \sigma_2 \exp\left(-w_2/kT\right) \tag{3.15}$$

where the pre-exponential factor $\sigma_2 \leq \sigma_1$ and the activation energy of hops w_2 equals approximately half of the localized states zone width in the middle of the gap ($w_2 = (E_1 - E_2)/2$).

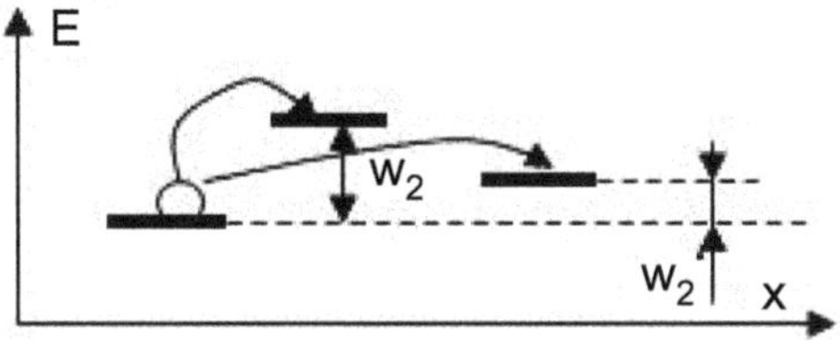

Figure 3.21 Schematic illustration of nearest-neighbor and variable-range hopping.

At low temperature, the probability of nearest-neighbor hopping decreases. In this situation, hopping between two sites whose transfer integral is smaller, but activation energy (w_2') is smaller as well, becomes probable (Fig. 3.21). The conductivity temperature dependence for this case, given by Mott, is as follows:

$$\sigma = \sigma_3 \exp[-(T_0/T)^{1/4}] \tag{3.16}$$

where σ_3 and T_0 are constants.

The total conductivity of material is determined as an integral extended along all electron states. In the case of Boltzmann distribution of electrons above the Fermi level, it is given by

$$\sigma = e\int N(E)f(E)\mu(E)dE \tag{3.17}$$

where $f(E)$ is the Boltzmann function and e is the electron charge. The energy dependencies of equation terms (3.17) are shown in Fig. 3.22. The Boltzmann function is given for different temperatures $T_1 > T_2 > T_3$. At temperature T_1, charge carriers are excited to delocalized states ($E > E_C$). The density of states $N(E)$ and carrier mobility $\mu(E)$ are large enough in this case and conductivity $\sigma(E)_{T1}$ is high as well. At lower temperature T_2, there are no carriers in delocalized states, but there are carriers in the conduction band tail ($E_C > E > E_A$). However, the density of states and carrier mobility are significantly less in this case and engendered by these carriers, conductivity ($\sigma(E)_{T2}$) is lower too. At the lowermost temperature T_3, conductivity $\sigma(E)_{T3}$ is carried out by carriers at the states near the Fermi level.

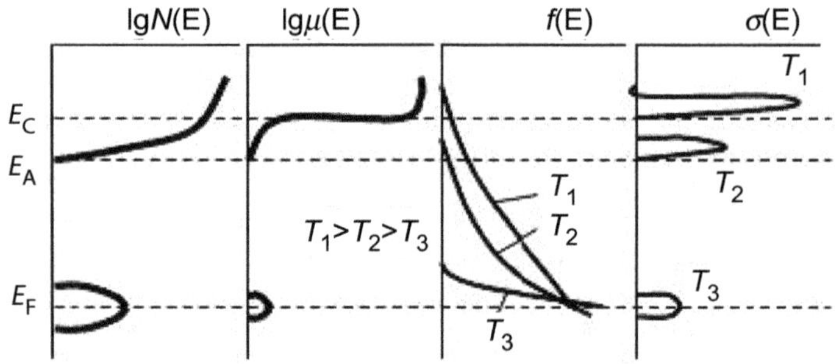

Figure 3.22 Energy dependencies of functions from Eq. (3.17).

The temperature dependence of non-crystalline semiconductor conductivity in a wide temperature range is shown in Fig. 3.23. There are four regions on this curve:

1. At high temperature, conductivity in extended states is observed (Eqs. 3.7 and 3.8).
2. The next region is the hopping conduction in the band tails (Eqs. 3.13 and 3.14).
3. The third region corresponds to nearest-neighbor hopping (Eq. 3.15).
4. The last one is variable-range hopping (Eq. 3.16).

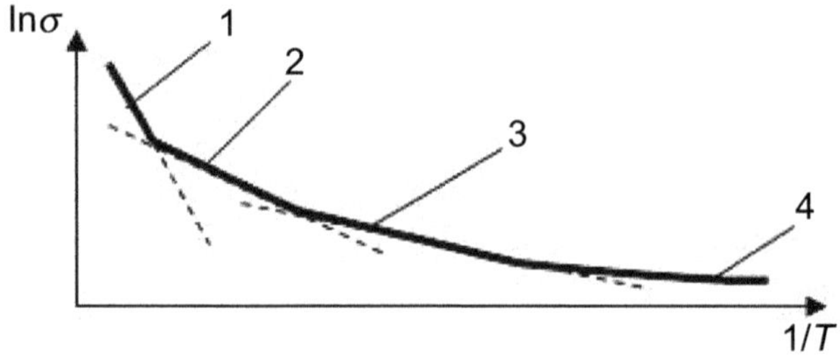

Figure 3.23 Schematic illustration of electrical conductivity temperature dependence of non-crystalline semiconductors (1: conductivity in extended states; 2: hopping conduction in the band tails; 3: nearest-neighbor hopping; and 4: variable-range hopping).

The types of electron transport described above determine features of conductivity frequency dependence of non-crystalline semiconductors. In the case of carrier transport in extended states (energy more than E_C and less than E_V), the frequency dependence of conductivity is described as follows:

$$\sigma(\omega) = \sigma(0)/(1 - \omega^2\tau^2) \tag{3.18}$$

where $\sigma(0)$ is the direct current conductivity, ω is the angular frequency, and τ is the relaxation time. The relaxation time in covalent-bonded materials equals to 10^{-15} s approximately because of the electronic nature of polarization. It means that there is no conductivity dependence on frequency in the radio frequency band in this case.

Logically, hopping conduction in the band tails (energy between E_C and E_A and between E_V and E_B) and in localized states near the Fermi level (energy between E_1 and E_2) has to depend on frequency. In the case of direct current, the carrier has to pass through the sample from one electrode to the other. The direct current resistance is determined by the biggest barrier value between neighbor's localized states. An increase in frequency leads to a decrease in the number of localized states and barriers, which the carrier passes during half-cycle. This is way the frequency dependence of hopping conductivity is given by

$$\sigma(\omega) = C\omega^{-S} \tag{3.19}$$

where C is a constant and $S = 0.7 - 1.0$.

The frequency dependencies of conductivity in the case of different types of electron transport are shown in Fig. 3.24.

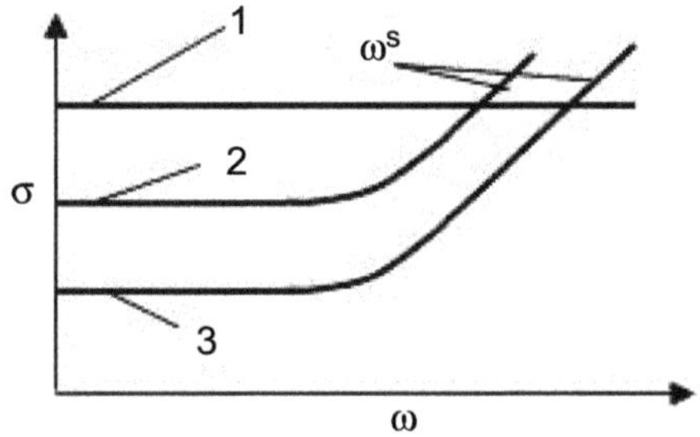

Figure 3.24 Schematic illustration of conductivity frequency dependences in the case of carrier transport in extended states (1), band tails (2), and near the Fermi level (3).

In practice, the type of electron transport providing the largest value of conductivity at a given temperature will dominate.

3.2.2 Thermoelectric Power

Thermoelectric power (S) is determined as the ratio of the voltage value (ΔU) between two points of a semiconductor at different temperatures and the temperature difference (ΔT) of these points. In the case of n-type crystalline semiconductors, thermoelectric power is given by

$$S = \Delta U/\Delta T = (k/e)[(E_C - E_F)/(kT) + A] \tag{3.20}$$

where k is the Boltzmann's constant, A is a coefficient depended on the carrier scattering mechanism (A = 2–4 for crystalline semiconductors), and e is the electron charge. Thermoelectric power of n-type semiconductors is negative because electron charge is negative. In the case of p-type crystal, it is necessary to change ($E_C - E_F$) to ($E_F - E_V$) and to change the negative electron charge to positive hole charge in Eq. (3.20). As a result, thermoelectric power of p-type semiconductors is positive. Therefore, the thermoelectric power polarity sign gives information about the conduction type of a semiconductor.

In the case of carrier transport in extended states of non-crystalline semiconductors, the value of coefficient A in Eq. (3.20) changes because carrier scattering increases as a result of absence of long-range order. As a rule, A = 1 in this case, and thermoelectric powers for semiconductors of n- and p-type are given by

$$S = (k/e)\ [(E_C - E_F)/(kT) + 1] \tag{3.21}$$

$$S = -(k/e)\ [(E_F - E_V)/(kT) + 1] \tag{3.22}$$

Transition to carrier transport in band tails leads to further lowering of coefficient A value (A < 1) owing to a decrease in the density of states. On the other hand, $(E_C - E_F)/(kT) >> 1$ and $(E_F - E_V)/(kT) >> 1$ because the Fermi level lies near the middle of the gap. Therefore, it is possible to ignore coefficient A. Besides, energies E_C and E_V in Eqs. (3.21) and (3.22) have to change to E_A and E_B (Fig. 3.20). As a result, we obtain for n- and p-type conduction

$$S = (k/e)(E_A - E_F)/(kT) \tag{3.23}$$

$$S = -(k/e)(E_F - E_B)/(kT) \tag{3.24}$$

Equations (3.21)–(3.24) show that the temperature dependency of thermoelectric power is determined by the temperature dependency of carrier concentration both for carrier transport in extended states ($E_C - E_F$ and $E_F - E_V$) and for carrier transport in band tails ($E_A - E_F$ and $E_F - E_B$). In the case of electrical conductivity (Eqs. 3.7, 3.8, 3.13, and 3.14), this statement is correct only for carrier transport in extended states ($E_C - E_F$ and $E_F - E_V$). If carrier transport is in band tails, conductivity activation energy equals the sum of carrier concentration and carrier mobility activation energies ($E_F - E_V + w_1$ and $E_A - E_F + w_1$). Therefore, it is possible to get information about the carrier-transport mechanism from the comparison of conductivity and thermoelectric power temperature dependencies. If dependencies ln $\sigma(1/T)$ and $S(1/T)$ have the same absolute values of slope, it means that carrier transport in extended states has taken place. In the case of different absolute values of slope, carrier transport is in band tails and slope difference gives the hop activation energy w_1.

In the case of hopping conductivity in localized states at the Fermi level, thermoelectric power may be calculated by the expression used for metals (because electrons move inside energy interval kT near the Fermi level in both cases):

$$S = \frac{\pi^2}{3} \cdot \frac{k^2 T}{e} \frac{d(\ln \sigma)}{dE}\bigg|_{E=E_F} \qquad (3.25)$$

3.2.3 Hall Effect Anomaly

The Hall effect consists of electron trajectory deflection in the transverse magnetic field due to the Lorentz force and the appearance of potential differences in the sample:

$$U = R_H(BI/d) \qquad (3.26)$$

where B is the magnetic induction, I is the electric current, d is the sample thickness, and R_H is the Hall coefficient, given by

$$R_H = r/en \qquad (3.27)$$

where r is a factor depending on the scattering mechanism ($r \approx 1$), e is the electron charge, and n is the carrier concentration. The Hall effect measurement gives extensive information about electrical transport in crystalline semiconductors: type and concentration of charge carriers (Hall coefficient sign is negative for n-type and positive for p-type crystals).

However, the Hall effect measurement results show a number of anomalies caused by the random structural network. The Hall coefficient sign is negative in chalcogenide glassy semiconductors despite thermopower measurements of these materials indicating p-type of conductivity. Moreover, double reversal of R_H is observed in hydrogenated amorphous silicon. The Hall coefficient is positive in phosphorus-doped (n-type) a–Si:H and negative in boron-doped (p-type) material. A number of models have been proposed to explain these anomalies (see, e.g., Elliott, 1984; Okamoto, Hattori, and Hamakawa, 1993). But we just point out here that measurement of the Hall effect is not a good tool for determination of the charge sign of carriers in non-crystalline semiconductors.

3.2.4 Time of Flight Method

An effective method of determination of localized states density distribution in the band tails of non-crystalline semiconductors is the drift mobility measurement by the time of flight experiment. The diagram of the experiment is shown in Fig. 3.25. A voltage pulse is applied to the sample with two blocking electrodes (one of which is a semi-transparent electrode). As a result, an electric field is induced in the sample. The sample is exposed to a short pulse of strong absorbed light through the semi-transparent electrode. Electrons and holes are created in the sample near the semi-transparent electrode due to light absorption. Positive or negative carrier packet (depending on voltage polarity) drifts across the sample under electric field action.

Electric current flows in the external circuit during carrier packet drift. Drift mobility is expressed by

$$\mu_D = d/F \cdot t_D \tag{3.28}$$

where d is the sample thickness, F is the electric field intensity, and t_D is the time of carrier flight. In the ideal case of Gaussian conduction, carrier packet reaches the opposite electrode in time t_D and the electric current in the external circuit drops to zero sharply.

In the case of non-crystalline semiconductors, localized states in band tail act as carrier traps, which capture carriers. In turn, some carriers excite back from the traps to the band. The probability of the last process depends on the localized state energy and measurement temperature. The processes of carrier capture and release lead to the spread of carrier packet. The time dependence of carrier concentration is expressed by

$$n_D(t) = n_D(0)\cdot\exp(-t/\tau_D) \tag{3.29}$$

where $n_D(0)$ is the carrier concentration at $t = 0$ and τ_D is the average carrier lifetime. The time dependence of drift current in this case is shown in Fig. 3.26a. One may see that it is impossible to determine the time of flight from this curve. However, it has been shown by Pfister and Scher (1978) that in the case of log–log plot, this dependence has two linear parts (Fig. 3.26b). The transit time of carrier packet's center of gravity is determined by the point of intersection of these linear parts. This value can be used as a parameter t_D in Eq. (3.28). As a result, carrier drift mobility may be determined from expression (3.28).

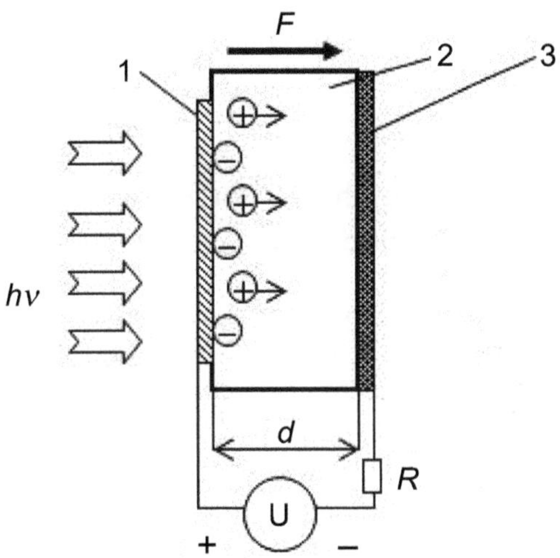

Figure 3.25 Diagram of the time of flight experiment (1 is semi-transparent electrode, 2 is semiconductor film, 3 is electrode, *d* is film thickness, and *F* is electric field intensity).

On the other hand, carrier drift mobility may be expressed by

$$\mu_D = \mu_\sigma(N_C/N_t)\exp[-(E_C - E_t)/kT] \tag{3.30}$$

where μ_σ is Hall mobility, N_C and N_t are density of states at E_C and E_t levels. Hence, it is possible to determine the energy position of localized states ($E_C - E_t$) from the slope of ln $\mu_D(1/T)$ dependence. The point of intersection of the curve and the axis of ordinates gives the concentration of traps at the given energy. Therefore, application of the time of flight method in a wide temperature range gives information about the localized state density distribution in the band tail.

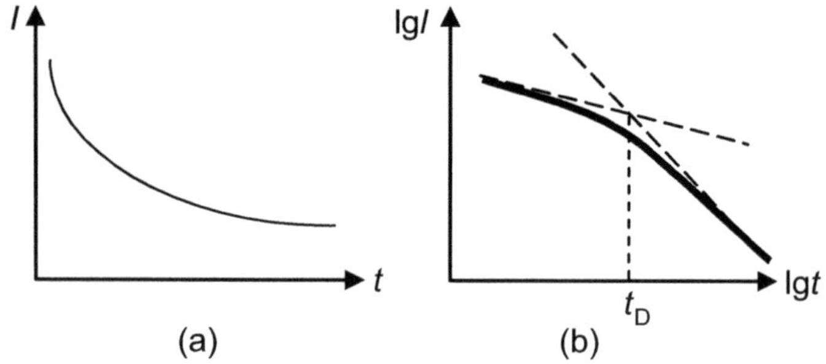

Figure 3.26 Schematic illustration of drift current dependence on time in the case of linear (a) and log–log (b) plots.

3.2.5 Features of the Charge Carrier Transport in Organic Semiconductors

As noted in Section 2.5.5, electrons in organic semiconductors are localized on the molecules. Therefore, if charge carriers move along the delocalized zone within the molecule, mobility should be 1–10 cm^2/V·s at the temperature of 300 K. But in fact mobility in organic semiconductors is typically less than 1 cm^2/V·s. It can be concluded that the mobility of charge carriers and hence the electrical conductivity in organic semiconductors are determined by the overlap of molecular orbitals of the neighboring molecules. In this case, the charge transport process is described by either hopping or tunneling mechanisms (Coropceanu et al., 2007) and the mobility of charge carriers can be expressed as the sum of two parts:

$$\mu = \mu_{tun} + \mu_{hop} \qquad (3.31)$$

where the first term is due to electron tunneling and dominates transport at low temperatures and the second term is related to hopping process and becomes dominant at higher temperatures.

For the majority of organic semiconductors, the temperature dependence of conductivity is described by the usual exponential expression

$$\sigma = \sigma_0 \cdot \exp(-\Delta E/kT) \quad (3.32)$$

where the activation energy is the sum of the activation energies of concentration (ΔE_n) and mobility (ΔE_μ) of charge carriers:

$$\Delta E = \Delta E_n + \Delta E_\mu$$

which corresponds to the hopping conductivity mechanism.

The activation energy of mobility is equal to the energy of hops as it is in inorganic semiconductors (see Eq. 3.12). At the same time, the activation energy of concentration of charge carriers, ΔE_n, in organic semiconductors differs from ΔE_n in inorganic materials.

It is known that an exciton may be formed at the generation of electron–hole pairs due to the Coulomb interaction between electrons and holes. The energy of the Coulomb interaction between the two charges is described by

$$E_{ex} = q_1 \cdot q_2 / 4\pi \cdot \varepsilon \cdot \varepsilon_0 \quad (3.33)$$

where q_1 and q_2 are values of charges and ε is the dielectric constant of the material.

The value of dielectric constant is large enough in inorganic semiconductors. As a result, the binding energy of the formed exciton (Wannier exciton) is small (~10 meV) and the exciton spontaneously destroys, forming free electrons and holes. The dielectric constant value is considerably lower in organic materials. Consequently, Frenkel excitons with binding energies up to 1 eV appear and need additional energy E_{ex} to create free charge carriers. Thus, in the case of organic semiconductors, the activation energy of charge carrier concentration is determined by (for p-type material)

$$\Delta E_n = E_F - E_V + E_{ex} \quad (3.34)$$

Therefore, it is necessary to take into account the existence of excitons in the description of charge transport processes in organic semiconductors.

3.3 Optical Properties of Disordered Semiconductors

In the vast spectrum of electromagnetic waves (from angstroms to kilometers), light or the visible part of electromagnetic radiation takes up only a small part from 0.40 to 0.75 microns. It corresponds to frequencies (7.5 ÷ 4.0) × 10^{14} Hz and photon energies from 3.1 to 1.7 eV. The interaction of light flux (I_0) with a homogeneous solid includes reflection (I_0R where R is the reflection factor) and absorption (A). As a result, the light flux intensity may be written as follows:

$$I = I_0(1 - R) \exp[-\alpha(\hbar\omega)x] \tag{3.35}$$

where $\alpha(\hbar\omega)$ is the absorption coefficient, $\hbar\omega$ is the photon energy, $\hbar$ is Planck's constant, and ω is the frequency. Thus, the light flux is expressed by

$$I_0 = I_0R + I_0A + I_0T \tag{3.36}$$

where T is a part of the flux transited through a sample, or

$$R + A + T = 1 \tag{3.37}$$

Interaction of the light flux with the solid may be described through the refractive index as well. The complex refractive index is given by

$$n(\hbar\omega) = n(\hbar\omega) + ik(\hbar\omega) \tag{3.38}$$

where $n(\hbar\omega) = c/[v(\hbar\omega)]$ is the real part of the refractive index, $c = 3 \times 10^5$ km/s is the light velocity in vacuum, v is the light velocity in the solid, and $k(\hbar\omega)$ is the imaginary part of the refractive index. It correlates with the absorption coefficient as follows:

$$\alpha(\hbar\omega) = 4\pi k(\hbar\omega)/\lambda \tag{3.39}$$

where λ is the wavelength.

The main type of optical absorption in semiconductor materials is fundamental absorption. It leads to electron excitation from the valence band into the conduction band. The probability of photon absorption is determined by photon energy (it has to be no less than the optical gap value $\hbar\omega \geq E_{opt}$) and by the density of states distribution in the valence band ($N(E)_V \sim E^p$) and in the conduction band ($N(E)_C \sim E^q$). In the case of parabolic density of states,

$p = q = 1$, and the dependence of absorption coefficient on photon energy is given by

$$\alpha n\hbar\omega = P(\hbar\omega - E_{opt})^m \tag{3.40}$$

where n is the refractive index and P is a constant. The index of a power m = ½ in the case of direct transitions and m = 2 in the case of indirect transitions (transitions that are accompanied by phonon emission). Indirect transitions predominate at the optical absorption edge when photon energy is close to the gap value. Thus, the absorption coefficient of crystalline semiconductors depends on photon energy as square-law characteristic in this case. The $(\alpha n\hbar\omega)^{1/2} - (\hbar\omega)$ plot yields a straight line extrapolation, which, at its intersection with the abscissa axis, gives the value of the optical gap because $(\hbar\omega) = E_{opt}$ at $(\alpha n\hbar\omega)^{1/2} = 0$.

In contrast to crystals, dependence of disordered semiconductor absorption coefficient on photon energy has four segments (Fig. 3.27). Portion A is the region of fundamental absorption. Portion B is the optical absorption edge where α depends on photon energy as square-low characteristic. These regions are similar to crystals. However, in contrast to sharp subsequent decrease in absorption coefficient in crystals (dotted line in Fig. 3.27), the square-low characteristic substitutes for the exponential one in disordered semiconductors

$$\alpha = P\exp[-(\gamma/T^*)(E_{opt} - \hbar\omega)] \tag{3.41}$$

where P and γ are constants and parameter T^* is defined as follows:

$$T^* = T \text{ if } T > T_{cr}$$

$$T^* = T_{cr} \text{ if } T \leq T_{cr}$$

where T_{cr} is some characteristic temperature of the material. As a result, the exponential tail of the absorption edge arises (portion C), which is called the Urbach tail.

The exponential tail of the absorption edge is generally observed in non-crystalline semiconductors and in some crystals as well (e.g., in crystals of trigonal selenium and cadmium sulfide). The origin of the Urbach tail does not have a single explanation. In the case of non-crystalline semiconductors, it is tempting to explain the Urbach tail by electron transitions between the extended states of one band and the localized states of another band tail due to the exponential dependence of localized state density in the tail. But it was pointed out by Mott and Davis (1979) that this simple explanation is hardly probable. The optical absorption spectra of eight non-crystalline semiconductors at room temperature are shown in Fig. 3.28. As one can see from the figure, curve slopes are approximately the same in the case of essentially different materials. It is difficult to assume that band tails are identical to such a degree in these materials. So there has to be a more general reason for exponential tail origin. As a rule, the Urbach tail is explained by the presence of exciton absorption or by the influence of internal electric field on absorption.

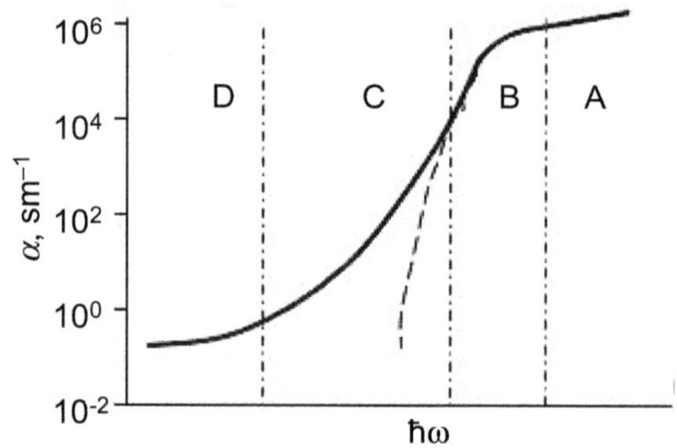

Figure 3.27 Schematic illustration of absorption coefficient dependence on photon energy.

Portion *D* in the dependence of absorption coefficient on photon energy (Fig. 3.27) is caused by optical transitions between localized states in the middle of the gap and extended states in the conduction or valence band.

As mentioned above, the value of the optical gap in the case of crystalline semiconductors may be estimated by the extrapolation of the $(\alpha n\hbar\omega)^{1/2}$–$(\hbar\omega)$ plot to the abscissa axis. It is possible to use the same procedure in the case of non-crystalline semiconductors. But the absorption edge length (portion *B*) is rather small in this case due to the presence of the Urbach tail. That is why extrapolation may produce a considerable error. If the exponential tail is observed in the optical absorption spectrum, the value of the optical gap may be found by the so-called Urbach focus point position. A schematic illustration of the absorption coefficient spectral dependence in the

Urbach tail region at different temperatures is shown in Fig. 3.29. Taking into account the linear temperature dependence of the gap value, expression (3.41) at $T > T_{cr}$ may be given as follows:

$$\alpha = P\exp\left[-\{\gamma(E_{opt}(0) - \hbar\omega - \beta T)/T\}\right] \tag{3.42}$$

where $E_{opt}(0)$ is the optical gap value at 0 K and β is the temperature coefficient of gap value.

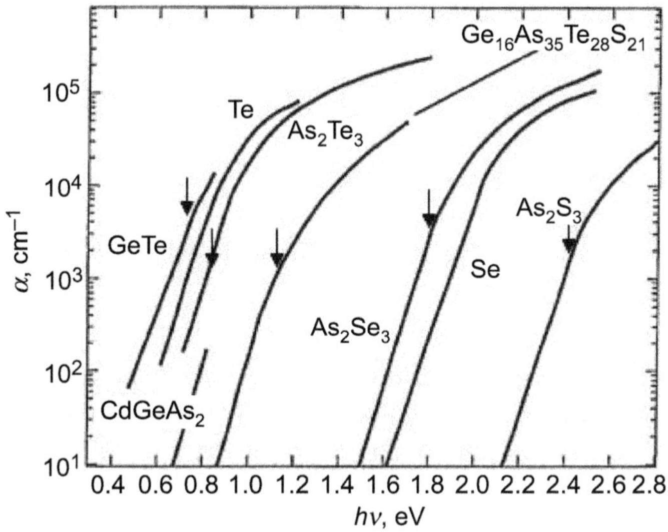

Figure 3.28 Optical absorption exponential tail of non-crystalline semiconductors at room temperature. The arrows indicate doubled activation energies of electrical conductivity (Mott and Davis, 1979).

In the case of $\hbar\omega = E_{opt}(0)$, absorption coefficient $[\alpha = P\exp(\gamma\beta)]$ does not depend on temperature (focus point A in Fig. 3.29), and the position of this point gives the value of the optical gap. This method is applicable to crystals and amorphous silicon. But in the case of chalcogenide glassy semiconductors, the characteristic temperature T_{cr} as a rule is more than the glass transition temperature T_g. As a result, expression (3.41) transforms to

$$\alpha = P\exp\left[-\delta(E_{opt} - \hbar\omega)\right] \tag{3.43}$$

where $\delta = \gamma/T_{cr}$ is constant and the absorption coefficient spectrum in the Urbach tail region does not depend on the temperature for solid materials. This does not allow to find the Urbach focus point and the value of the optical gap.

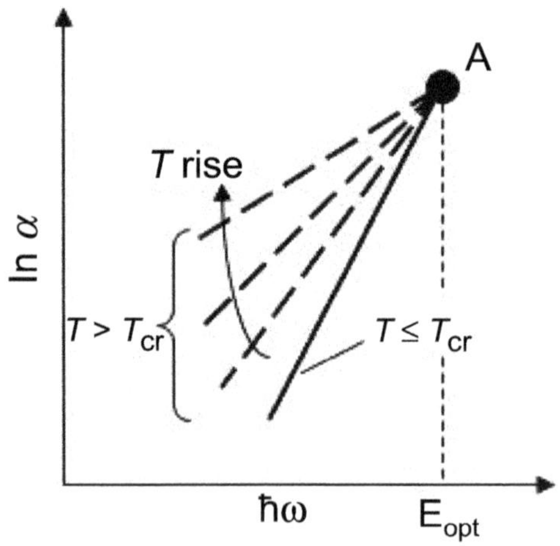

Figure 3.29 Schematic illustration of the absorption coefficient spectral dependence in the Urbach tail region at different temperatures (*A* is the Urbach focus point).

There is another method to estimate the optical gap value in non-crystalline semiconductors. It is based on determining the optical absorption coefficient value for photon energy equaling the value of the mobility gap. The latter is determined by electrical measurements. It is known that the Fermi level lies nearly in the middle of the mobility gap in non-crystalline semiconductors. Consequently, the activation energy of conductivity in extended states is equal to approximately half of the mobility gap value ($E_C - E_F = E_\sigma$). It means that the optical gap value has to correspond to photon energy equal to twice conductivity activation energy ($\hbar\omega = 2E_\sigma$). The optical absorption coefficient at room temperature is equal to approximately 10^3 cm^{-1} at this photon energy for many non-crystalline semiconductors (Fig. 3.28). So the value of the optical gap is determined as the photon energy corresponding to the optical absorption coefficient $\alpha = 10^3$ cm^{-1}. The optical band gap determined in this manner is larger by 0.1–0.2 eV than the value found from the approximation of the parabolic portion of the spectral dependence of the optical absorption coefficient. This is probably due to the fact that the temperature dependence of the mobility gap is not taken into account. One can see that this method is rather imprecise as well.

3.4 Photoelectrical Properties of Disordered Semiconductors

The photoelectrical properties of semiconductors are based on electric conductivity changes under the direct influence of light. If photon energy corresponds to regions *A* or *B* of the absorption coefficient spectral dependence (Fig. 3.27), absorption of photons leads to the generation of electron–hole pairs. As a result, charge carrier concentration and the electric conductivity of the semiconductor increase. The total electric conductivity of the semiconductor under illumination is given as follows:

$$\sigma = \sigma_{\mathrm{d}} + \sigma_{\mathrm{ph}} = e\mu_{\mathrm{n}}(n + \Delta n) + e\mu_{\mathrm{p}}(p + \Delta p) \qquad (3.44)$$

where $\sigma_{\mathrm{d}} = e\mu_{\mathrm{n}}n + e\mu_{\mathrm{p}}p$ is dark conductivity, $\sigma_{\mathrm{ph}} = e\mu_{\mathrm{n}}\Delta n + e\mu_{\mathrm{p}}\Delta p$ is photoconductivity, and Δn and Δp are excess charge carrier concentrations.

The time dependence of electron concentration is determined by the continuity equation

$$\frac{\partial n}{\partial t} = -\mathrm{div}\left(\frac{j_n}{e}\right) + G_n - \frac{\Delta n}{\tau_n} \qquad (3.45)$$

where G_n is the carrier-generation rate and j_n is the electron current density. The right part of Eq. (3.45) shows electron concentration changes owing to electric current (the first term), charge carrier generation (the second term), and charge carrier recombination (the third term). Where electric current equals zero ($j_n = 0$), Eq. (3.45) transforms to

$$\frac{\partial n}{\partial t} = G_n - \frac{\Delta n}{\tau_n} \qquad (3.46)$$

And in the case of constant conditions $\left(\frac{\partial n}{\partial t} = 0\right)$, Eq. (3.46) transforms to

$$G_n = \Delta n/\tau_n \qquad (3.47)$$

The excess charge carrier concentration may be written as follows:

$$\Delta n = G_n\tau_n \qquad (3.48)$$

The photocarrier-generation rate is expressed by

$$G_n = I\alpha\eta \qquad (3.49)$$

where $I = I_0(1 - R)$ is the light flux inside the sample, I_0 is the incident light flux, α is the absorption coefficient, and η is the quantum efficiency. The quantum efficiency is determined as a ratio between the number of generated photocarriers and the number of absorbed photons and has value less than one usually. Taking into account expression (3.49), excess charge carrier concentration is given by

$$\Delta n = I\alpha\eta\tau \tag{3.50}$$

and photocurrent density is expressed as follows:

$$j_{ph} = e\Delta n\mu E = eI\alpha\eta\tau\mu E \tag{3.51}$$

where μ is the carrier mobility and E is the electric field intensity. On the other hand, the photocurrent density may be expressed from Ohm's law by

$$j_{ph} = \sigma_{ph}E \tag{3.52}$$

Photoconductivity may be determined from Eqs. (3.51) and (3.52) as follows:

$$\sigma_{ph} = eI\alpha\eta\tau\mu. \tag{3.53}$$

As one can see from Eq. (3.53), photoconductivity depends on four factors: light flux intensity [$\sigma_{ph} = f(I)$], light flux spectral characteristic [$\sigma_{ph} = f(\alpha)$, $\alpha = f(\hbar\omega)$], temperature [$\sigma_{ph} = f(\alpha\mu\tau)$, α, μ, $\tau = f(T)$], and electric field intensity [$\sigma_{ph} = f(\eta)$, $\eta = f(E)$]. Let us examine these dependencies.

3.4.1 Dependence on Light Flux Intensity

It has been shown from experiment that in the case of relatively small light flux intensity, photoconductivity varies directly with it in accordance with Eq. (3.53). But at high flux intensity, photoconductivity rise becomes slower and the dependence is given by

$$\sigma_{ph} = AI^n \tag{3.54}$$

where A is constant and $n = 0.5$–1.0. The photocurrent dependence on light flux intensity is shown in Fig. 3.30. The reason for deviation from linear dependence is in the increase in excess charge carrier concentration. At high flux intensity, it becomes considerably higher in comparison with equilibrium carrier concentration ($\Delta n >> n$). It leads to a rise in the photocarrier recombination rate and to a decrease in their lifetime. As a result, the lifetime in Eq. (3.53) becomes a function of flux intensity too [$\alpha = f(I)$].

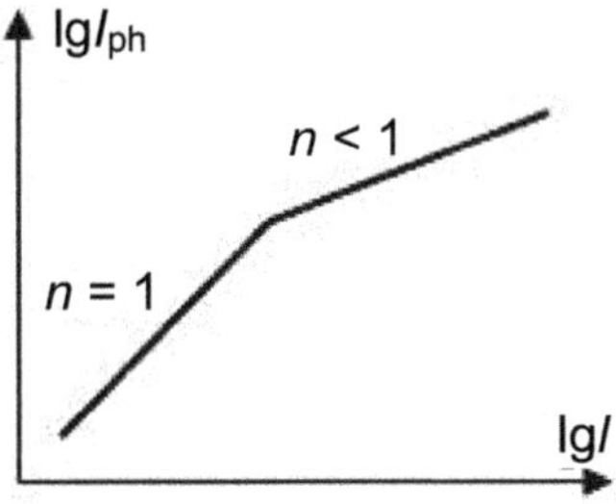

Figure 3.30 Schematic illustration of photocurrent dependence on light flux intensity.

3.4.2 Dependence on Spectral Characteristics

The dependence of photoconductivity on photon energy and the spectral dependence of absorption coefficient are shown in Fig. 3.31.

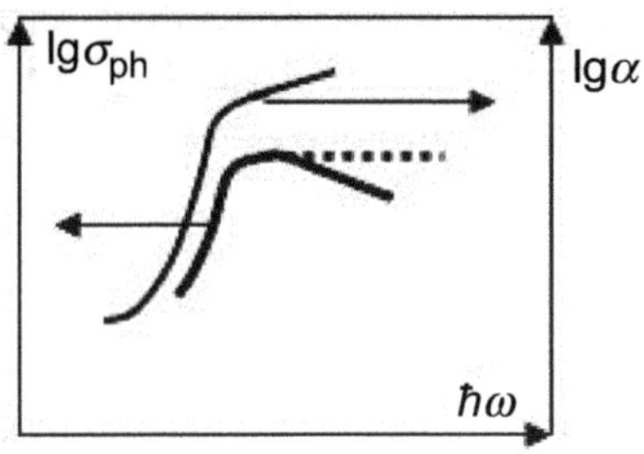

Figure 3.31 Schematic illustration of photoconductivity and absorption coefficient spectral dependences.

One can see from the figure that the dependence $\lg\sigma_{\mathrm{ph}}(\hbar\omega)$ has two considerably different parts. In the region of optical absorption, edge photoconductivity repeats the spectral dependence of absorption coefficient in accordance with Eq. (3.53). However, all photons are absorbed in the sample at a certain value of absorption coefficient and further rise in photon energy does not lead to photoconductivity augmentation in spite of absorption coefficient increase. It is possible to assume that photoconductivity has to be constant in this region of photon energy (dotted line in Fig. 3.31). But experiment indicates reduction in photoconductivity.

The decrease in photoconductivity is connected with surface effects. The increase in the value of absorption coefficient results in a decrease in the photon absorption free path, so absorption occurs nearer the surface, which has higher concentration of defect states. It leads to a rise in the photocarrier recombination rate. The effect is observed in both crystalline and non-crystalline semiconductors, and it is more apparent in crystals because of lesser concentration of defect states in the volume.

3.4.3 Dependence on Temperature

The temperature dependence of photoconductivity is shown in Fig. 3.32. One can see that the dependence has two different regions. At low temperatures when $\sigma_{ph} > \sigma_d$, photoconductivity increases with rising temperature (region I). However, at high temperatures when dark conductivity exceeds photoconductivity ($\sigma_{ph} < \sigma_d$), the latter decreases with rising temperature (region II).

There are three temperature-dependent terms in Eq. (3.53): optical absorption coefficient (α), carrier lifetime (τ), and carrier mobility (μ). It is necessary to note that carrier drift mobility (μ_D, expression 3.28) has to be analyzed in the presence of a strong electric field.

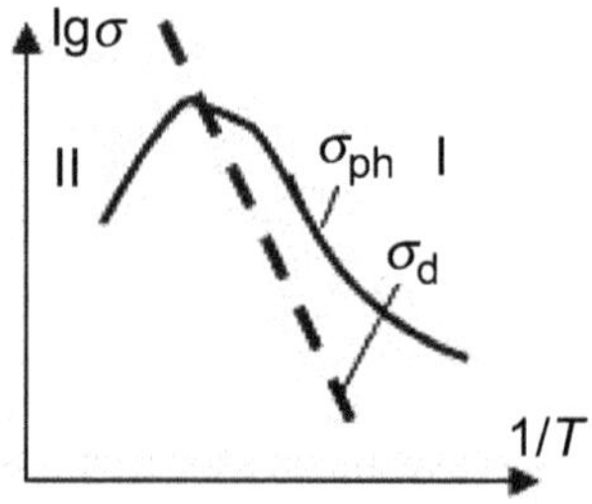

Figure 3.32 Schematic illustration of photoconductivity temperature dependence of non-crystalline semiconductors.

Among the characteristics mentioned above, the carrier drift mobility has the strongest temperature dependence. It is determined by trapping the processes of photocarriers on band tail localized states, and it increases with rising temperature. In the case of $\sigma_{ph} > \sigma_d$, the recombination of photoexcited electrons and holes takes place through deep recombination levels, and photoconductivity temperature dependence is determined by the temperature dependence of drift mobility. However, when dark conductivity exceeds photoconductivity ($\sigma_{ph} < \sigma_d$), equilibrium carrier concentration exceeds the concentration of photocarriers and recombination takes place between photoexited and thermoexited carriers. Equilibrium carrier concentration increases with rising temperature. It leads to a decrease in photocarrier lifetime (τ) and to a decrease in photoconductivity in accordance with Eq. (3.53).

3.4.4 Dependence on Electric Field Intensity

The dependence of photoconductivity on electric field intensity is determined by the dependence of quantum efficiency (η) on electric field. Because quantum efficiency is the ratio between the number of generated photocarriers and the number of absorbed photons and η is usually < 1, the following question arises: why is the number of created electron–hole pairs contributing to photoconductivity is less than the number of absorbed photons? To answer this question, let us examine the process of optical generation of the electron–hole pair.

The absorption of photon with energy $\hbar\omega \geq E_{opt}$ leads to electron transition from the valence band to the conduction band and the creation of the electron–hole pair with energy difference $E \approx \hbar\omega$ (Fig. 3.33). There are two possible ways of subsequent development. Electron and hole may separate under the influence of electric field or concentration gradient (solid arrows in the figure) and become free charge carriers. Alternatively, charge carriers may lose part of their energy by phonon scattering, couple through Coulomb attractive interaction, and recombine with each other (double arrow in the figure).

Because electric field is conducive to charge carrier separation, an increase in electric field intensity leads to a rise in quantum efficiency. The value of quantum efficiency becomes close to 1 in the case of an electric field intensity of 10^5 V/cm (Fig. 3.34, solid line). This explains the experimental electric field dependence of photoconductivity usually observed in non-crystalline semiconductors.

However, Juska, Arlauskas, and Montrimas (1987) have shown that quantum efficiency in non-crystalline selenium films exceeds 1 in the case of an electric field intensity of more than 10^5 V/cm (Fig. 3.34, dashed lines). In other words, the absorption of one photon leads to the formation of several electron–hole pairs.

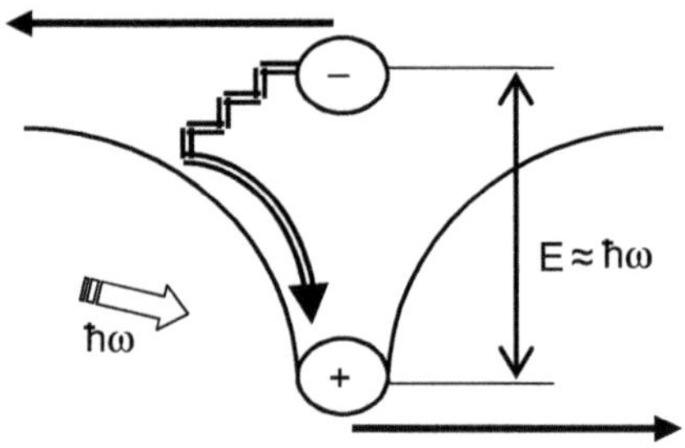

Figure 3.33 Illustration of optical generation and recombination processes (see text).

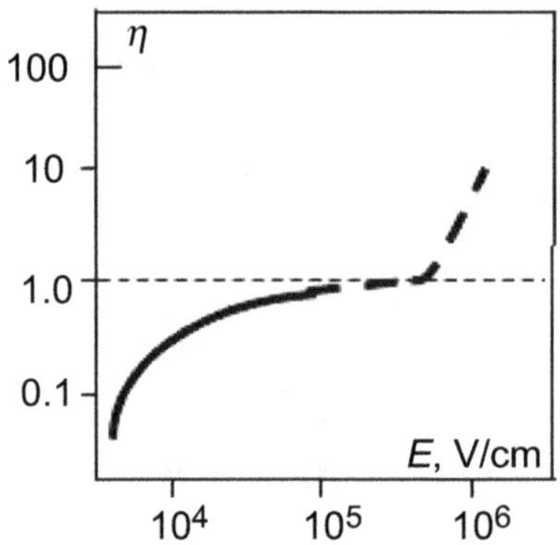

Figure 3.34 Field dependence of quantum efficiency in non-crystalline selenium.

This phenomenon may be explained by avalanche carrier multiplication in electric field. In this case, an increase in film thickness at constant electric field intensity has to produce a rise in quantum efficiency because of the greater number of carrier ionization acts. The experiment confirms this assumption (Kasap et al., 2009): an increase in film thickness from 1 to 35 μm causes an increase in the quantum efficiency by three orders of magnitude.

The experimental evidence mentioned above has been received in the early experiments (Juska, Arlauskas, and Montrimas, 1987) and allowed to increase sensitivity of TV camera tubes (vidicons) by dozens of times (see Chapter 6). However, the existence of avalanche carrier multiplication in non-crystalline semiconductor was a rather unexpected result because the absence of long-range order in atomic arrangement leads to small carrier free path. Explanation for the avalanche multiplication mechanism in non-crystalline semiconductors has been provided by the modified "lucky-drift" model only 25 years later (Jandieri et al., 2008). This model is based on the assumption that there is elastic scattering (from the potential fluctuations) and inelastic scattering (from optical phonons) in non-crystalline semiconductors. Elastic collisions are much more frequent and not all the carrier energy is lost upon inelastic scattering. Therefore, carriers can gain sufficient energy during drift to cause impact ionization.

Chapter 4

Methods for Controlling Properties of Disordered Semiconductors

A distinctive feature of crystalline semiconductors is the strong dependence of electrical conductivity on the presence of impurities. An admixture of a thousandth of a fraction per cent impurity can lead to a change in electrical conductivity by orders of magnitude and in the conductivity type. It is precisely this availability of a simple method for controlling electrical property by doping that ensures the progress of modern solid-state electronics and microelectronics.

The excess electron of a donor (phosphor atom in silicon, for example) transfers to the anti-bonding orbital and forms a donor center under the bottom of the conduction band in the case of donor doping. It leads to a shift in the Fermi level toward the conduction band and to an increase in free electron concentration. A similar situation occurs in the case of acceptor doping. However, the doping of non-crystalline semiconductors does not lead to noticeable changes in their electrical properties. The insensitivity of non-crystalline semiconductors to impurities was discovered by Kolomiets (1960) and has become a characteristic feature of these materials.

Let us consider the reasons for this phenomenon. In the case of chalcogenide glassy materials with flexible disordered structural network, each foreign atom can satisfy its valency requirements and will not act as an electrically active center. Even if a number of foreign atoms remain electrically active, their impact is negligible.

Disordered Semiconductors: Physics and Applications (2nd Edition)
Anatoly Popov

ISBN 978-981-4774-37-6 (Hardcover), 978-1-315-10042-5 (eBook)
www.panstanford.com

Such effect originates in the relaxation of structural network, which leads to a variation in the ratio between negatively (C_1^-) and positively (C_3^+) charged intrinsic defect states (valence-alternation pair states, see Chapter 3). Concentrations of these states are same in the undoped material (Fig. 4.1a). A rise in the concentration of electrically active impurity centers (donors, for example) leads to electron redistribution between intrinsic and extrinsic defect states. The concentration of C_3^+ centers decreases, and the concentration of C_1^- centers increases (Fig. 4.1b). As a result, the Fermi-level position and electrical conductivity remain unchanged.

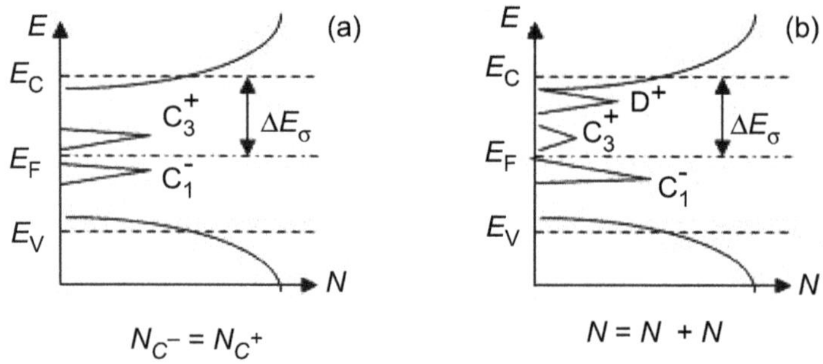

Figure 4.1 Density of charged states in the gap of undoped (a) and doped (b) chalcogenide glassy semiconductor (ΔE_σ is conductivity activation energy).

The aforementioned reasoning cannot be applied to amorphous silicon, which has a rigid fourfold coordinated structural network. A rigid structural network does not allow easy satisfaction of the valency requirements for foreign atoms. On the other hand, a rigid structural network leads to a high concentration of intrinsic defects. The concentration of foreign atoms proved to be less than the concentration of intrinsic defects (Fig. 3.17a), and impurities do not influence the Fermi-level position and the electrical conductivity value.

The insensitivity of non-crystalline semiconductors to impurities creates huge difficulties in device design. That is why some methods have been proposed for controlling the properties of these materials:

- Doping of hydrogenated amorphous silicon
- Chemical modification of chalcogenide films

- Inversion of conductivity type in balk glassy chalcogenide
- Structural modification of non-crystalline semiconductors

4.1 Doping of Hydrogenated Amorphous Silicon

The possibility of purposeful material property changes by means of doping is the feature that distinguishes hydrogenated amorphous silicon from other non-crystalline semiconductors. This possibility was demonstrated by Spear and Le Comber in 1975 and allowed the widespread practical application of a–Si:H observed today (Madan, 2006).

As mentioned in Chapter 3, the incorporation of hydrogen atoms in amorphous silicon sharply decreases the concentration of intrinsic defects and the density of localized states in the gap. As a result, the donor (or acceptor) level density exceeds the density of intrinsic defect levels (Fig. 3.15a) after doping. It leads to a shift in the Fermi-level position and allows to change the electrical conductivity value by up to 6–10 orders of magnitude and to change the type of conductivity as well.

Because the generally used method of preparation of amorphous hydrogenated silicon films is silane (SiH_4) glow discharge decomposition (see Chapter 5), doping is carried out by the addition of gases containing donor or acceptor atoms to silane. A gas mixture of silane, hydrogen, and phosphine (PH_3) is used in the case of phosphorus doping, and a gas mixture of silane, hydrogen, and diborane (B_2H_6) is used in the case of boron doping. As a rule, the doping level is controlled by changing the PH_3/SiH_4 or B_2H_6/SiH_4 ratio in the range from 10^{-7} to 10^{-1}.

The dependences of a–Si:H electrical conductivity at room temperature and the position of the Fermi level on the level of doping by phosphorus and boron are shown in Fig. 4.2 (Spear and Le Comber, 1975). The conductivity of undoped a–Si:H equals 10^{-8}–10^{-9} $\Omega^{-1}cm^{-1}$. Increasing the phosphorus concentration leads to a smooth rise in conductivity up to 10^{-2} $\Omega^{-1}cm^{-1}$ and the Fermi level shift to the conduction band. At the same time, the activation energy of conductivity decreases from 0.6 to 0.2 eV. It is necessary to note that small phosphine additives to silane (PH_3/SiH_4 = 10^{-5}–10^{-4}) lead to a sharp conductivity increase in comparison with undoped material. It indicates relatively low localized state density in the middle of the gap. At higher levels of phosphine content, the Fermi level draws nearer to the conduction band. The movement of the Fermi level slows down because a high localized state density in the conduction band tail and conductivity saturation is observed at $\sigma_{T=300\text{K}} \approx 10^{-2}\ \Omega^{-1}\text{cm}^{-1}$ and $E_\sigma \approx 0.2$ eV. In the case of boron admixture, the situation is different. Small diborane additives to silane lead to a decrease in the conductivity of a–Si:H by four orders of magnitude to the value $10^{-12}\ \Omega^{-1}\text{cm}^{-1}$. The activation energy of conductivity increases to 0.8 eV. When the gas ratio B_2H_6/SiH_4 exceeds the value of 10^{-5}–10^{-4}, conductivity begins to rise together with the increase in diborane concentration. The maximum conductivity value is approximately the same as in the case of phosphorus doping. The rise in conductivity is accompanied by a decrease in the activation energy of conductivity from 0.8 to 0.2 eV (Spear and Le Comber, 1975).

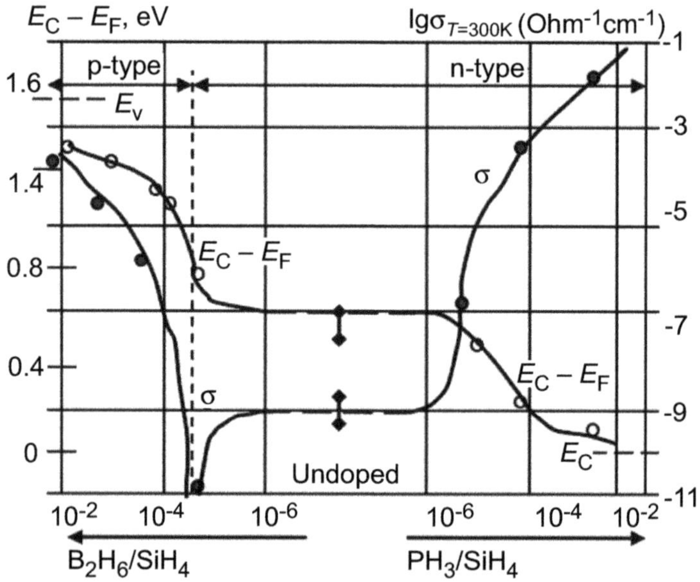

Figure 4.2 The effect of phosphorous and boron doping on the conductivity and the position of the Fermi level in a–Si:H. On the right, these quantities are plotted against the phosphorous concentration in the gas mixture used for the specimen preparation and on the left against the boron concentration. Reprinted from Spear and Le Comber (1975), with permission from Elsevier.

The dependences described above may be explained under the assumption that undoped hydrogenated amorphous silicon is not an intrinsic semiconductor. It has weak n-type conductivity, and the Fermi level lies a little bit above the middle of the gap. In the case of phosphorus doping, the shift in the Fermi level up to the conduction band leads to a decrease in the activation energy of conductivity and to an increase in conductivity. In the case of boron doping, the Fermi level shifts down to the valence band, but it leads to an increase in the activation energy of conductivity and to a decrease in conductivity because the n-type of conductivity still remains. At the gas ratio of approximately 10^{-4}, the Fermi level reaches the middle of the gap and the conductivity activation energy gains the maximum value. The increase in diborane concentration after this point yields to p-type conductivity and to a lower activation energy of conductivity.

Doping efficiency depends on film preparation conditions. However, a comparison of phosphorus (or boron) atom concentration with the value of the electrical conductivity of the film shows that only one-third of admixture atoms affect the Fermi-level position at best (Spear and Le Comber, 1984). There are two reasons for this. First, only a part of phosphorus (boron) atoms is fourfold coordinated and acts as donors (acceptors) in a disordered structural network. It has been shown (Stutzmann et al., 1987) that the concentration of fourfold coordinated phosphorus atoms is proportional to the square root of the total incorporated atom concentration. Second, the doping mechanism in hydrogenated amorphous silicon differs from that in a single crystal. In the former case, the excess electrons of donor atoms occupy empty localized states near the Fermi level. It leads to a shift in the Fermi level toward the conduction band. The value of the shift (ΔE_F) depends on the ratio of donor concentration (N_D) to localized state density near the Fermi level [$g(E_F)$]:

$$\Delta E_F \sim N_D/g(E_F) \tag{4.1}$$

In the case of fixed N_D, the value of the Fermi-level shift is inversely proportional to $g(E_F)$. Because the localized state density near the Fermi level in non-hydrogenated amorphous silicon reaches 10^{20} cm^{-3}, it is clear why doping does not influence the Fermi-level position and properties of this material. The positions of the phosphorus donor level and the boron acceptor level in hydrogenated amorphous silicon are E_C – 0.2 eV and E_V + 0.3 eV, respectively.

4.2 Chemical Modification of Chalcogenide Glassy Semiconductor Film Properties

As mentioned above, chalcogenide glassy semiconductors are not sensitive to impurities because a foreign atom can satisfy its valency requirements and it does not act as an electrically active center due to the relaxation of the structural network in a material lacking long-range order. Therefore, it is necessary to prevent structural relaxation during material preparation in order that foreign atoms remain electrically active centers. Thus, from the point of view of physics, the idea of chemical modification of chalcogenide glassy films consists in the incorporation of impurity atoms into the growing chalcogenide film under conditions where relaxation of the atomic structure is not probable. These conditions exist below the glass transition temperature (Ovshinsky, 1977).

This idea was realized by the simultaneous radio-frequency sputtering of chalcogenide glass and the desired amount of additive. The additive concentration in the film is varied by changing the number of small additive disks attached to the face of the sputtering target of the host material (Flasck et al., 1977). The conductivity temperature dependences of $Ge_{32}Te_{32}Se_{32}As_4$ thin films modified by nickel are shown in Fig. 4.3. The electrical characteristics of films with various nickel content are listed in Table 4.1.

As one can see from the figure and the table, the conductivity activation energy of unmodified film is constant throughout the temperature range and equals approximately half of the chalcogenide glass gap value. Nickel modification leads to an increase in the conductivity value by up to eight orders of magnitude. Besides that, the region of extrinsic conduction appears on the $\sigma(1/T)$ curves. The rise in nickel concentration results in a decrease in the conductivity activation energy from 0.74 to 0.18 eV. At the same time, the value of the optical gap changes only insignificantly. In addition, this modification causes changes in the thermoelectric power (Seebeck coefficient) from positive to negative value (Table 4.1). It means that transition takes place from p-type conductivity to n-type conductivity of the material. These results confirm that the reason for electrical characteristic changes is a shift in the Fermi level. Similar effects have been observed with different transition metal modifiers in a number of chalcogenide glasses (Flasck et al., 1977).

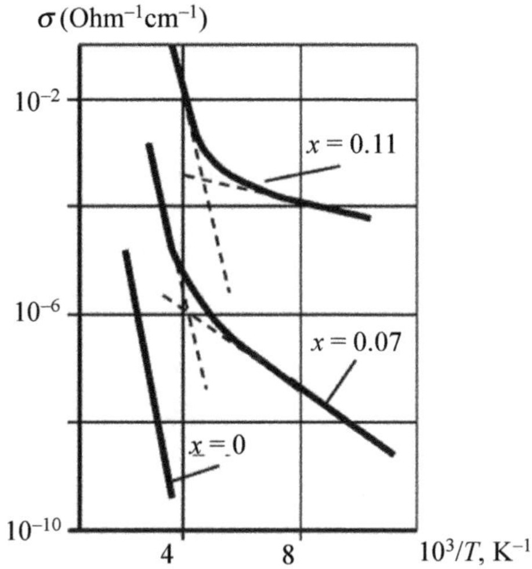

Figure 4.3 Conductivity dependences of $(Ge_{32}Te_{32}Se_{32}As_4)_{1-x}Ni_x$ films on temperature.

Table 4.1 Characteristics of chemical modified films $(Ge_{32}Te_{32}Se_{32}As_4)_{100-x}Ni_x$

X (at.%)	E_σ (eV)	σ_{293K} ($\Omega^{-1}cm^{-1}$)	*S* (μV/K)
0	0.74	4×10^{-10}	+2000
7	0.33	1.4×10^{-5}	+54
9.1	0.25	6.5×10^{-4}	−82
11.4	0.18	3.8×10^{-2}	−84

Thus, the chemical modification of chalcogenide glassy films leads to effects similar to the doping of crystalline semiconductors and hydrogenated amorphous silicon.

4.3 Conductivity Type Inversion in Bulk Glassy Chalcogenide

The results of thermoelectric power measurements indicate p-type conductivity for all bulk chalcogenide glasses. The possibility of changing conductivity type in chalcogenide glassy films by chemical modification (see Section 4.2) gives rise to attempts to prepare bulk chalcogenide glass with n-type conductivity. Usually, it is done by adding metals to the melt before quenching. A lot of materials were tested. However, in the case of a small amount of added metal (within the limits of a few percent), glasses retain p-type conductivity. An increase in metal content in the melt leads to crystallization of the material during cooling. For example, a concentration of less than 1 at.% of nickel or chrome can be realized in glassy As_2Se_3 without crystallization of the material. In the case of glassy selenium, the highest possible bismuth concentration is 2 at.%. Substitution of selenium for the As–Se system allows to increase bismuth concentration to up to 4 at.%, but the material retains p-type conductivity.

The largest amount of metal can exist in glasses of Ge–S and Ge–Se systems. The glass formation regions for the Ge–Se–Bi system at various cooling rates are shown in Fig. 4.4 (Pazin and Borisova, 1969). It is possible to introduce up to 14 at.% of bismuth to the $GeSe_{3.5}$ and $GeS_{3.5}$ glasses without crystallization of the material.

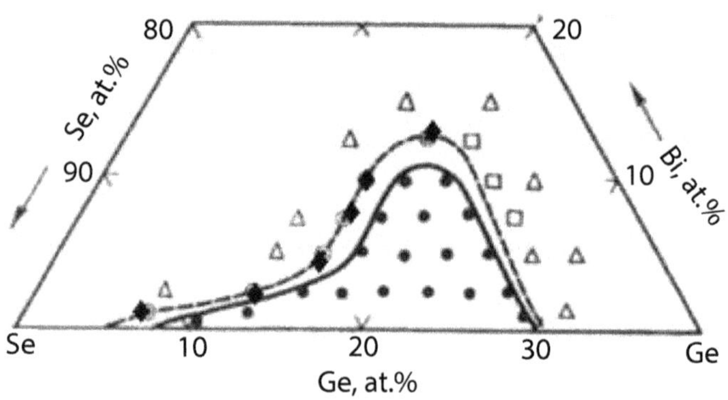

Figure 4.4 Glass formation regions for the Ge–Se–Bi system (• – slow cooling, ♦ – cooling in the ice water, □ – glassy-crystalline phase, Δ – crystalline phase).

The temperature dependences of electrical conductivity of bulk glasses $(GeSe_{3.5})_{100-x}Bi_x$ are shown in Fig. 4.5 (Aivasov et al., 1995). An increase in bismuth concentration leads to a rise in the absolute conductivity values by up to six orders of magnitude and to a decrease in the conductivity activation energy (the slope of curves lg $\sigma(1/T)$ decreases). Besides that, the Seebeck coefficient (thermoelectric power) changes from positive to negative sign at a bismuth content of about 8 at.%. This indicates a change in conductivity from p-type to n-type. As a rule, these facts are interpreted as the Fermi-level shift toward the conduction band.

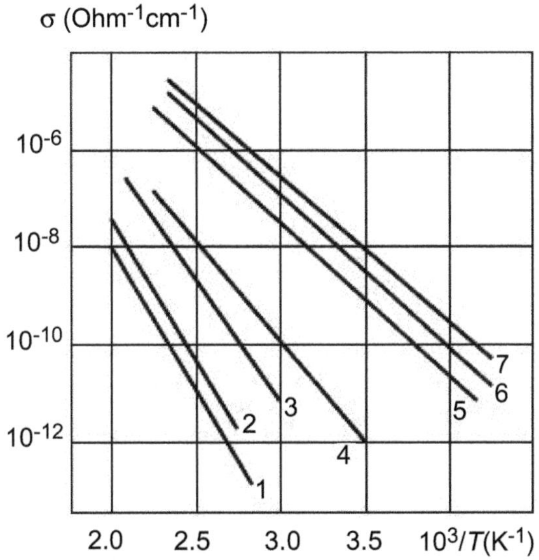

Figure 4.5 Conductivity temperature dependence of $(GeSe_{3.5})_{1-x}Bi_x$ glasses for: $x = 0$ (1), $x = 0.04$ (2), $x = 0.06$ (3), $x = 0.08$ (4), $x = 0.10$ (5), $x = 0.12$ (6), and $x = 0.14$ (7).

However, in contrast to the chemical modification of glassy chalcogenide films (see Section 4.2), there is no region of extrinsic conduction on the conductivity temperature dependence curves in this case. Dependences lg $\sigma(1/T)$ are straight lines with constant slope throughout the temperature range. It contradicts the assumption about the Fermi-level shift. An increase in the absolute value of conductivity and a decrease in the conductivity activation energy may arise from both the Fermi-level shift and the change in the forbidden gap value. Measurements of the optical gap value in $(GeSe_{3.5})_{1-x}Bi_x$ glasses show that the optical gap value decreases from 1.95 to 1.20 eV when the bismuth concentration rises from 0 to 14 at.%. Moreover, the concentration dependences of the conductivity activation energy of half of optical gap have analogous characteristics and similar absolute values (Fig. 4.6; Aivasov et al., 1995). It indicates that the observed changes in electrical properties are due to the value of forbidden gap lowering without the Fermi-level shift.

However, it is difficult to explain changes in the Seebeck coefficient sign and conduction type in samples with bismuth content of 8 at.% and more based on this assumption. Is the change in conduction type possible when the Fermi-level position does not change?

It is known that hole conduction is equal to electron conduction in the case of intrinsic semiconductors. Because conductivity $\sigma = en\mu_n + ep\mu_p$, the condition of intrinsic conduction is the equality $n\mu_n = p\mu_p$. Hole mobility in chalcogenide glassy semiconductors exceeds electron mobility. It means that $p\mu_p > n\mu_n$, and thermoelectric power has positive sign even if the Fermi level is situated at the middle of the gap. Bismuth creates deep traps for holes in chalcogenide glasses but has weak influence on electron mobility. As a result, hole mobility becomes less than electron mobility at certain bismuth concentration and the sign of thermoelectric power changes in spite of constant position of the Fermi level.

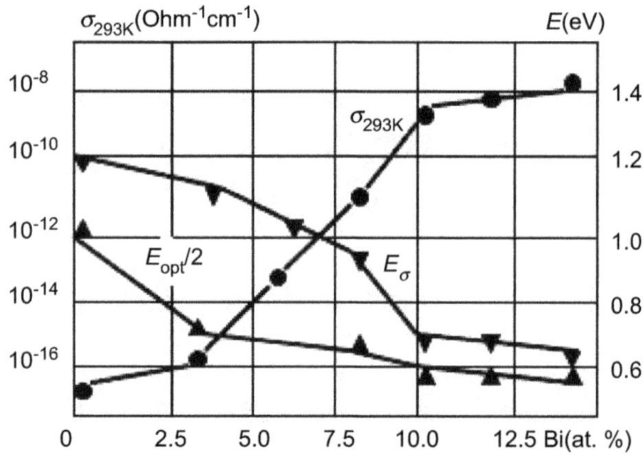

Figure 4.6 Dependences of room temperature conductivity (σ_{293K}), activation energy of conductivity (E_σ), and half of optical gap value ($E_{opt}/2$) on bismuth concentration in the $(GeSe_{3.5})_{1-x}Bi_x$ glasses.

4.4 Structural Modification of Disordered Semiconductors Properties

As mentioned above, the most essential feature of disordered semiconductors is their weak sensitivity to impurities. The problems of control over the properties and reproducible synthesis of these materials with prescribed properties still remain pressing in spite of some solutions being found (see Sections 4.1–4.3).

As an alternative to controlling semiconductor properties by doping, a method of structural modification of the properties of disordered semiconductors has been proposed. The method allows to control properties by changing the material's structure without changing its chemical composition (Popov, Michalev, and Shemetova, 1983). The physical basis of the method is the fact that the free energy of disordered materials has several minima corresponding to various metastable states of the system (see Chapter 1).

It has been shown in Chapter 2 that four levels of structural characteristics are necessary for the common description of the structure of disordered solids, namely: short-range order in atomic disposition, medium-range order in atomic disposition, morphology, and subsystem of defects. Structural changes are possible at all the levels of structural characteristics mentioned above. So one can conclude that, on the whole, four levels of structural modification differ by the type of change in material structure, namely: level of short-range order (level 1), level of medium-range order (level 2), level of morphology and heterogeneity (level 3), and level of defect subsystem (level 4) (Table 4.2; Popov, Vorontsov, and Popov, 2001).

4.4.1 Structural Modification at the Level of Short-Range Order

Structural changes at the level of short-range order lead to variations in all the basic properties of a material. For example, polymorphic crystalline modifications of carbon (diamond, graphite, and carbine) possess fundamentally different physicochemical properties because of the different hybridizations of electron orbitals and different atomic structures at the short-range-order level. Amorphous carbon films incorporate structural units of different allotropic modifications, with the relative content of these units determined by film growth modes and varying widely for the same preparation method. Correspondingly, the coordination of atoms varies (between 2 and 4) together with other parameters of the first coordination sphere (Vasil'eva and Popov, 1995). When films of amorphous hydrogenated carbon (a–C:H) are obtained by

Table 4.2 Levels of structural modification

Level	Structural changes	Method of treatment	Characterization of sensitive properties	Groups sensitive properties	Examples of sensitive properties
1	Short-range order	Various methods and modes of preparation	All properties	All properties	All properties
2	Medium-range order	External factor treatment during preparation or thermal treatments	Properties associated with rearrangement of structural units	Mechanical properties, phase transitions	Viscosity, hardness, Young modulus, photo-contraction of films, temperature and activation energy of crystallization
3	Morphology and heterogeneity	Changes in the preparation and treatment modes	Properties dependent on heterogeneities	Electrical, optical	AC conductivity
4	Defect subsystem	Changes in preparation modes, treatments affecting the defect subsystem	Properties dependent on the distribution of localized states density and on the Fermi-level position	Electrical, photo-electric	Field-dependent conductivity

rf-ion-plasma sputtering in an argon–hydrogen atmosphere, merely changing the substrate temperature and discharge power may give films (Popov et al., 1995), in which the optical gap varies by two orders of magnitude (between 0.02 eV for graphite-like films and 1.85 eV for films with predominance of the diamond-like phase) and the dark conductivity by more than 10 orders of magnitude (between 7 and 2×10^{-10} Ω^{-1}cm^{-1}). The dependences of resistivity and the value of the optical gap of a–C:H on substrate temperature and the charge power are represented in Fig. 4.7 (Popov, Vorontsov, and Popov, 2001).

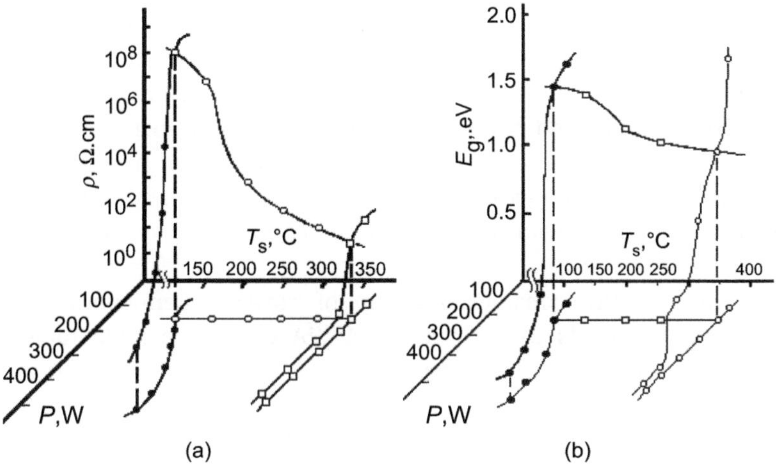

Figure 4.7 Resistivity ρ (a) and optical gap E_g (b) of a–C:H films in relation to their preparation modes (T_s is the substrate temperature, P is the rf discharge power). Reprinted from Popov et al. (2001), with permission from Pleiades Publishing.

Thus, the first level of structural modification involves pronounced changes in the short-range order of atomic arrangement, that is, changes in the hybridization of electron orbitals of all (or most) atoms constituting the sample. Carbon has long been considered the only (and unique in this respect) element of those considered here that exists in allotropic crystalline modifications of diamond (sp^3-hybridization) and graphite (sp^2-hybridization). The discovery of carbines (sp-hybridization) in 1960 (Sladcov et al., 1972) provided more evidence of the unique properties of carbon. At the same time, there have been reports that the short-range order may change substantially under certain conditions for other elements as well. For example, cubic α and β modifications of crystalline selenium with atom coordination numbers 4 and 6, respectively, were obtained (Andrievsky, Nabitovich, and Krinykovich, 1959; Andrievsky and Nabitovich, 1960) in electron-beam-induced crystallization of thin films. However, these reports failed to attract due attention of the scientific community at that time.

The monopoly of carbon on the possibility of existence of forms with different hybridizations of electron orbitals and, consequently, with different atom coordination numbers was radically broken up by a series of investigations (Khohlov, Mashin, and Khohlov, 1998; Mashin and Khohlov, 1999). New forms of silicon that appeared under certain conditions in films of amorphous silicon and in a–Si:H were discovered in these studies—silicine with *sp*-hybridization of electron orbitals and atom coordination of 2 and a form with sp^2-hybridization and atom coordination of 3. However, it should be noted that in these investigations devoted to silicon and in studies of selenium, the new forms were obtained only under certain, rather specific conditions.

Thus, in the considered group of covalent semiconducting materials (periods 2–5, Groups IVA–VIA of the periodic table), structural changes at the short-range-order level are observed for elements belonging to Group IVA—periods 2 and 3 (C, Si)—and Group VIA—period 4 (Se) (Table 4.3), i.e., for three elements out of the considered nine. We should also note the increase in the first coordination number from 2 to 3 in tellurium melt (at 600°C), with the covalent nature of chemical bonds being preserved (Fig. 4.8; Poltavzev, 1984), and the changes in the short-range order (bond length) in glassy GeS_2 under pressure (Miyauchi et al., 2001). These elements show no fundamental distinctions in the electron shell structure or other parameters from the rest of the considered elements. In view of the above, it seems reasonable to assume that structural changes at the short-range-order level are not unique to carbon, being characteristic to all the considered covalent semiconducting materials. Experimental evidence in favor of this assumption has been obtained for carbon, silicon, and selenium, and obtaining it for other elements is only a matter of time and attention devoted to the problem.

Table 4.3 Short-range-order level

	Columns | Coordination number (N_c)								
Periods	**IVA**		N_c	**VA**		N_c	**VIA**		N_c
	6		2						
2		**C**	3						
	12		4						
	14		2	15			16		2
3		**Si**	3		**P**	3		**S**	
	28		4	31			32		
	32			33			34		2
4		**Ge**	4		**As**	3		**Se**	4
	72,5			75			79		6
				51			52		2
5					**Sb**	3		**Te**	3[a]
				122			128		

a: melt

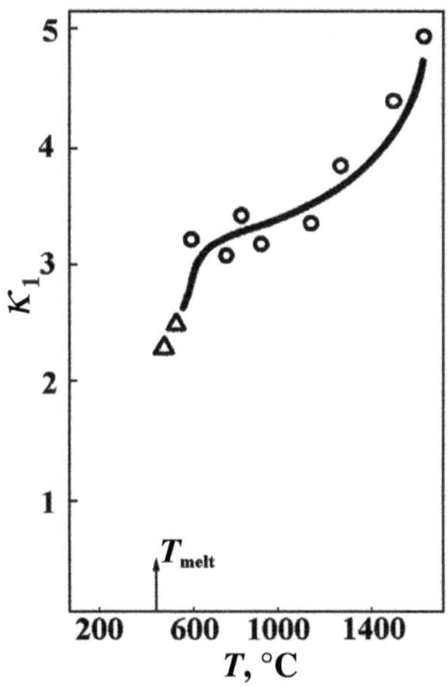

Figure 4.8 Temperature dependence of the liquid tellurium first coordination number (Poltavzev, 1984).

Structural modification at the level of short-range order not only determines considerable changes in the properties of a material but also influences upon other groups of structural characteristics, namely: the medium-range order, the morphology, and the subsystem of defects. It was shown (Miyauchi et al., 2001) that the changes in the tetrahedron structure of glassy GeS_2 caused by high pressure, that is, changes in the short-range order lead to a change in the mutual packing of tetrahedrons, and so to the change in the medium-range order. And more significant changes of all the structural characteristics should be expected at the transition from the tetrahedral coordinated network to the layer structure or to the linear polymers.

4.4.2 Structural Modification at the Medium-Range-Order Level

This level of structural modification implies changes in the dihedral angle distribution, degree of molecule polymerization, extent of disorder in alloys, etc. The magnitude of the dihedral angle dispersion determines the mutual arrangement of the tetrahedron, the probability of formation of cells with five-membered, six-membered, or seven-membered rings in the amorphous silicon. Changing the medium-range order without changing the short-range order mainly affects the macroscopic properties of the material (viscosity, microhardness, Young modulus, photo-contraction of films; see Table 4.2). The properties governed by the electronic structure of the material, which are dependent mainly on the short-range order, should change relatively weakly. However, Sarsembinov et al. (2001a, b) showed that changes in the medium-range order in the films of the As–Se system do influence their optical and electrical properties.

In the case of modification at the level of medium-range order, the efficiency of property control is determined by the flexibility of the structural network. The latter decreases with an increasing degree of covalent coupling of the structural network (the average number of covalent bonds per atom is $N/2$, where N is the average coordination number) and increases with the increasing average number of lone-pair electrons per atom ($[NVE - N]/2$, where NVE is the average number of electrons in the outermost shells of the atoms). Thus, the flexibility of the structural covalent-bonded network varies directly as the ratio ($[NVE - N]/N$). At the same time, the rise in chemical bond ionicity or the degree of bond metallization decreases the directional character of chemical bonds and increases the structural network flexibility. Thus, it is necessary to establish the proportion of ionic and metallic bond components.

The ionicity of chemical bond can be determined by the electronegativity difference of atoms as follows:

$$S = 1 - \exp(-\Delta x^2/4) \tag{4.2}$$

where $\Delta x = x_A - x_B$ is the electronegativity difference of atoms A and B. In the case of multi-component material, the average coefficient of bond ionicity is given by

$$I_C = \sum S_n N_n \tag{4.3}$$

where n is the number of various chemical bonds in the material, S_n is the ionicity of bond n, and N_n is the fraction of n bonds in the material. The metallicity of the bonds increases with the increase in atomic number in the periodic table. However, it is rather difficult to estimate metallicity in terms of this quantity. As a rule, the degree of bond metallization (M) is taken to be proportional to the average number of electrons per atom. Thus, the covalent coupling of the material structural network is characterized by the parameter $N(1 - I_C - M)/2$ and the flexibility of the structural network varies in proportion to $(NVE - N)/N(1 - I_c - M)$.

As a rule, the structural modification at the medium-range-order level is realized at a temperature range above the glass transition temperature. Therefore, the efficiency of property control depends not only on the obtained value of structural changes but also on the preservation of these changes during the cooling process of the material. At an infinitely high cooling rate, the number of various structural units and their ratio in the material are preserved the same as in the melt before cooling. However, the structural relaxation of the material can take place during the cooling process at the finite cooling rates and the rate of structural relaxation increases with increasing melt temperature. It is necessary to increase the cooling rate to neutralize the structural relaxation process. The boundary cooling rate needed to avoid structural relaxation depends on temperature as follows (Dembovsky and Chechetkina, 1981a):

$$V_b = \nu_0(RT^2/\Delta G^*)\exp[-\Delta G^*/(RT)] \quad (4.4)$$

where ν_0 is the vibration frequency of molecular groups, R is the universal gas constant, and ΔG^* is the free energy of the initial reorientation of molecular groups (Dembovsky and Chechetkina, 1981b).

The real cooling rate depends on the mass, shape, and heat capacity of the sample and on cooling conditions:

$$V_q = T/\tau = (T_0/\tau)\exp(-t/\tau) \quad (4.5)$$

where T_0 is the initial melt temperature, t is the cooling time, and τ is the cooling time constant

$$\tau = mc/s\lambda \quad (4.6)$$

where m and s are the mass and surface area of the cooling sample, c is the material specific heat, and λ is the convective heat transfer factor.

The condition of structural relaxation absence is given by

$$V_q \geq V_b \quad (4.7)$$

The maximum value of the cooling time constant needed to avoid structural relaxation may be obtained from the equality of Eqs. (4.4) and (4.5):

$$\tau_{max} = (\Delta G^*/\nu_0 RT)\exp[\Delta G^*/(RT)] \quad (4.8)$$

Equation (4.8) allows to formulate demands to preserve the structure of the melt in the solid sample after quenching.

The value of maximum possible change in the structure and, therefore, in the properties of the material can be written as

$$\Delta\Phi = f\{(NVE - N)/[N(1 - I_C - M)]\}_{Vq \geq Vb} \quad (4.9)$$

where f is the unspecified function and $(NVE - N)/[N(1 - I_C - M)]$ is the criterion of structural modification efficiency (CSME) at the medium-range-order level. The physical meaning of Eq. (4.9) is expressed by the fact that the possibility of obtaining various structural configurations increases with an increase in the degree of freedom of structural units and of the number of lone-pair electrons capable of forming unusual structural units in a given material.

The dependence of microhardness changes for a number of chalcogenide glasses as a consequence of changes in the thermal histories of materials as a function of CSME is shown in Fig. 4.9 (Popov, Michalev, and Shemetova, 1983).

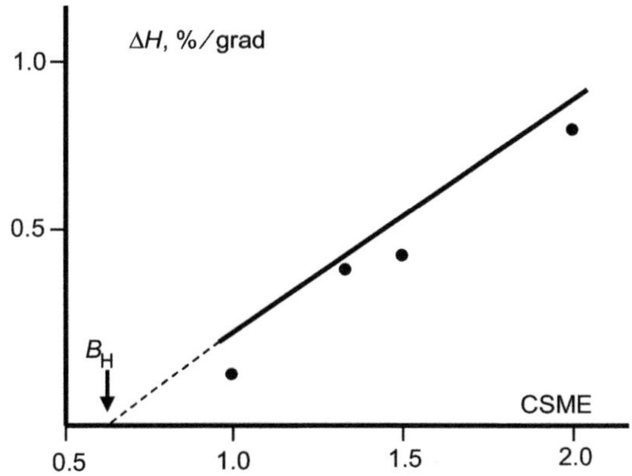

Figure 4.9 Microhardness changes as a function of CSME for chalcogenide glasses with various thermal histories of materials (see text).

Various thermal histories were obtained by changing the melt temperatures before cooling in the temperature range satisfying the requirements (4.7). Data are provided in percentages per one degree of the melt temperature change.

One can see from Fig. 4.9 that the value of microhardness change increases linearly with increasing CSME:

$$\Delta H = \left[\frac{NVE - N}{N(1 - M - I_C)} - B_H \right] A_H \tag{4.10}$$

where A_H is the proportionality factor and B_H is the boundary CSME value below which it becomes impossible to change the material property by structural modification at this level. The question arising in connection with this is "what determines the applicability limit of the structural modification at the medium-range-order level and where does it lay?"

Qualitatively, it may be stated that this limit is determined by the rigidity of the structural network: for changes in the medium-range order to occur, the network must exhibit certain flexibility and flexible bonds are to be present. In view of the above, it seems reasonable to suggest that the limit in question corresponds to the rigidity threshold of the structural network (Philips, 1979a, b, 1980) at which the average number of force constants per atom (N_{const}) becomes equal to the number of degrees of freedom ($N_d = 3$):

$$N_{const} = N_d = 3 \tag{4.11}$$

The number of force constants depends on the type of chemical bonds. In the case of non-directed bonds, stretch force constants are present only and $N_{const} = N/2$ (where N is the coordination number). Since bond-bending modes are also present in covalent-bonded materials, the number of force constants is given by

$$N_{const} = N/2 + [N(N-1)]/2 = N^2/2 \tag{4.12}$$

and the critical coordination number (N_C) corresponding to the rigidity threshold of the covalent-bonded structural network is defined by

$$N_C^2/2 = 3 \text{ or } N_C = 2.4 \tag{4.13}$$

The values of average coordination number for materials of systems $A_x^{IV}(A_x^{V})B_{1-x}^{VI}$ are shown in Table 4.4. The rigidity threshold ($N_C = 2.4$) is shown by the dashed line. The materials with flexible structural network are on the right-hand side of that line. Structural modification at the medium-range-order level is possible in these materials. To the left of the dashed line, there are materials for which structural modification at this level is impossible because of their rigid structural network. However, experimental studies of these systems (Yun et al., 1989; Aitken, 2001) have shown that the rigidity threshold actually lies at higher average coordination numbers. It was demonstrated (Popov, 1994) that the reason for the discrepancy between the calculated and experimental values of the rigidity threshold comes from the ionic component and metallization of chemical bonds being neglected. The rigidity threshold obtained for the aforementioned systems taking into account the ionicity of chemical bonds is shown in the table by the solid line. In accordance with this suggestion, the line is the applicability limit of structural modification at the medium-range-order level.

It is necessary to note that a change in the material's thermal history is not the only way of modifying the structure at the middle-range order level. The combined effect of thermal and other external factors (electric and electromagnetic fields, ultrasonic vibrations, and others) increases the structural modification efficiency at the middle-range-order level.

Table 4.4 Medium-range order and morphology levels

Columns		$\overset{IV}{A}_x(\overset{V}{A}_x)\overset{VI}{B}_{1-x}$ system Value of x							Column
IV A	VA	0.60	0.50	0.40	0.33	0.25	0.20	0.11	VI A
Ge 4 *4*					GeS_2 2.66 *2.40*	GeS_3 2.50 *2.30*	GeS_4 2.40 *2.24*	GeS_8 2.22 *2.12*	Se 2 2
Ge 4 *4*			GeS 3.00 *2.73*	Ge_2S_3 2.80 *2.49*	GeS_2 2.66 *2.30*	GeS_3 2.50 *2.23*	GeS_4 2.40 *2.19*	GeS_8 2.22 *2.10*	S 2 2
	As 3 *3*	As_3Se_2 2.60 *2.53*	AsSe 2.50 *2.42*	As_2Se_3 2.40 *2.31*	$AsSe_2$ 2.33 *2.25*	$AsSe_3$ 2.25 *2.19*	$AsSe_4$ 2.20 *2.16*	$AsSe_8$ 2.11 *2.09*	Se 2 2
	As 3 *3*	As_3S_2 2.60 *2.49*	AsS 2.50 *2.38*	As_2S_3 2.40 *2.26*	AsS_2 2.33 *2.22*	AsS_3 2.25 *2.17*	AsS_4 2.20 *2.14*	AsS_8 2.11 *2.08*	S 2 2

– – – – (dashed line) rigidity threshold in the case of purely covalent bonds;
——— (solid line) rigidity threshold in the case of ionicity is taken into account;
2.25 average coordination number, N;
2.25 (italics) $N(1 - I_c)$, where I_c = mean coefficient of bond ionicity.

4.4.3 Structural Modification at the Morphology/Heterogeneity Level

The necessary condition for obtaining any solid non-crystalline material is the thermodynamically non-equilibrium process of its synthesis. In conformity with the basic concepts of the theory of self-organization (synergetics), the non-equilibrium conditions of material formation result in the appearance of heterogeneities. Thus, general considerations suggest the presence of heterogeneities in all of the non-crystalline semiconductors in question.

The presence of macroheterogeneities (columns, globules, cones, etc.) on a scale of tens of nanometers and more in non-crystalline material films is considered film morphology. Certain morphology (columns structure) was observed in hydrogenated amorphous silicon films by Knights and Lujan in 1979. It was demonstrated later that certain morphology is a characteristic feature of all amorphous tetrahedral semiconductors. At the same time, a "structureless" smooth surface is commonly observed in vitreous materials. Considering that the presence of heterogeneity is in the nature of non-crystalline solids themselves, one can conclude that glassy materials contain microheterogeneities. It points to the manifestation of not the morphology, but medium-range order (Popov, 2002). Thus, the heterogeneity is an internal characteristic of non-crystalline solids and can manifest itself both by the appearance of certain morphology and by areas of ordered regions of atomic structure outside the elementary cells (medium-range order). However, such a conclusion implies the answer to a number of questions, namely: Why are the two classes of non-crystalline semiconductors characterized by various kinds of heterogeneity? What factors cause the separation between them and where does such boundary passes? To answer these questions, let us return to the medium-range order of atomic arrangement.

Medium-range order is linked with the correlation between atom positions in the range of up to one or a few nanometers. Models of medium-range order (see Section 2.2) can be divided into two groups (Table 4.5). In the first group, the ordered regions determining the medium-range order are the part of the surrounding matrix (or they are linked with each other) without breaking the continuous network. In the second group, between the ordered regions there are boundaries at which the structural network breaks. The possibility for the first or the second situation to be realized in a specific non-crystalline material depends on the flexibility of its structural network. The limit in question corresponds to the rigidity threshold of the structural network at which the average number of force constants per atom becomes equal to the number of degrees of freedom (see Section 4.4.2). In the first case, the network must exhibit certain flexibility. The materials in which the formation of the ordered regions determining the medium-range order (microheterogeneities) is made possible without breaking the continuous covalent network are placed on the right-hand side of the solid line in Table 4.4. As stated above, structural modification at the level of medium-range order is possible in these materials.

Table 4.5 Models describing medium-range order of atomic arrangement

Authors and year	Model	Presence of boundary
Lebedev A., 1921	Microcrystals	Yes
Leadbetter A. and Apling A., 1974	Microcrystals	Yes
Phillips J., 1981	Clusters	Yes and no
Busse L., 1984	Layers	No
Lucovsky G., 1987	Regular distribution of dihedral angles	No
Elliot S., 1991	Void correlation	No
Golicova O. and Kazanin M., 2000	Nanoclusters	Yes
Golicova O., 2001		
Nakhmanson S. et al., 2001 Voyles P. et al., 2001	Grains embedded in a disordered matrix	Yes (with low concentration of defects)
Popescu M. and Bradaczek H., 2001	New paracrystallite model	No

The materials with a rigid structural network are on the left-hand side of the solid line. Formation of the ordered regions in these materials breaks the continuous covalent network and forms limits between microheterogeneities. While synthesizing such materials, some microheterogeneities can reach the size of macroheterogeneities, forming the specific morphology. Hence in the considered materials, structural modification is made possible at the level of morphology. Morphology changes affect the properties sensitive to macroheterogeneities (see Table 4.2).

Thus, the limit between medium-range order and morphology levels of structural modification is the rigidity threshold of the structural network. The morphology level of structural modification can be applied only to materials with structural network rigidity higher than the aforementioned rigidity threshold.

4.4.4 Structural Modification at the Defect Subsystem Level

This level of structural modification is associated with changes in the defect subsystem under the influence of either sample fabrication conditions or various external factors and is manifested by changes in the spectrum of localized states in the gap, which in turn lead to a shift in the Fermi-level position and modify the properties related to the electron subsystem.

Selection of the influence factor is important in this case because it has to interact with defects but not with the structural network on the whole. It has been shown (Popov, 1996) that treating hydrogenated amorphous silicon films with ultraviolet radiation leads to a significant change in the spectrum of localized states in the material, resulting from a change in the relative content of Si–H, Si–H_2, and Si–H_3 complexes. Dependences of infrared absorption peak areas corresponding to various complexes Si–H_n in a–Si:H on ultraviolet radiation dose are shown in Fig. 4.10.

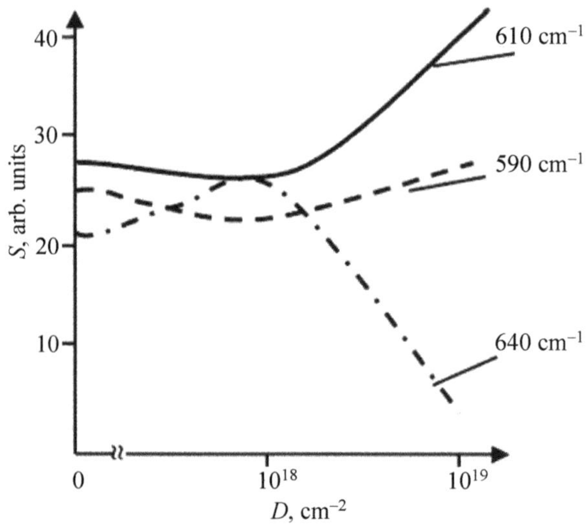

Figure 4.10 Dependences of infrared absorption peak areas corresponding to various complexes Si–H_n in a–Si:H on ultraviolet radiation dose.

The area of peak at 640 cm^{-1} corresponding to Si–H_3 complexes decreases, while the area of peak at 610 cm^{-1} corresponding to Si–H complexes increases with increasing ultraviolet radiation dose. The area of peak at 590 cm^{-1} corresponding to Si–H_2 complexes changes slightly. Alteration in the relative content of Si–H_n complexes leads to changes in the localized state distribution in the gap (Fig. 4.11). One can see that the length of valence and conduction band tails decreases and the density of localized states in the middle of the gap increases under the influence of ultraviolet radiation. In turn, this leads to enhanced stability of a–Si:H films and to improvement in thin-film transistor parameters.

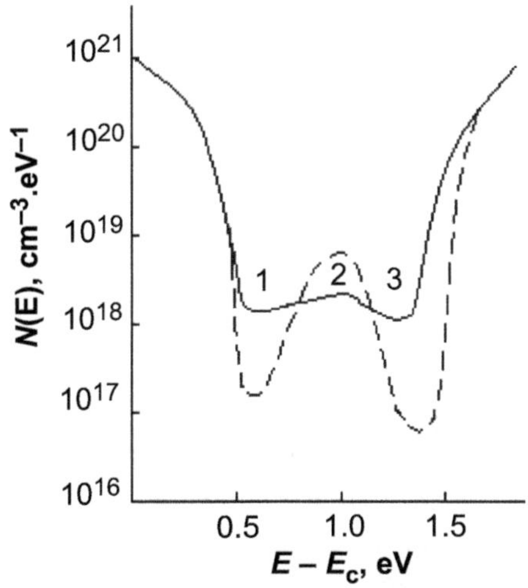

Figure 4.11 Localized state distribution in the gap of a–Si:H before (solid line) and after (dashed line) exposure to ultraviolet radiation (dose 10^{19} cm^{-1}) calculated from the results of constant photocurrent method (1), conductivity temperature dependence (2), and space-charge-limited currents (3) (Popov, 1996).

The effect of amorphous silicon pseudo-doping that varied with film preparation conditions has been observed by Golicova (1991). Dembovskii, Chechetkina, and Kozukhin (1985) observed the effect of weak electric and magnetic fields on quasi-molecular defects (see Chapter 3) and properties of vitreous selenium, arsenic triselenide, and materials of the selenium–tellurium system. Dembovsky, Zyubin, and Grigor'ev (1998) extended these results to sulfur and arsenic trisulfide. Thus, the defect subsystem level of structural modification is observed experimentally both in a tetrahedral material with a rigid covalent structural network (a–Si:H) and in vitreous materials of Group VI and V–VI chalcogenide glasses (Table 4.6). The above suggests that structural modification at this level is inherent, as is the structural modification at the short-range-order level, in all the considered non-crystalline semiconducting materials.

Table 4.6 Defect subsystem level

Columns		Systems		Column
IIV A	V A	$A_2^V B_3^{VI}$	$A^{VI}B^{VI}$	VI A
C				
Si	P			
Ge	As	As_2S_3	S–Se	S
		As_2Se_3	Se–Te	Se
		As_2Te_3		Te

It can thus be concluded that there exist four levels of structural modification of the properties of non-crystalline semiconductors that differ in the resulting structural changes. Control over the properties of non-crystalline semiconductors by changing the structure at the short-range-order level is, in principle, possible for all of the considered materials. Control over properties by treating the defect subsystem is also applicable to all these materials. Control over properties by changing the medium-range order is possible for vitreous semiconductors that have high values of the criterion of structural modification efficiency, with the applicability limit of this level corresponding to the rigidity threshold of the structural network, calculated with consideration of ionicity and metallization of chemical bonds. On the other side of this boundary (materials with a rigid structural network) lies the area of applicability of structural modification, involving changes in the morphology of materials.

4.4.5 Correlation between Structural Modification and Stability of Material Properties and Device Parameters

Structural modification (changes in structure) can take place in the production process as well as in the process of affecting material or devices on the basis of different factors (Table 4.2). Moreover, atomic structure and consequently properties of non-crystalline semiconductors can change in the process of using such devices because their operation is subjected, as a rule, to the influence on the material by electric or electromagnetic fields, high temperature, and so on. Two conclusions follow from the above, and these must be taken into consideration when designing devices on the basis of the observed materials.

The first conclusion consists of the possibility to improve the parameters of such devices by purposeful effect on the structure of the material during or after their production. The present approach has been put into practice during the creation of electrophotographic photoreceptors on the basis of glassy selenium (thermal treatment and electromagnetic radiation), integrated circuit of reprogramming memory matrix on the basis of chalcogenide glassy semiconductors (γ-emission), photo-thermoplastic optical information memory units on the basis of chalcogenide glasses (thermal treatment), and thin-film transistors on the basis of hydrogenated amorphous silicon (ultraviolet emission) (Popov, 2004).

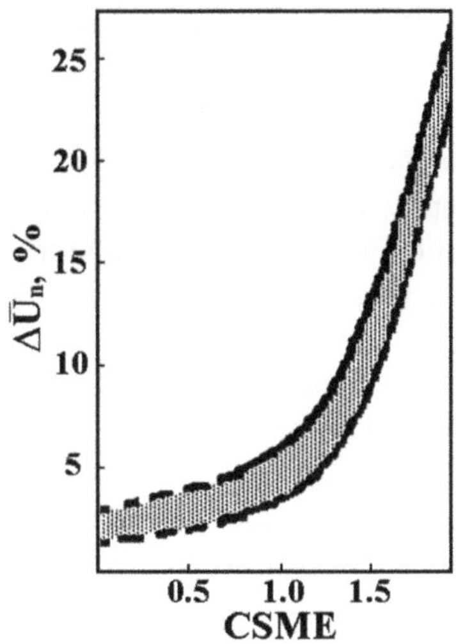

Figure 4.12 Statistical variations of threshold voltage in various chalcogenide glasses as a function of the criterion for structural modification efficiency.

The second conclusion is that the possibility of changes in the non-crystalline semiconductor structure during the device operation leads to statistical variations of device characteristics that cannot be avoided or eliminated. The value of these variations will grow in line with the level of structural changes possible in non-crystalline semiconductors used at the given levels of external factors. The dependence of statistical variations of threshold voltage in various chalcogenide glassy semiconductor switches on the value of the criterion of structural modification efficiency is shown in Fig. 4.12 (Popov, Michalev, and Shemetova, 1983). As seen in this figure, experimental data of statistical variations of threshold voltage map well to power law dependence on the value of criterion efficiency of structural modification.

Thus, in many cases, the structural modification of the properties of non-crystalline semiconductors is an effective method of device parameter control. But it is also necessary to take into consideration that the possibility of structural changes in non-crystalline semiconductors in the period of device operation defines the value of unavoidable variation of device parameters.

4.5 Chemical Modification of the Atomic Structure of Disordered Semiconductors

It was noted earlier that doping of non-crystalline semiconductors with donor or acceptor impurities usually does not lead to a shift in the Fermi level and, therefore, to any significant changes in their electrical properties (see introducing to Chapter 4). At the same time, the addition of impurities may lead to changes in the atomic structure (chemical modification of the structure) and in the defect subsystem and thus influence the properties of the material. Let us consider chemical modification of the structure and defect subsystem on the example of glassy selenium.

Admixtures in selenium are usually divided into three groups in accordance with their influence on the structure (Popov, 2004): isoelectron admixtures (oxygen, sulfur, tellurium), branching admixtures (elements of the fourth and fifth groups), and univalent admixtures (hydrogen, halogens, alkaline metals, thallium).

A similar construction of the outermost electron shell of atoms of the sixth group most likely prevents the formation of new structural units under the doping of selenium by isoelectron admixtures. The mixture of tellurium with selenium leads to the formation of mixed-up molecules and a strong decrease in the level of chain molecule polymerization. An additional peak appears on the RDF curve at the distance 2.8 Å equal to the bond length Te–Te. The decrease in the degree of polymerization leads to a decrease in the crystallization activation energy of selenium with the admixture of tellurium. The lower potential of tellurium atom ionization promotes the expansion of the photoconductivity spectrum of doped selenium in a long wave range. Resistivity and activation energy of conductivity both decrease consistently with the growth of tellurium concentration.

With the addition of sulfur, a peak at 355 cm^{-1} appears in the Raman scattering spectrum of selenium and the intensity of this peak grows with an increase in the content of sulfur. This peak is linked with the formation of mixed-up rings of sulfur and selenium. Thus, when doped with sulfur, the number of mixed-ring molecules increases and the level of polymerization of molecules Se_n decreases.

In contrast to tellurium and sulfur, small concentrations of oxygen have a significant influence on the electrical properties of glassy selenium. So the resistivity of glassy selenium decreases by six orders of magnitude with the addition of 5×10^{-3} at.% oxygen (Fig. 4.13) (Lacourse, Twadell, and MacKenzie, 1970). The presence of oxygen significantly changes the type of frequency dependence of conductivity (Baratov and Popov, 1990). The dependence of the electrical conductivity of the undoped selenium on the frequency at room temperature (Fig. 4.14, curve 1) in the frequency range of 10^0–10^4 Hz corresponds to the frequency dependence of hopping conductivity

$$\sigma = k \cdot \omega^n \tag{4.14}$$

with the value of $n = 0.64$. When 2.5×10^{-2} at.% oxygen is added to selenium (Fig. 4.14, curve 2), the n value drops to less than 0.1 in the frequency range 10^2–10^4 Hz. This indicates transition from the hopping conductivity in undoped selenium to the conduction through delocalized states in selenium doped with oxygen.

It is interesting to note that the addition of arsenic in selenium doped with oxygen returns the frequency dependence of its electrical conductivity to values close to undoped selenium (Fig. 4.14, curve 3). This is probably a result of the interaction of arsenic and oxygen atoms to form an arsenic oxide.

The unusually strong influence of small oxygen concentrations on the electrical properties of non-crystalline selenium is explained (Popov, 2004) by the interaction between oxygen atoms and the subsystem of intrinsic charged defects in selenium. Due to the significantly higher electronegativity (3.5 for oxygen and 2.4 for selenium), oxygen atoms create O_1^- defects, which leads to an increase in the hole concentration in the material.

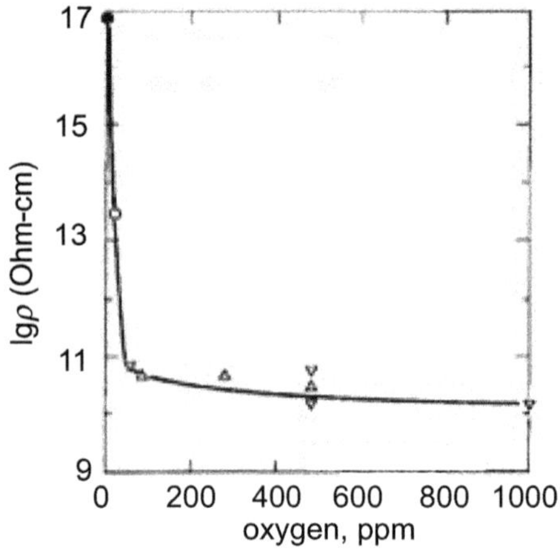

Figure 4.13 Dependence of non-crystalline selenium resistivity on oxygen concentration. Reprinted from Lacourse et al. (1970), with permission from Elsevier.

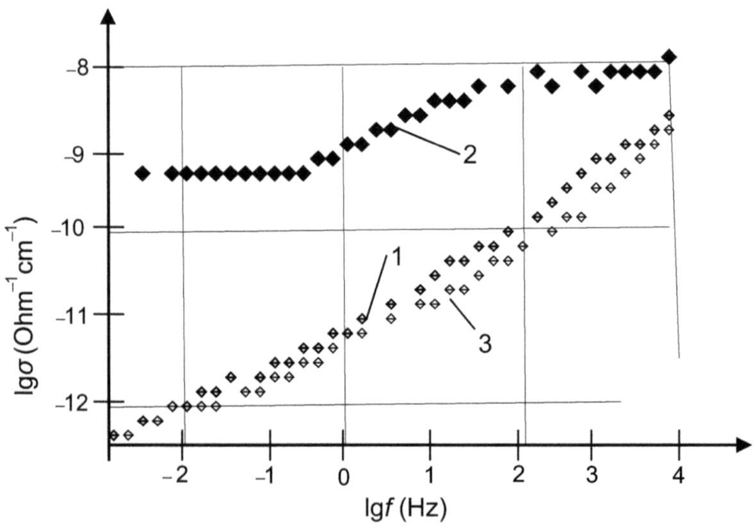

Figure 4.14 Dependence of non-crystalline selenium conductivity on frequency at room temperature (1: undoped selenium, 2: selenium doped with oxygen, 3: selenium doped with arsenic and oxygen).

Among the different univalent admixtures, the influence of halogens on selenium has been studied more extensively. Atoms of halogens have the electron configuration s^2p^5 and can terminate the ends of polymer molecules Se_n (Fig. 4.15a)

$$Se_1^0 + X_0^0 \rightarrow Se_2^0 + X_1^0 \tag{4.15}$$

or interact with the selenium atom inside a molecule and form covalent bond. In this case, the coordination of the selenium atom increases and an Se_3^+ center is formed (Fig. 4.15b):

$$Se_2^0 + X_0^0 \rightarrow Se_3^0 + X_1^0 \rightarrow Se_3^+ + X_1^0 + e \tag{4.16}$$

The released electron can form an Se_1^- center with an atom of selenium at the end of the polymer molecule (see Section 3.1.3):

$$e + Se_1^0 \rightarrow Se_1^- \tag{4.17}$$

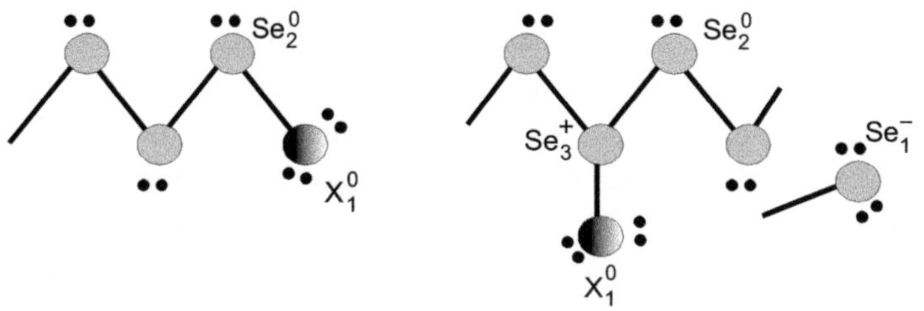

Figure 4.15 Bonding configurations of halogen atom (X) in non-crystalline selenium.

The aforementioned kinds of interactions of halogen atoms with selenium will influence the molecular structure of the latter in the following manner. The termination of chain-ends with atoms of halogens will lead to a decrease in the degree of polymerization of chain molecules and must bring decreasing viscosity and activation energy for the crystallization of matter. The formation of Se_3^+ centers connected to the halogen atoms reduces the number of Se_3^+ centers that bond the molecules of selenium with each other. The reduction in the amount of bridges between molecules must also lead to decreasing viscosity and activation energy for crystallization of matter even at a constant level of polymerization. The results of experimental investigation of the influence that the halogen (bromine) admixture has on the structure of selenium confirm the above conclusions (Popov et al., 1980).

The addition of arsenic (branching admixture) to selenium decreases the concentration of ring molecules and is proved by the results of Raman spectra investigation. Besides, arsenic atoms bond separate molecules of selenium with covalent bonds that leads to rising glass transition temperature and activation energy for the crystallization of selenium in line with increasing arsenic concentration. An admixture of germanium affects selenium in a similar way (Mott and Davis, 1979).

4.6 Structural, Chemical, and Phase Modification of Amorphous Diamond-Like Silicon–Carbon Films

It was noted in Section 2.5.4 that diamond-like amorphous silicon–carbon films (ASCFs) exhibit high stability. This makes it possible to change their properties over a wide range using chemical, structural, and phase modification. Let us consider these possibilities in the case of ASCFs and metal-containing nanocomposites based on ASCFs produced by the simultaneous decomposition of polyphenylmethylsiloxane (PPMS) and magnetron sputtering of metal.

4.6.1 Structural Modification of ASCFs

In order to achieve structural modification of ASCF properties, two methods have been used (Barinov et al., 2015a):

1. Changes in the magnitude of negative bias voltage applied to the substrate holder during film growth
2. Presence and partial pressure of argon in the chamber during film growth

The value of the negative bias voltage applied to the substrate holder determines the kinetic energy of positively charged PPMS particles generated in plasma. Increasing the voltage leads to an increase in the mobility of the particles on the substrate, an increase in their packing density, and a decrease in the concentration of dangling bonds. This increase in the packing density of particles in the film alongside the increase in bias voltage is confirmed by the results of surface morphology study using atomic force microscopy (Fig. 4.16; Barinov, 2014). As can be seen from the dependence of the waviness of the film surface on the voltage applied to the substrate holder, waviness is reduced threefold when the bias voltage is increased from 100 to 1000 V.

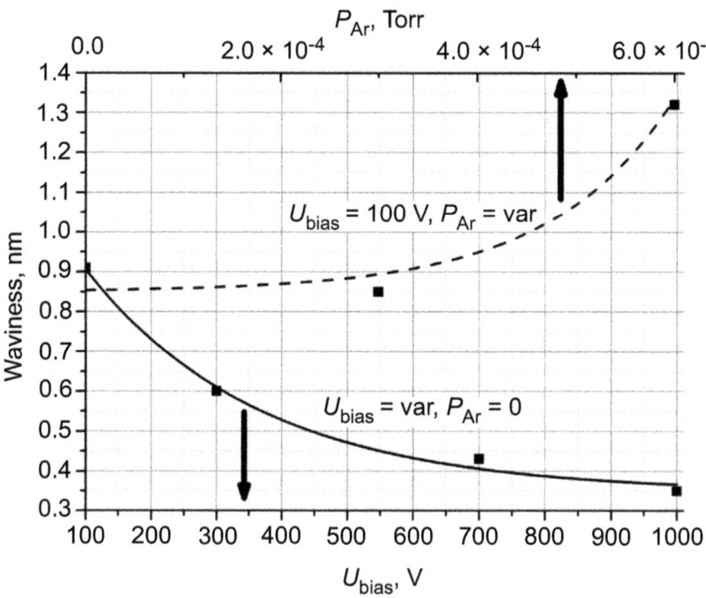

Figure 4.16 Dependence of waviness of the film surface on the bias voltage applied to the substrate holder (U_{bias}) and on the partial pressure of argon in the reactor.

The aforementioned changes in the structure of the films induce changes in their electrical properties. An increase in bias voltage leads to a decrease in the absolute values of electrical conductivity and to an increase in the activation energy of conductivity. Electrical conductivity in diamond-like silicon–carbon films is defined by the hopping mechanism in the tails of the localized states (Barinov et al., 2014)

$$\sigma = \sigma_0 \exp\left(-\frac{E_a - E_F + w}{kT}\right) \quad (4.18)$$

where E_a is the tail of the conduction band edge, E_f is the Fermi-level position, and w is the activation energy of the charge carrier mobility in the tail of localized states. An increase in packing density leads to a reduction in the concentration of dangling bonds and to a decrease in the density of localized states in the band tails as well as to a reduction in the length of band tails themselves. The reduction in dangling bond concentration leads to an increase in the barriers between localized states and, as a consequence, to an increase in the activation energy of charge carrier mobility (w). The reduction in band tail length leads to an increase in the activation energy of the charge carrier concentration ($E_a - E_f$). The combined result of these two factors is reduction in the conductivity of the material. It should be noted that Shustin et al. (2013) observed a similar decrease in the electrical conductivity of carbon films prepared by plasma-chemical deposition method, when increasing the bias voltage applied to the substrate.

Adding argon to the vacuum chamber (in the presence of a negative bias voltage on the substrate holder) leads to the bombardment of the growing film by argon ions. The intensity of the bombardment increases with an increase in the partial pressure of argon. Ion bombardment of the film during its growth should lead to an increase in the concentration of defects in the form of dangling bonds. Consequently, there should be effects opposite to those observed with increasing negative bias voltage applied to the substrate. Experiments confirm the expectations. The waviness of the film surface rises with the increase in the partial pressure of argon (Fig. 4.16). The results of deep-level transient spectroscopy (Q-DLTS) of films (Fig. 4.17) show that the height of the peak in the case of films prepared in the presence of argon and hence the concentration of localized states are substantially higher than those of films produced in the absence of argon.

In the case of hopping conduction in band tails, an increase in the density of localized states leads to an increase in electrical conductivity. This is consistent with the character of electrical conductivity temperature dependence of silicon–carbon films prepared in the absence of argon in the chamber and in the presence of argon with various partial pressures. The change in the partial pressure of argon causes a change in conductivity by nearly six orders of magnitude (from 3×10^{-12} to 10^{-6} $\Omega^{-1}cm^{-1}$ at room temperature).

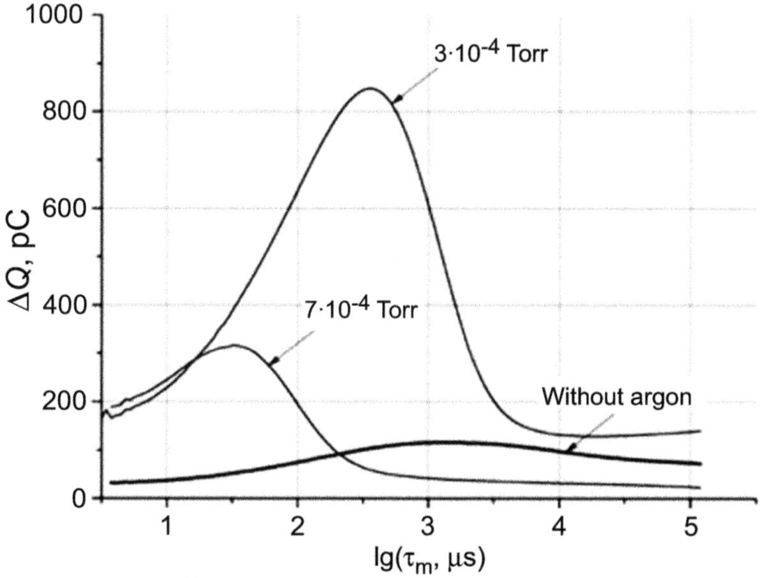

Figure 4.17 Q-DLTS spectra of films prepared at various argon partial pressures.

An increase in electrical conductivity is accompanied by a decrease in its activation energy (from 0.53 to 0.24 eV), which is also consistent with an increase in the density of localized states in the band tails. Dependences of conductivity on the magnitude of negative bias voltage applied to the substrate holder during film growth and on the partial pressure of argon in the chamber during the film growth are shown in Fig. 4.18.

4.6.2 Chemical and Phase Modification of ASCF

The high stability of amorphous silicon–carbon matrix makes it possible to add high concentrations of metals to the material in order to control its properties. In other words, it allows for chemical modification of its properties. The addition of metal results in the formation of nanocrystalline phase in the amorphous matrix, that is, a change in the phase composition of the material. In most cases, the nanophase consists of carbide metal nanocrystals (see Fig. 2.19). It should be noted that the carbon for forming metal carbide is extracted from the silicon–carbon matrix. Thus, the addition of metal leads not only to a change in the chemical and phase composition of the sample as a whole (due to the formation of metal carbide nanocrystals), but also to changes in the chemical composition of the silicon–carbon matrix, due to the depletion of its carbon atoms. Consequently, changes in the properties of the material caused by the addition of metals may be conditioned by two factors: formation of a new phase (metal carbide) and the change in the chemical composition of the initial silicon–carbon matrix.

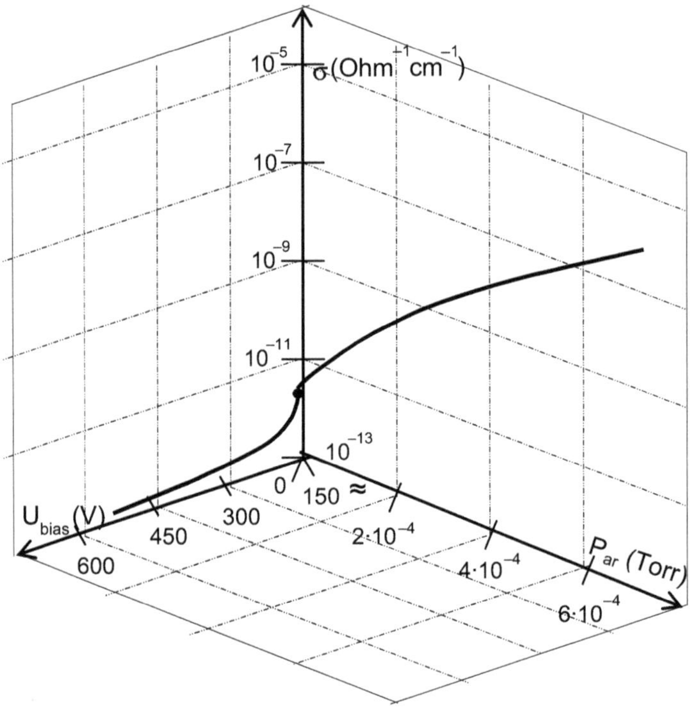

Figure 4.18 Conductivity dependences of silicon–carbon films on technological factors (U_{bias} is negative bias voltage applied to the substrate holder and P_{ar} is the partial pressure of argon in the chamber).

Consider the dependences of some properties of diamond-like silicon–carbon films on the added metal concentration. The dependences of electrical conductivity of films on the concentration of the metal added to them are shown in Fig. 4.19 (Barinov et al., 2015b).

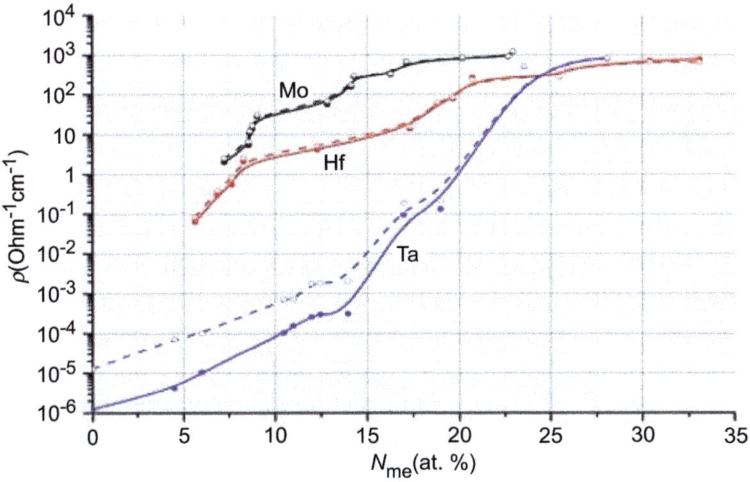

Figure 4.19 Dependence of silicon–carbon film electrical conductivity (solid line at $T = 300$ K, the dotted lines at $T = 400$ K) on the added metal concentration.

As can be seen from the figure, in the case of tantalum-containing films, the dependence has three clearly expressed areas. This kind of dependence is typical for percolation systems (Zolotukhin, Kalinin, and Sitnikov, 2006). In the first area when the metal concentration is in the range 0–14 at.%, electrical conductivity increases smoothly from the value of 10^{-6} Ω^{-1}cm^{-1} to 3×10^{-4} Ω^{-1}cm^{-1} at $T = 300$ K, that is, over two orders of magnitude. In the second area where the metal concentration varies between 14 and 22 at.%, there is a sharp growth in electrical conductivity by six orders of magnitude. Finally, in the third area (where the metal concentration exceeds 22 at.%), conductivity shows virtually no dependence on metal concentration. Thus, the addition of metal allows to change electric conductivity by nine orders of magnitude. The differences in temperature dependences and, consequently, also in the mechanisms of electrical conductivity in the three areas mentioned above should not be overlooked. In the first area, there is a strong dependence of electrical conductivity on temperature. This indicates its activation nature with activation energy 0.2–0.3 eV. In that range of metal concentrations, tantalum carbide nanocrystals do not have contact with each other and electric conductivity is determined by the transport of charge carriers in the amorphous silicon–carbon matrix. Hopping conductivity via localized states caused by defects in the amorphous material is typical for such systems. At the same time, the increase in electrical conductivity in this area by more than two orders of magnitude indicates a significant increase in the density of localized states in the amorphous silicon–carbon matrix with the rise in metal concentration, leading to a decrease in the activation energy of both concentration and mobility of charge carriers (see Eq. 4.18). In turn, the increase in the density of localized states can be caused by a decrease in the concentration of carbon atoms in the silicon–carbon matrix with the formation of metal carbide nanocrystals.

In the third area, the temperature coefficient of conductivity is close to zero. This indicates the metallic nature of transport of charge carriers through the channels formed by nanocrystals of tantalum carbide. Finally, the second area is a transitional area from the conductivity in the amorphous matrix to the percolation conduction through the nanocrystals. This is evidenced by a sharp increase in the absolute values of electrical conductivity, and a sharp decline in its temperature dependence.

Silicon–carbon films with molybdenum and hafnium are characterized by the shift in the boundaries between the three aforementioned areas of concentration dependences of conductivity in the direction of lower metal concentrations. In this case, judging by the nature of the concentration dependences as well as by the magnitude of the temperature coefficient of electrical conductivity, the area with conductivity through channels formed by nanocrystals occurs when the metal concentration reaches 10–15 at.%. The reason for this may lie either in a different geometry of the nanocrystalline carbide phase or different chemical affinity for oxygen of the added metals.

Changes in the chemical composition of the silicon–carbon matrix (and hence in its structure) due to the formation of metal carbides, as well as the formation of the nanocrystalline phase, should influence the mechanical properties of the material. In this connection, it is interesting to analyze the dependence of film hardness and elastic modulus on metals concentration (Fig. 4.20, Barinov et al., 2016).

An increase in the concentration of metal carbides' nanocrystals that have high values of mechanical properties should lead to an increase in the hardness and modulus of elasticity of the samples with higher metal content. It can be seen from the figure that just such dependences are observed in the case of tungsten-containing samples.

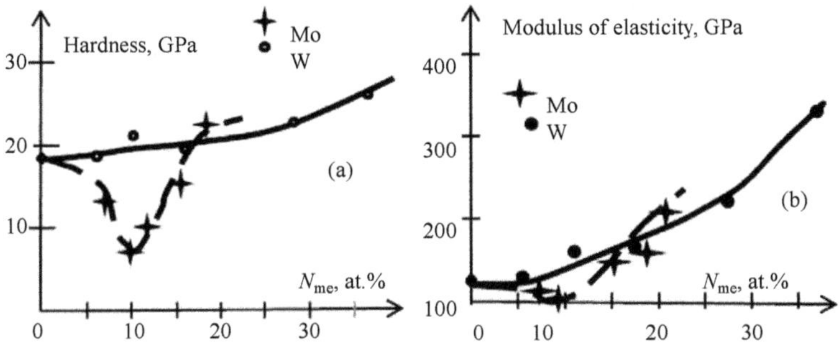

Figure 4.20 The dependences of hardness (a) and elastic modulus (b) of silicon–carbon films on metal concentration.

However, in the case of molybdenum-containing films, there is a different kind of concentration dependences. Hardness and the modulus of elasticity decrease with an increasing concentration of the metal to a minimum value at 9 at.% molybdenum, and then rise, reaching the undoped sample value at 15–17 at.% molybdenum. Moreover, the magnitude of decrease in hardness at the minimum reaches 60% of the hardness of the undoped sample. This type of concentration dependences for samples with molybdenum at low metal concentrations can be explained by a decrease in the mechanical properties of the silicon–carbon matrix caused by the decline in the number of carbon atoms with sp^3-hybridization of the electron orbitals and insufficient concentration of carbide metal nanocrystals for the compensation of this decline.

However, this prompts the question of why the type of property concentration dependences is different in the case of tungsten-containing and molybdenum-containing films. Investigations performed by high-resolution transmission electron microscopy have shown that molybdenum carbide nanocrystals are larger in size than tungsten carbide nanocrystals with the same metal content. An increase in the size of nanocrystals leads to a reduction in their number and to an increase in the distance between them. As a result, mechanical properties at low concentrations of molybdenum are mainly determined by the properties of the silicon–carbon matrix that change, as noted above, due to a decline in the number of carbon atoms with sp^3-hybridization of the electron orbitals.

4.6.3 Nanostructuring of ASCF

It was shown above that the chemical modification of diamond-like carbon–silicon films by introducing metals leads to the formation of a nanocrystalline phase and significant changes in the physical and chemical properties of the material. In this case, the nanocrystalline phase is distributed uniformly over the sample volume. At the same time, an effective method of modifying the properties is to create a specified distribution of the nanocrystalline phase or nanostructuring of the material.

The creation of a certain distribution of the nanocrystalline phase over the sample thickness is possible by the controlled variation in the relation between the growth rate of the silicon–carbon matrix and the metal sputtering rate (Popov et al., 2014a). A general view of the transverse cross section and the plot of the tantalum distribution over the thickness of the film prepared using this method are shown in Fig. 4.21. Five different layers were formed in the film during its fabrication: The first layer does not contain tantalum, and layers from the second to the fifth have different concentrations of tantalum. The microphotograph is taken in the mode of scanning transmission electron microscopy with the recording of electrons scattered at large angles, which makes it possible to obtain images with contrast sensitive to the atomic number Z or with the so-called Z contrast. The lighter regions in this image correspond to a higher tantalum concentration. The concentrations of tantalum and other elements in these layers are shown in Table 4.7.

Figure 4.21 and Table 4.7 show that with a decrease in the average growth rate of the silicon–carbon matrix (V), the average tantalum concentration in the sample increases from zero (in the first layer at the switched-off magnetron) to 80 wt.% in the fifth layer. The silicon–carbon matrix in all layers remains amorphous regardless of the metal concentration, which is confirmed by the high-resolution electron microscopy pictures and the character of Fourier images obtained from these pictures. At the same time, nanocrystals are observed in microphotographs in all layers in which tantalum is present. The two-dimensional Fourier image indicates that they are tantalum carbide (TaC) nanocrystals.

Table 4.7 Average values of film growth rate and values of element concentrations in layers

Layer no.	V, nm/min	Carbon, wt%	Silicon, wt%	Oxygen, wt%	Tantalum, wt%
1	43	58	32	10	0
2	43	31	16	7	46
3	28	26	15	4	55
4	8.75	10	13	2	75
5	5.3	6	12	2	80

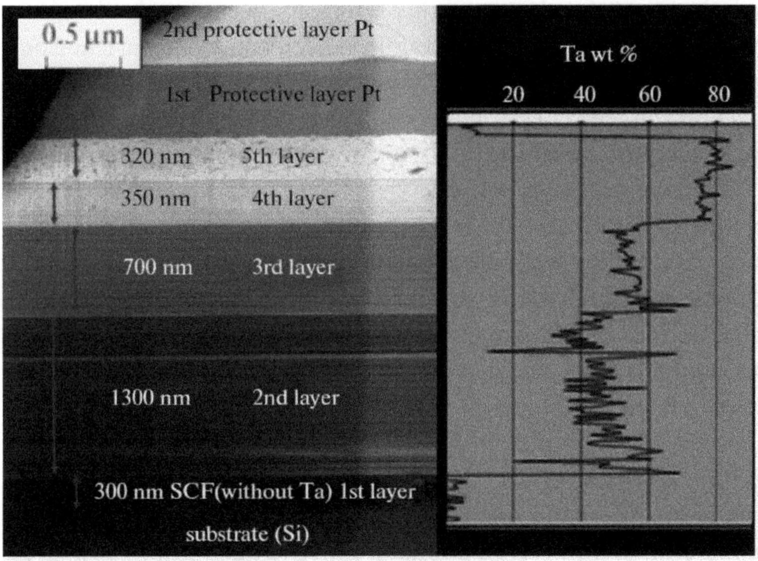

Figure 4.21 General view of the transverse cross section (dark-field image) and tantalum distribution over the film thickness. Reprinted from Popov et al. (2014a), with permission from Springer.

Figure 4.21 also shows a plot of the tantalum distribution in each of the layers formed under different technological modes. In the first layer prepared on the switched-off magnetron, tantalum is absent and the film is an amorphous silicon–carbon matrix. In the second layer, polyhedral tantalum carbide nanocrystals are observed with diameters of 1–2 nm. Nanocrystals are distributed inhomogeneously over the layer's thickness. The regions of their increased concentration (the tantalum concentration in these regions is 45 wt.%) with thicknesses of about 5 nm alternate with the regions with decreased concentrations of the nanocrystalline phase (the tantalum concentration of 25 wt.%) with thicknesses of 15–20 nm. The tantalum concentration in these regions differs, on average, 1.5–2 times. In the third and fourth layers, the character of the nanophase distribution (polyhedral tantalum carbide nanocrystals) has the same regularities as those in the previous layers. However, with the increase in the average tantalum concentration, the concentration of tantalum carbide nanocrystals and the thickness of regions enriched with nanocrystals increase and the distances between these regions decrease. At the same time, the difference in the tantalum concentrations in enriched and depleted regions decreases. Finally, in the fifth layer, at a tantalum concentration of 80 wt.%, nanocrystals are distributed over the layer thickness quite uniformly.

The dependences of the geometric characteristics of the nanophase distribution over the layer thickness and the dependence of the ratios of the tantalum concentrations in enriched and depleted regions on the ratio of the tantalum sputtering rate to the growth rate of the silicon–carbon matrix are shown in Fig. 4.22. It can be seen in this figure that the thickness of regions enriched with nanocrystals changes from 5 nm to more than 30 nm, and the distances between enriched layers (depleted layers) decrease from 15–20 nm to 2 nm with an increase in the tantalum sputtering rate. The ratio of the tantalum content in enriched and depleted regions decreases with an increase in the average tantalum concentration from 250% at an average tantalum concentration of 46 wt.% to 20% at an average tantalum concentration of 80 wt.%.

Thus, it is possible to control the nanophase distribution over the thickness of the films of metal-containing silicon–carbon composites in a wide range by the controlled variation of the ratio of the tantalum sputtering rate to the growth rate of the silicon–carbon matrix.

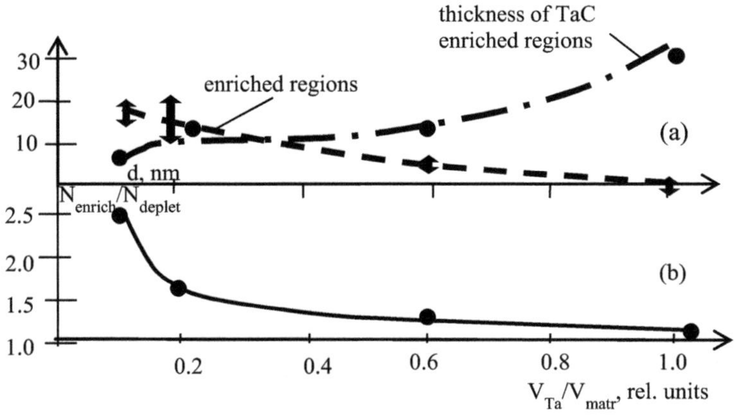

Figure 4.22 Dependences of the geometric characteristics of the nanophase distribution over the thickness of layers (a) and the ratios of tantalum concentrations in enriched and depleted regions (b) on the ratio of the tantalum sputtering rate to the growth rate of the silicon–carbon matrix. Reprinted from Popov et al. (2014a), with permission from Springer.

Chapter 5

Preparation Methods of Disordered Semiconductor Films

The use of disordered semiconductors in a device is usually limited to thin films. Such films must have predefined properties, and the preparation methods must guarantee the reproducibility of these properties. Not surprisingly, different methods are used for the preparation of different disordered semiconductors. For example, thin films of chalcogenide glassy semiconductors are manufactured by thermal vacuum evaporation and radio frequency sputtering methods, while methods of radio frequency or DC glow discharge gas decomposition, thermal or catalytic chemical vapor deposition, and reactive sputtering are usually used for the preparation of hydrogenated amorphous silicon films.

5.1 Technological Distinctions of Chalcogenide Glassy Film Preparation

The most common method of thin-film preparation is thermal vacuum evaporation. In the case of chalcogenide glassy films, the distinctions in preparation consist of evaporation methods and evaporator designs. The simplest construction is an open evaporator with direct resistive heating. A boat-type evaporator made of a high-melting-temperature metal (tantalum or tungsten) with an ingot of

Disordered Semiconductors: Physics and Applications (2nd Edition)
Anatoly Popov

ISBN 978-981-4774-37-6 (Hardcover), 978-1-315-10042-5 (eBook)
www.panstanford.com

material to be evaporated is heated by electric current in a chamber in vacuum below 10^{-5} Torr. The ingot of material melts under the action of Joule heat, turns into a vapor, and deposits on substrate. Film thickness is controlled by the time of deposition with the use of a shutter in front of substrate. This type of evaporator is used (a) in the case of elementary materials that have relatively low melting temperature (e.g., selenium) and (b) in the case of stable chemical compositions (e.g., stoichiometric compositions As_2S_3 and As_2Se_3).

The majority of chalcogenide glassy semiconductors are two- or multi-component materials. These materials are inclined to thermal dissociation. Moreover, components of these materials have different values of vapor pressure and, therefore, different evaporation rates. As a result, using an open evaporator with direct resistive heating leads to a quantitative change in chemical composition along the thickness of the film. (Instead of the initial *AB* composition in the evaporator, the film may be "a sandwich" of *A* and *B* components.) Yet, in some cases, it is possible to prepare a film of two-component material by using this type of evaporator. Various chemical elements have different slopes of curves of saturated vapor pressure on temperature dependencies. If these curves cross at acceptable temperature value, then the evaporation rates of components are equal at this temperature. However, the technological process is intrinsically unstable in this case. Furthermore, questions on the meaning of "acceptable" temperature and determination of the evaporation temperature of chalcogenide film preparation process remain.

The evaporation temperature defines the vapor pressure and thereby the film growth rate. First, the film growth rate should not be too slow or too fast. Second, the thermal dissociation probability of a multi-component material depends on the evaporation temperature as well. Considering the above, a decrease in the evaporation temperature seems to be advisable. On the other hand, the kinetic energy of vapor particles increases with increasing evaporation temperature. This results in improved adhesion of film, which is important from the point of view device reliability. The listed considerations conflict with each other even for simple compositions. And very often it is impossible to find an acceptable compromise for multi-component chalcogenide glassy film preparation. Therefore, special evaporator designs and evaporation methods are employed.

The *methods of flash evaporation* are based on very rapid (momentary) heating and evaporation of a material before the process of thermal dissociation takes place. There are various ways to realize this idea. In one case (large current pulse evaporation), capacitor *C* is charged by voltage source *U* and closed to evaporator (Fig. 5.1a). A large current pulse heats the evaporator instantly up to high temperature. However, in the case of an open-boat evaporator, rapid heating leads to splash out of the material due to "vapor cushion" formation between the evaporator walls and the evaporated material. To prevent this, two covers with displaced outlets cover the evaporator (Fig. 5.1b). This method can only produce films with limited thickness, which is an obvious shortcoming.

An evaporator with a vibrating bin is free from this shortcoming. In this case, the vibrating bin feeds small portions of the material on the evaporator surface heated to high temperature. These portions evaporate instantly on contact with the heated surface. There is no limitation of film thickness. However, the determination of temperature conditions is a rather complicated task in this case. Underheating of the evaporator leads to decomposition of the material, whereas overheating leads to the splash effect.

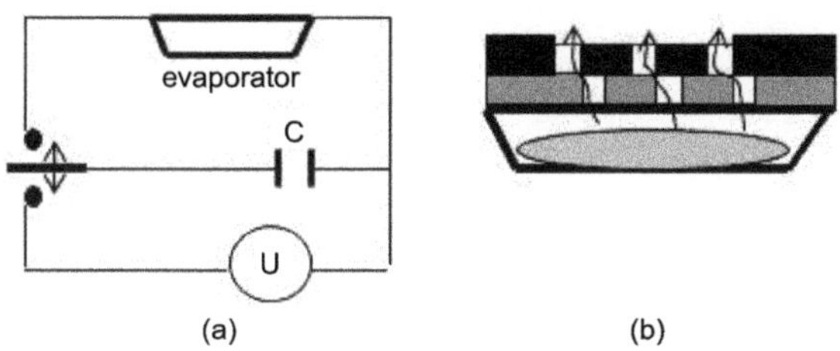

Figure 5.1 Large current pulse evaporation: diagram (a) and evaporator design (b).

A solution to these constraints in the case of chalcogenide glassy films is provided by an evaporator with a continuous feeding of material to be evaporated (Fig. 5.2) (Popov, 2008). The heated part of the evaporator (1) is a tantalum or tungsten band with holes. A piston (9) presses the evaporated material (2) to the heated part of the evaporator. Vapor passes through the holes in the evaporator band.

The evaporator body (4) has a water-cooling system (6) and so only the upper part of the evaporated material reaches the evaporation temperature. Under these conditions, vapor composition does not depend on the evaporation time or on the amount of evaporated material. As a result, the chemical composition of the film remains constant in the process of preparation.

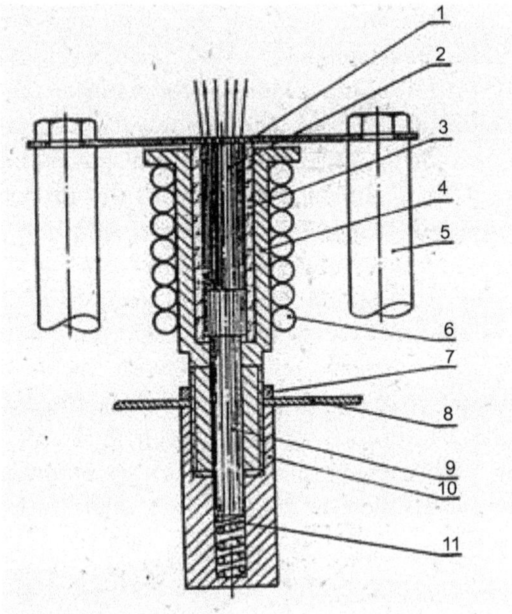

Figure 5.2 An evaporator with continuous feeding of the evaporated material for chalcogenide film evaporation (1 is the heated part of the evaporator, 2 is the evaporated material, 3 is a quartz tube, 4 is the upper part of body, 5 is the current lead, 6 is the water-cooling system, 7 is nut, 8 is basis, 9 is piston, 10 is the lower part of body, and 11 is spring).

The other group of flash evaporation methods is laser-induced evaporation and electron beam evaporation. The essence of the former is the evaporation of a material by short high-intensity laser pulses. A small amount of the material is heated up to 10^6 °C during a laser pulse time of 10^{-7}–10^{-8} s. This method is characterized by

- Identical chemical composition of the vapor and the material being evaporated

- Ability to obtain high rates of film growth (up to 6 μm/min) with exact control of film thickness by the number of laser pulses
- Possibility to obtain super thin films with a single pulse or to produce a super lattice
- Conversion of the evaporated material into plasma due to interaction with electromagnetic radiation and high temperature and deposition of film from plasma
- Absence of heating of vacuum chamber parts and minimal film pollution

The laser-induced evaporation method retains the shortcoming of the splash effect. The characteristics of the electron beam evaporation method are similar to those of the laser-induced evaporation method. However, technical realization is more complicated in this case because the electron gun has to be inside the vacuum chamber.

One of the ways to solve the contradiction between the necessity to decrease the evaporation temperature for minimizing the multi-component material dissociation process and the necessity to increase it for adhesion improvement is the *evaporation from quasi-closed volume method*. The concept behind this method is illustrated in Fig. 5.3.

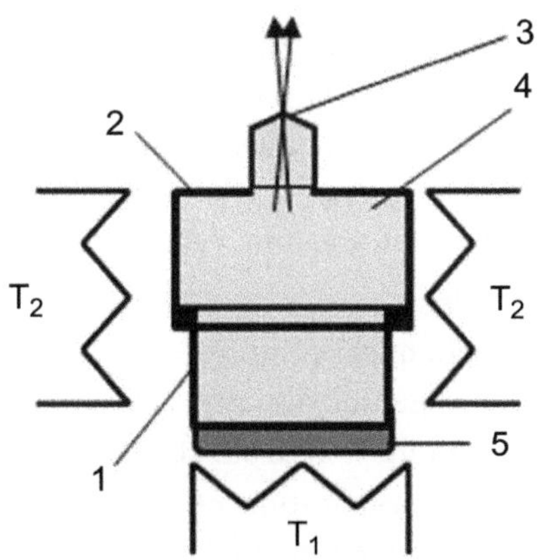

Figure 5.3 Schematic illustration of an evaporator with quasi-closed volume (1 is the evaporator body, 2 is vacuum-tight cover, 3 is capillary, 4 is vapor, 5 is melt, and T_1 and T_2 are heaters).

The evaporator consists of the lower part 1 and cover 2 with a capillary hole 3 on the top. The bottom of the evaporator is heated to temperature $\boldsymbol{T}_1$, which is enough for material melting only. The vapor goes out from the evaporator through the capillary; that is why the vapor pressure increases in the evaporator space over the melt. This part (4) of the evaporator is heated to temperature $\boldsymbol{T}_2 > \boldsymbol{T}_1$. It leads to a rise in the kinetic energy of vapor particles and to an increase in film adhesion.

The methods described above have the same algorithm. At first, bulk chalcogenide glass of the required chemical composition is synthesized. After that various methods are used to prevent dissociation of the synthesized material during the evaporation process. On the other hand, if these materials are inclined to dissociation, it may be better to exclude the process of synthesis at all. This idea was realized in the *method of co-evaporation from different sources*. Let us consider the preparation of a chalcogenide film with A_xB_y composition by this method. Components A and B are loaded in two evaporators. The evaporator temperatures are selected from the condition that the ratio of component evaporation rates (V) is equal to the required ratio of component concentration in the film:

$$V_A(T_A)/[V_B(T_B)] = x/y \tag{5.1}$$

The condensed film composition is equal to A_xB_y under this condition. It is necessary to note that it is possible to prepare films with a determined chemical composition profile along the thickness of the film by temperature changes of one or both evaporators during the evaporation process.

On the whole, thermal vacuum evaporation methods have a number of shortcomings:

- Changes in the chemical composition in the process of multi-component chalcogenide film preparation
- Rather poor film adhesion to substrate
- Difficulties in the preparation of large-area films with even thickness because vapor source has point or linear form
- Sluggishness of process in the case of resistive heated evaporators

It is possible to solve all these problems, but it leads to significantly more complex technological equipment. To a certain degree, sputtering methods are free from these shortcomings.

Sputtering methods include the following modifications:

- Direct current (cathode) sputtering
- Magnetron sputtering
- Radio frequency sputtering
- Reactive sputtering

A schematic illustration of diode-type direct current sputtering is shown in Fig. 5.4a. Cathode (1) and anode (2) are located in the vacuum chamber (3). The target of sputtered material (4) is placed on the cathode, and substrates (5) are placed on the anode. After air evacuation, argon leaks in the vacuum chamber up to a pressure of 10^{-2} Torr, approximately. A high voltage applied to the electrodes leads to electron emission from the cathode. Electrons ionize argon atoms, and glow discharge (6) is initiated between the electrodes. Positive argon ions are accelerated in the electrical field and bombard the cathode target. As a result, the target surface sputters and material particles deposit on the substrates.

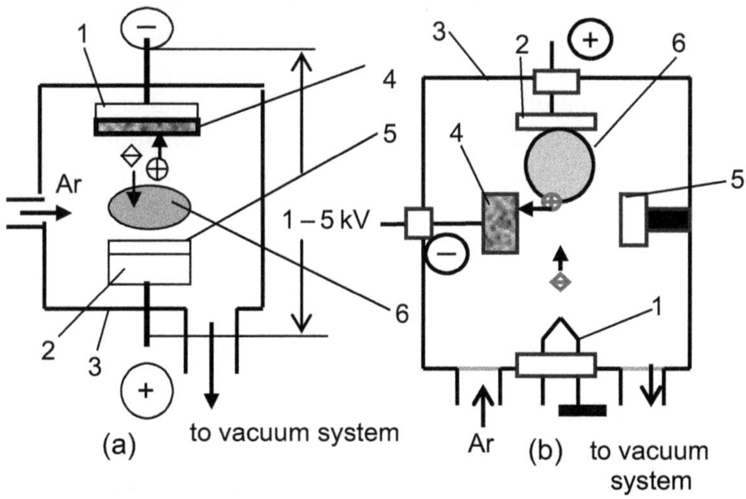

Figure 5.4 Schematic illustration of diode-type (a) and triode-type (b) sputtering reactors (1 is cathode, 2 is anode, 3 is vacuum chamber, 4 is target of sputtered material, 5 are substrates, and 6 is glow discharge).

The sputtering method has the following features:

- Because the energy of bombarded ions (keV) is considerably higher than the energy of chemical bonds (eV), various materials have close sputtering rates. In the case of a diode-type system, sputtering rates are 100–300 Å/min.
- The same reason leads to the high energy of sputtered particles and the high adhesion of the film to substrate.
- Relatively low sputtering temperature prevents dissociation of the material and ensures low chemical composition changes in the film in comparison with the target of the sputtered material.
- It is easier to solve the problem of preparation of large-area films with even thickness because the vapor source has flat form.
- The addition of chemically active gases (oxygen, nitrogen, etc.) to argon in the vacuum chamber during elementary material sputtering allows to prepare compound films with these gases (oxides, nitrides, carbides, etc.). This method is called *the reactive sputtering method.*

A diode-type direct current sputtering reactor has relatively slow rate of film growth. The sputtering rate is determined by the concentration of the bombarded target ions. The latter depends on the concentration of the cathode-emitted electrons and on the gas pressure in the reactor. Because electron emission from the cold cathode is small enough, it is necessary to grow the gas pressure up to 10^{-2}–10^{-1} Torr. On the other hand, increasing the gas pressure leads to a rise in sputtered particle scattering by gas atoms and to a decrease in film growth. Thus, it is not possible to increase the film growth rate within the limits of diode-type direct current sputtering.

A triode-type direct current sputtering reactor (Fig. 5.4b) has been developed to solve this problem. The main difference consists in the division of cathode and target functions. In other words, the third electrode—thermo-emission cathode—is added to the chamber. It increases electron and, consequently, argon ion concentrations and allows to decrease the gas pressure to approximately 10^{-3} Torr. Moreover, it allows to bring substrates nearer the target and to increase the film growth rate as well. As a result, the film growth rate rises up to 1000 Å/min.

However, it is possible to increase the argon ion concentration without increasing the electron concentration by extending electron tracks from the cathode to the anode and appropriately increasing the number of ionization acts. It is known that electrons spiral in the axial magnetic field. Thus, the superposition of electrical and magnetic fields leads to an increase in the gas ionization efficiency and the film growth rate. Application of cross-wise electrical and magnetic fields in the reactor is the basis of *magnetron sputtering*.

Direct current sputtering is an effective method of film preparation in the case of materials with high electrical conductivity. However, it does not give possibility to spurt dielectrics and high-resistivity semiconductors. The reason is that the ions bombarding the target accumulate on the target's surface and form a positive charge that compensates the applied electrical field. It leads to close sputtering process. Therefore, *radio frequency sputtering* is used to prepare films of materials with high resistivity. A high-frequency voltage (10–13 MHz) is applied to the target in this case. Thus, the target is bombarded by positively charged gas ions during the time of half-period and is bombarded by electrons that neutralize the ion positive charge during the time of the next half-period. In fact, the target is bombarded by positively charged gas ions for the duration of time more than the half-period as ion mobility is less than electron mobility and the effect of target self-bias takes place. Radio frequency sputtering allows preparation of high-resistivity multi-component chalcogenide glassy films, hydrogenated amorphous silicon films, silicon dioxide, silicon nitrides films, etc.

The methods described above ensure the preparation of films with a given constant chemical composition and with a given profile of chemical composition along the film thickness. One could think that the problem of preparation of films with the required properties is solved. But such conclusion is premature in the case of disordered semiconductor films. The absence of long-range order in the atomic arrangement leads to various atomic structures at constant chemical composition (see Chapter 4) and to different film properties. Therefore, the properties of disordered material films depend strongly on the conditions of film preparation even in the case of invariable chemical composition. The main factors that determine the chemical composition, atomic structure, and consequently properties of disordered material films are shown in Fig. 5.5.

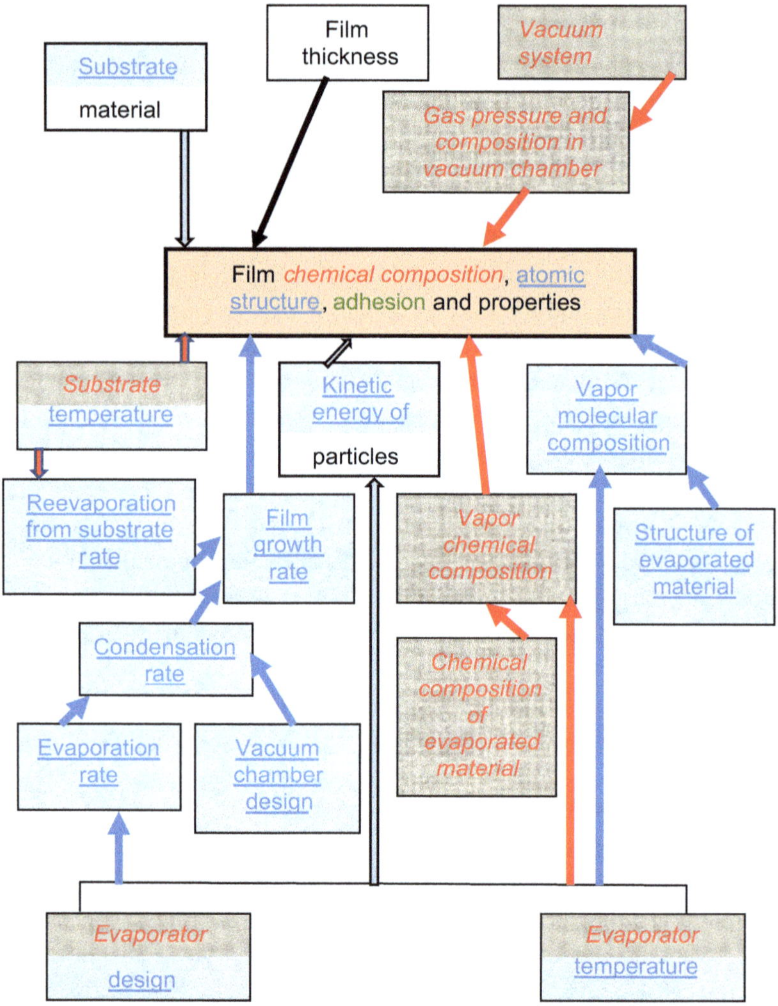

Figure 5.5 Determined film characteristic factors.

The chemical composition of film (red italics in the figure) is determined by the vapor chemical composition. The latter depends on the following factors:

- Chemical composition of the evaporated material
- Evaporator temperature

- Evaporator design
- Substrate temperature
- Gas pressure and composition in the vacuum chamber

The film atomic structure (blue underlined text in the figure) depends to a large extent on the mobility of condensed particles on the substrate. The latter is determined by:

- Substrate temperature
- Kinetic energy of condensed particles, which in turn depends on evaporator temperature and design
- Film growth rate, which depends on the rate of particle condensation and, therefore, on the geometrical characteristics of the vacuum chamber and evaporation rate. The latter depends on evaporator temperature and design as well. Besides that, the film growth rate depends on substrate temperature, because re-evaporation from the substrate is possible at high temperatures.
- Vapor molecular composition, which is determined by evaporator temperature and design and by the structure of evaporated material
- Substrate material

Film adhesion (black bold text in the figure) is determined by the kinetic energy of condensed particles, material and surface conditions of the substrate. These properties also depend on film thickness.

It is necessary to take into account all the factors mentioned above when considering device manufacture. Let us consider a solution to these problems in the production of selenium electrophotographic (xerographic) drums (see Chapter 6). In this case, duralumin drum substrates are used. The drum height has to be no less than the width of the duplicated original (300, 420, and 660 mm as a rule), and drum diameters lie within the limits of 25–250 mm. Because an elementary material (selenium) is used in this case, we can exclude the film's chemical composition changes.

It is known from experiments that selenium evaporation on cold (room temperature) substrate leads to the formation of films that have low value of crystallization activation energy. Spontaneous crystallization takes place even at room temperature in these films. On the contrary, selenium films prepared at the substrate temperature of 70–80°C have high value of crystallization activation energy and remain in non-crystalline state at higher temperatures. In order to explain the facts mentioned above, let us consider the dependence of the concentration of nanocrystalline inclusions in amorphous selenium films on the substrate temperature during their preparation (Fig. 5.6; Montrimas and Petretis, 1974).

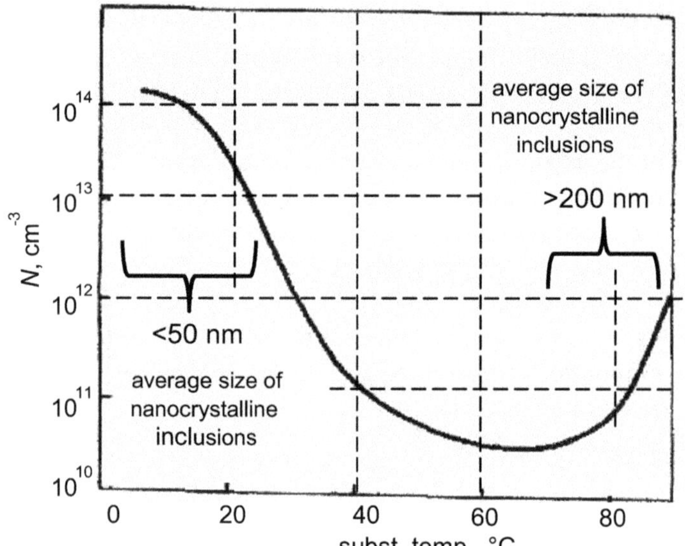

Figure 5.6 Dependence of nanocrystalline inclusions concentration in selenium films on substrate temperature. Reprinted from Montrimas and Petretis (1974), with permission from Elsevier.

As can be seen from the figure, the minimum value of the average concentration of nanocrystalline inclusions is observed in films of selenium fabricated at a substrate temperature of 60–80°C. When the substrate temperature lowers below 20°C, the concentration of inclusions increases by four orders of magnitude. Crystal sizes are less than 50 nm in this case. The increase in the substrate temperature to 90°C also leads to an increase in the concentration of inclusions by about an order of magnitude. Crystal sizes reach 200 nm in this case.

This dependence may be explained by the change in the thermodynamic conditions of film formation. In the case of films depositing on cold substrates ($T_{sub} < T_g$), short vapor species of selenium molecules impinging on the substrate have low mobility and freeze immediately on the film surface. Therefore, polymerization of separate molecular fragments into a long-chain molecule is hardly probable. As a result, this leads to a condensed film structure, which has a short-circuit selenium, high disorders, higher distance between chains, and low activation energy of crystallization (Belev and Kasap, 2009). Subsequently, the above structure of the films favors the process of selenium nanocrystal formation. Molecule mobility on the substrate increases with the substrate temperature. In the case of high substrate temperature, short vapor species of selenium arriving to the surface of the film can migrate and polymerize together to form long molecular chains in the film (see Chapter 2) and rearrangement of these molecules into crystalline lattice requires much more energy. As a result, the activation energy of crystallization increases, and the number of formed crystalline inclusions becomes lower.

Taking the above into consideration, the following requirements for the selenium drum preparation can be formulated:

- The substrate temperature has to be as high as possible without reaching the crystallization temperature. This requires fine control and adjustment of the substrate temperature in the vacuum chamber.
- The evaporation temperature has to be low to ensure sufficiently large molecular fragments in the vapor phase.
- The kinetic energy of molecular fragments in the vapor phase and, therefore, the evaporation temperature have to be high enough to ensure high film adhesion to the substrate.

One of the possible variants to solve these contradictory requirements is shown in Fig. 5.7 (Popov, 2008). The vacuum chamber (1) has cylindrical form. Ringed evaporators (2) are situated on the inside surface of the chamber wall and are heated by heaters (3). The drum substrate (4) is placed axially in the center of the vacuum chamber. Selenium is charged into the evaporators and thermally treated for transformation into crystalline trigonal phase, which consists of long-chain molecules (see Chapter 2). Selenium sublimation from solid state at the temperature 210–215°C is carried out (melt temperature of trigonal selenium is 218°C).

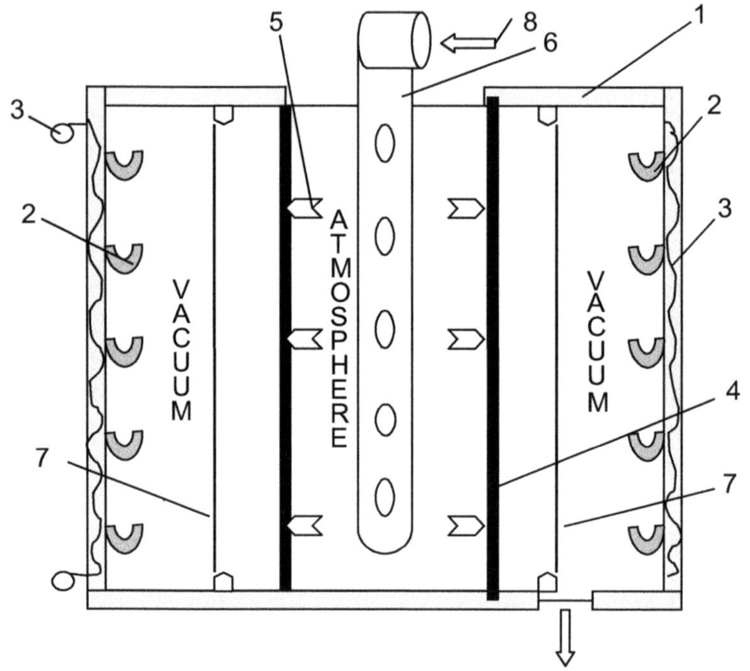

Figure 5.7 Schematic illustration of xerographic drum preparation system.

Substrate heating is fulfilled by the thermal radiation of evaporators. Therefore, the substrate temperature increases with evaporation time and reaches 80–85°C at the end of the film formation process. Sharp cooling of the substrate is necessary to prevent crystallization of the film in this situation. However, it is not easy to realize sharp cooling in vacuum. That is why, the drum substrate performs the function of vacuum chamber wall. The insides of the drum substrate are vented to the atmosphere. It allows accurate monitoring of the substrate temperature by thermocouples (5) and sharp cooling of the substrate by feeding water–air mixture (8) to the insides of drum substrate through a pipeline (6).

The last but not the least problem that needs to be addressed is the problem of film adhesion. Due to the low evaporator temperature (210–215°C), the kinetic energy of vapor particles is not high enough to guarantee good adhesion. It leads to insufficient adhesion of film to the substrate. To eliminate this problem, additional string-like heaters (7) are set in the vacuum chamber. These heaters are switched on at the beginning of evaporation for a short time only and heated up to 2000°C. As a result, the kinetic energy of vapor particles at the initial stage of film formation increases. It leads to a rise in film adhesion to the substrate. An electrophotographic drum manufacturing plant and selenium drums are shown in Fig. 5.8.

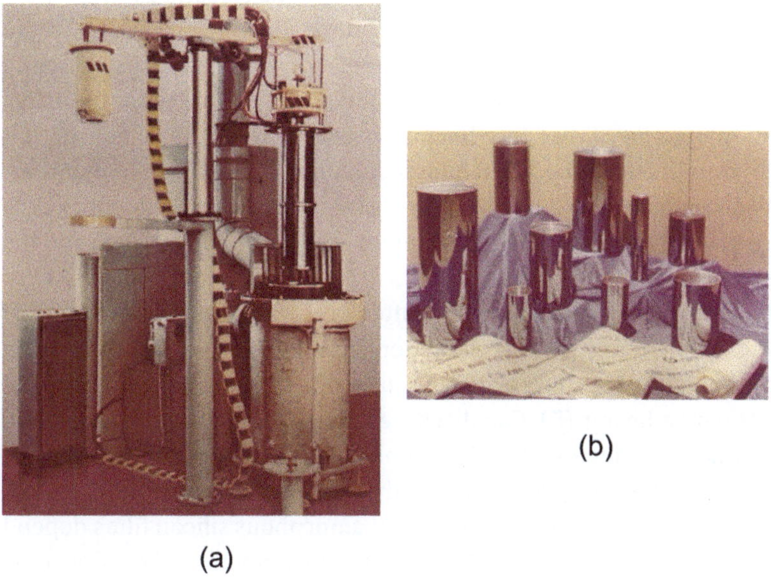

Figure 5.8 Electrophotographic drums manufacturing plant (a) and selenium drums (b).

5.2 Preparation of Hydrogenated Amorphous Silicon Films by Glow Discharge Decomposition Method

Glow discharge represents weakly ionized plasma. The concentration of electrons and positive ions is approximately 10^{10} cm^{-3}. This is 10^4 times less than the gas molecule concentration at pressure 10^{-1} Torr. Electron energy varies from 1 to 10 eV. This is comparable with the energy of chemical bonds. Owing to these characteristics, there is a wide set of different chemical units in the glow discharge plasma: atoms, ions, and various free radicals. These units have heightened chemical activity. That is why chemical reactions in glow discharge take place at lower temperatures in comparison with the normal conditions.

The method of radio frequency glow discharge (13.56 MHz) in silane (SiH_4) or in silane–hydrogen mixture (SiH_4 + H_2) is used for the preparation of hydrogenated amorphous silicon film. There are two types of reactors, which differ depending on the way of radio frequency power input to plasma: reactor with capacitive coupling and reactor with inductive coupling. A reactor with inductive coupling (Fig. 5.9a) has been developed first. The silane or gas mixture feeds to the fused quartz chamber (1) 5–10 cm in diameter. Vacuum pumping-out is used to create gas flow in the chamber. As a rule, the gas pressure in the chamber is equal to 0.1–1.0 Torr, and the gas flow rate lies within the limits of 0.1–10 cm^3/min. Silane is a fire hazard gas. It spontaneously ignites in the air at a concentration of 0.4%. Therefore, a gas reclamation facility (6) has to be mounted in front of the vacuum system. The substrate (3) is placed on a pedestal (4) with a heater (5). Gas-discharge plasma is activated by the radio frequency inductor (2). Silane molecules decompose in the plasma, and silicon atoms precipitate on the substrate.

The properties of hydrogenated amorphous silicon films depend strongly on plasma parameters, substrate temperature, gas flow rate, and geometrical characteristics of the chamber. It has been shown by experiment that the substrate temperature has to be within the limits of 250–350°C and the radio frequency power has to minimize to the value necessary to maintain glow discharge to obtain high-quality hydrogenated amorphous silicon films (films with low concentration of localized states in the gap). At lower temperatures, the mobility of atoms on the substrate surface decreases. This leads to increased defect concentration in the film. The increase in radio frequency power leads to the same effect due to the higher atom deposition rate. At higher temperatures, hydrogen effusion takes place from the a–Si:H film. It gives rise to a higher concentration of localized states in the gap.

Matsuda (1993) investigated the dependence of electron paramagnetic resonance (EPR) spin density in films deposited by the glow discharge method on the substrate temperature during their deposition. The EPR spin density characterizes the concentration of dangling bonds in a–Si:H films. It has been shown that a reduction in the substrate temperature from 250–300°C to room temperature causes an increase in the spin density by three orders of magnitude (from 5×10^{15} to 5×10^{18} cm^{-3}). A temperature increase above 250–300°C also increases the density of spins. However, it should be noted that the above parameters provide rather low film growth rate (100–200 Å/min).

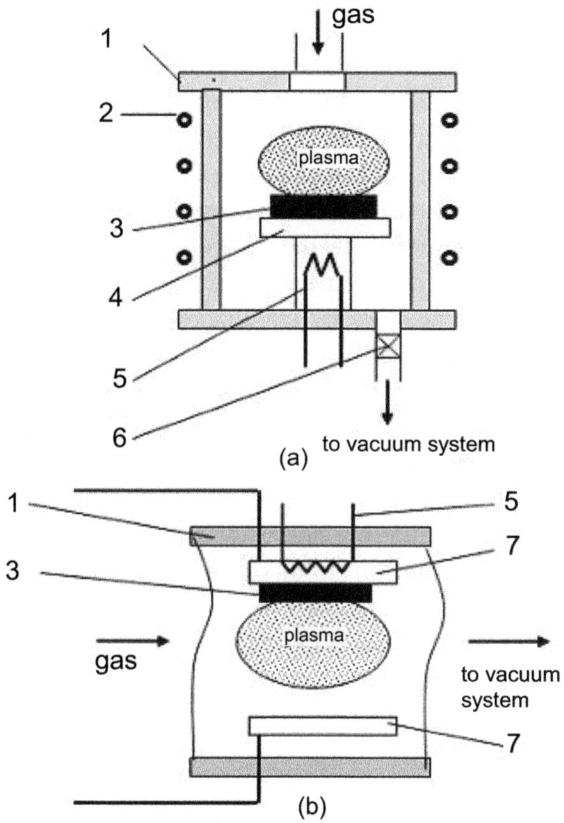

Figure 5.9 Schematic illustration of a reactor with inductive coupling (a) and a reactor with capacitive coupling (b): 1 is chamber, 2 is radio frequency inductor, 3 is substrate, 4 is pedestal, 5 is heater, 6 is gas reclamation facility, and 7 are electrodes.

The main shortcomings of a reactor with inductive coupling are non-uniformity of properties across film area and impossibility of preparing large-area films because increasing the reactor's diameter requires a sharp increase in the radio frequency generator power. That is why the other type of reactor is used in industrial production.

In the case of a reactor with capacitive coupling (Fig. 5.9b), the radio frequency power is applied to parallel flat electrodes (7) inside the chamber. One can see that it allows to increase electrode areas and, therefore, to prepare large-area uniform films. The parameters of deposition process (gas pressure, flow rate, substrate temperature, and radio frequency power) lay within the same limits as in the case of a reactor with inductive coupling.

The possibility of obtaining large-area uniform films is the main reason why reactors with capacitive coupling of radio frequency are applied in manufacturing devices based on hydrogenated amorphous silicon. Mass production requires increasing the production rate of the process equipment. In the case of reactors with capacitive coupling, there are various ways to solve this problem. An increased production rate is possible by simultaneous deposition of layers on several substrates. For example, 30 substrates with an area of about 1.5 m^2 each are placed simultaneously in the reactor for the deposition of hydrogenated silicon layers in the enterprise Hevel Solar, Russia (Fig. 5.10).

However, devices based on hydrogenated amorphous silicon, as a rule, consist of different layers with various types and values of conductivity. Manufacturing of such layers in the same reactor requires a thorough purification process after each deposition in order to avoid mutual contamination of layers. This problem can be avoided through the use of multi-chamber reactors (Hamakawa, 1993). It allows to increase the production rate and to avoid reciprocal contamination of layers. A multi-chamber reactor with a semi-continuous process of production of a–Si:H solar cells is shown schematically in Fig. 5.11 (Vikhrov and Bodygin, 1994).

There are a number of consecutive chambers with independent pumping-out systems. Chambers are separated from one another and from the external atmosphere by vacuum seals (1). Substrates (2) transfer from one chamber to the other by a roller conveyer (4), which simultaneously serves as the lower electrode of the radio frequency system.

Figure 5.10 A–Si:H and mc–Si:H layers manufacturing plant Hevel Solar, Russia. Image courtesy: E. I. Terukov.

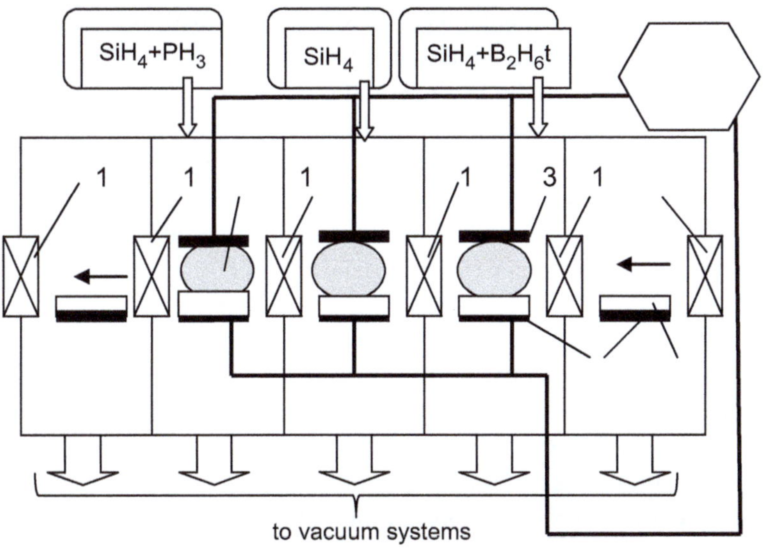

Figure 5.11 Schematic illustration of a multi-chamber reactor with semi-continuous process of *p-i-n* a–Si:H solar cell production (1 is vacuum seal, 2 is substrate, 3 is electrode, 4 is roller conveyer-electrode, and 5 is plasma).

First, the substrate is placed into the inlet chamber. The inlet vacuum seal shuts, and air evacuates from the chamber to create the ultimate vacuum. Then the next vacuum seal opens, and the substrate transfers to the second chamber, which is already evacuated to the ultimate vacuum. The vacuum seal shuts after substrate transfer, and the deposition process of a p-type a–Si:H layer starts after the gas mixture fills up the chamber and glow discharge initiates. At the same time, another substrate is placed into the inlet chamber. When the deposition process of the p-type layer is finished, the substrate transfers to the third chamber to deposit an *i*-type layer. The process continues in the same sequence up to the *p-i-n* structure preparation. It is necessary to note that industrial reactors have a larger number of vacuum chambers for the preparation of contact layers, protection layers, etc. Thus, the reactors allow to prepare finished solar cells.

A schematic illustration of a multi-chamber reactor with a continuous (roll-to-roll) process of production of *p-i-n* a–Si:H solar cells is shown in Fig. 5.12 (Popov, 2008). The reactor includes a number of consecutive chambers with independent pumping-out systems. A stainless steel foil (3) is used as the substrate. It rewinds continuously from the feeding roll (4) in the first chamber to the receiving roll (5) in the last one and passes through all chambers. Chambers are separated from one another by gas seals (1). The gas-feeding system (6) creates areas of higher pressure (8) in the gas seals. It prevents chambers from mutual contamination. Contact layers, a–Si:H layers with different types of conductivity, upper electrode and protection layers are deposited in the chambers, and finished solar cells are obtained on the receiving roll.

As mentioned above, the glow discharge decomposition method is characterized by relatively low film growth rate (about 1 μm/h). If the film thickness of less 1 μm is necessary for device application (thin-film transistors or solar cells, for example), such a growth rate does not hinder the technological process. But film thickness up to 30 μm is required for some applications (electrophotographic drums, for example). In this case, it is necessary to increase the growth rate by at least a factor of 10. It is certainly possible to increase the radio frequency power. Such a solution ensures the acceleration of film growth, but at the same time, it leads to an increase in the localized state concentration in the gap and to a sharp deterioration of film properties (photoconductivity decrease, in particular).

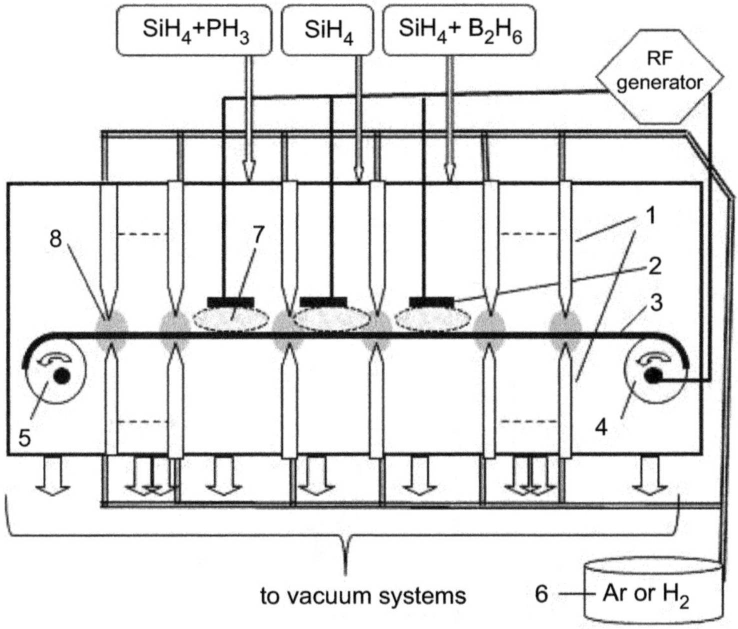

Figure 5.12 Schematic illustration of a multi-chamber reactor with continuous (roll-to-roll) process of *p-i-n* a–Si:H solar cell production (1 is gas seal, 2 is electrode, 3 is stainless steel substrate, 4 and 5 are feeding and receiving rolls, 6 is gas-feeding system, 7 is plasma, and 8 is the area of higher pressure).

The substitution of silane for polysilanes (Si_2H_6 or Si_3H_8) increases the growth rate by up to 20 times within the same technological conditions. Film properties remain practically the same as well (Spear and Le Comber, 1984). The dependencies of conductivity activation energy on the substrate temperature in the case of a–Si:H films prepared from SiH_4 and Si_2H_6 are shown in Fig. 5.13. The radio frequency power and preparation time are the same in both cases, but film thickness is 1 μm in the former and 20 μm in the latter. One can see from the figure that differences lay within the limits of experimental errors. The reason for the acceleration of a–Si:H film growth rate is the following. The limiting reaction stage that determines the film growth rate is the generation of SiH_2

radicals due to the decomposition of silane molecules. In the case of SiH_4 decomposition, reaction is given by

$$SiH_4 \rightarrow SiH_2 + H_2 \quad (5.2)$$

and in the case of disilane decomposition, reaction is given by

$$Si_2H_6 \rightarrow SiH_2 + SiH_4 \quad (5.3)$$

The activation energy of the latter is considerably less. For example, the rate of reaction (5.3) exceeds the reaction (5.2) rate by 100 times at the temperature of 400°C. Therefore, the higher film growth rate in the case of polysilane decomposition is explained by the increase in SiH_2 radical generation. Thus, the substitution of silane for polysilanes ensures acceleration of the a–Si:H film growth rate up to 20 μm/h.

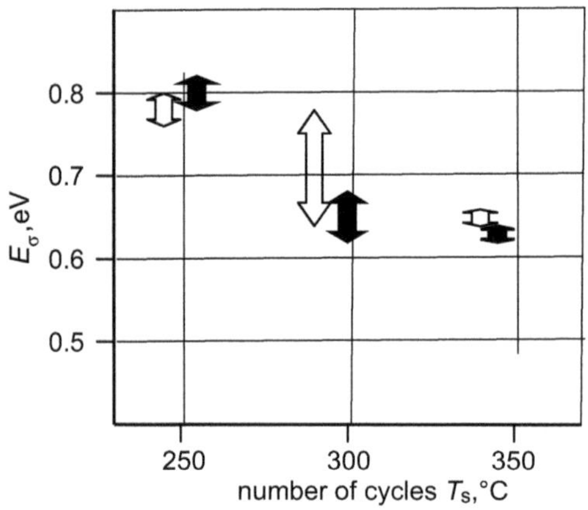

Figure 5.13 Dependences of the conductivity activation energy on the substrate temperature in the case of a–Si:H films prepared from SiH_4 (open marks, thickness is 1 μm) and Si_2H_6 (solid marks, thickness is 20 μm).

However, the cost of polysilanes is significantly higher than that of SiH_4. Moreover, polysilanes are unstable compounds. That is why, as a rule, polysilane is synthesized just before the film growth process, usually in the same reactor. This leads to more complexity in the production equipment and an increase in production costs.

The other problem of devices based on a–Si:H development is connected to the impossibility of low-resistance material preparation. Heavy doped low-resistance n^+ and p^+ layers are usually used to create ohmic (non-rectifying) contact in semiconductor devices.

However, as was mentioned in Section 4.1, it is impossible to prepare hydrogenated amorphous silicon with electrical conductivity more than $10^{-2}\ \Omega^{-1}cm^{-1}$ and conductivity activation energy less than 0.2 eV because of the high density of localized states in the band tails. Thin layers of hydrogenated microcrystalline silicon (μc–Si:H) can be used to create ohmic contacts in devices based on a–Si:H because the conductivity of μc–Si:H is higher than a–Si:H conductivity by three orders of magnitude. Hydrogenated microcrystalline silicon is prepared by the same method of glow discharge silane decomposition and in the same reactors by means of process characteristic changes.

In the general case, the reaction of silane decomposition in glow discharge is expressed by

$$SiH_4 \leftrightarrow SiH_x + (4 - x)H \leftrightarrow (Si + xH)_{sol} + (4 - x)H_{pl} \qquad (5.4)$$

where index "sol" means solid phase and index "pl" means plasma. Direct reaction (from left to right) leads to silane decomposition and a–Si:H film growth. Reversible reaction (from right to left) results in film erosion (etching of film in plasma). In the case of hydrogenated amorphous silicon preparation, direct reaction predominates because the system is far from chemical equilibrium. Proximity to chemical equilibrium leads to slower film growth rate and to increasing mobility of condensed atoms on the substrate. In turn, this creates conditions for crystalline nucleus formation.

Conditions close to chemical equilibrium are ensured by silane dilution with hydrogen up to 2–3% silane and increasing the glow discharge power by 5–10 times. As a result, the film growth rate falls by a factor of about 10. This is evidence of shift in reaction toward chemical equilibrium. The diffraction patterns of silicon films prepared under these conditions show amorphous-crystalline state of material (Spear and Le Comber, 1984). The content of crystalline phase in the film is 65–97% depending on preparation conditions (Fig. 5.14; Hata et al., 1981). Crystallite sizes range from 50 Å to 80 Å.

Investigations of hydrogenated microcrystalline silicon conductivity have shown that the conductivity of undoped μc–Si:H is approximately $10^{-2}\ \Omega^{-1}cm^{-1}$. It is higher than the conductivity of undoped a–Si:H by six orders of magnitude. The values of μc–Si:H film conductivity, $20\ \Omega^{-1}cm^{-1}$ and $8\ \Omega^{-1}cm^{-1}$, have been obtained by PH_3 and B_2H_6 doping, correspondingly (Spear and Le Comber,

1984). This allows to create ohmic contacts in devices based on hydrogenated amorphous silicon.

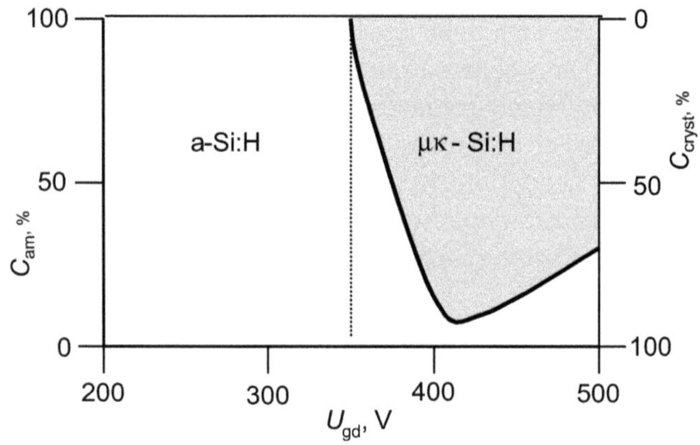

Figure 5.14 Dependence of amorphous (C_{am}) and crystalline (C_{cryst}) phases content in silicon film on glow discharge voltage (U_{gd}). Reprinted from Hata et al., Copyright 1981, The Japan Society of Applied Physics.

Investigations of hydrogenated amorphous silicon stimulated search for new materials with better properties. One of these attempts consisted in the substitution of hydrogen with fluorine because the energy of Si–F bond is significantly more than the energy of Si–H bond. It had hopes to decrease the localized state density in the gap of material and to minimize gas effusion at elevated temperatures. A series of investigations have been carried out with the substitution of SiH_4 with SiF_4 (Madan, 1984). Various gas combinations have been used: SiF_4, SiF_4 + Ar, SiF_4 + H_2, SiF_4 + SiH_4.

However, the total replacement of hydrogen with fluorine gives negative result. The localized state density in a–Si:F films increases in comparison with a–Si:H films. The reason may be in the big value of fluorine electronegativity (3.9). (Silicon electronegativity equals 1.8, and hydrogen electronegativity equals 2.1.) It leads to significant ionicity of Si–F chemical bonds, polarization of neighboring Si–Si chemical bonds, and growth of localized state density in the gap as a result.

At the same time, a combined application of fluorine- and hydrogen-containing gases brings positive result. Silicon films prepared from a SiF_4 and H_2 mixture (12–20% hydrogen) have lower localized state density in the upper part of the forbidden gap in comparison with a–Si:H films. The dependencies of electrical conductivity and the activation energy of conductivity on the doping level of a–Si:F:H films are shown in Fig. 5.15 (Madan and Ovshinsky, 1980). The curves of a–Si:H films are shown in the figure as well. The characteristic feature of fluorinated films is higher conductivity value in the case of heavy doped n-type samples. The following explanation may be provided for this fact. Films with high fluorine content are inclined to crystallization because of increasing chemical bond ionicity. This leads to silicon microcrystalline phase formation even at the normal regimes of film preparation and allows to create ohmic contacts to n-type a–Si:H.

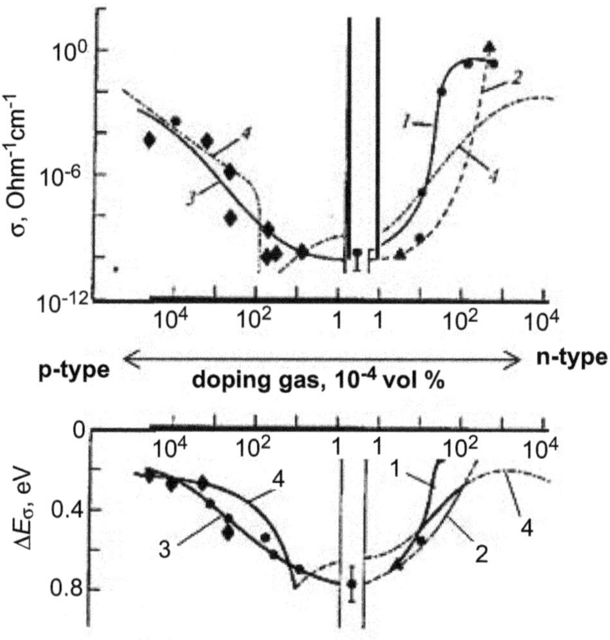

Figure 5.15 Dependences of electrical conductivity (σ) and activation energy of conductivity (ΔE_σ) on the doping level of a–Si:F:H films doped by arsenic (1), phosphorus (2), and boron (3) and a–Si:H films (4).

5.3 Preparation of $A^{IV}B^{IV}$ Alloys on the Base of Hydrogenated Amorphous Silicon

Combination of materials with different values of forbidden gap allows to create new highly effective optoelectronic and photoelectric devices. Compounds $A^{III}B^{V}$ and $A^{II}B^{VI}$ are used for this purpose in the case of crystalline semiconductors. Changes in the forbidden gap value in chalcogenide glassy semiconductors can be obtained by varying the material chemical composition. In the case of tetrahedral bonded amorphous semiconductor alloys, silicon and other elements of Group IV of the periodic table are used to vary the material's gap value.

Silicon forms solid solutions with other elements of Group IV (germanium, carbon, nitrogen). These elements or their compounds with silicon have the value of gap smaller (germanium) or bigger (SiC and SiN_4) than the value of the silicon gap. Therefore, the admixture of gaseous compounds of germanium, carbon, or nitrogen with hydrogen to silane allows to vary the value of the forbidden gap of the material within a wide range (Table 5.1).

Table 5.1 Forbidden gap values of A^{IV}–B^{IV} materials

Material	ΔE, eV
a–Si:H	1.8
a–$Si_{1-x}Ge_x$:H	1.1–1.8
a–$Si_{1-x}C_x$:H	1.8–2.6
a–$Si_{1-x}N_x$:H	1.8–4.0

A–$Si_{1-x}Ge_x$:H films with x value from 0 to 1 are prepared by the glow discharge decomposition of the $SiH_4 + GeH_4 + H_2$ gas mixture. Infrared spectroscopy shows in the films the presence of both Si–H and Ge–H bonds. However, the relative concentration of Si–H bonds exceeds the relative concentration of Ge–H bonds by approximately the factor of 3: [Si–H] / [Si] = 0.11 and [Ge–H] / [Ge] = 0.04. This indicates that hydrogen atoms are more effective in passivating the dangling bonds of silicon in comparison with germanium atoms. At the same time, the relative concentration of germanium dangling bonds (non-paired spin concentration [Ge↑]/[Ge]) decreases more quickly than the concentration of Ge–H bonds ([Ge–H]/[Ge]) grows. It means that hydrogen atoms not only passivate dangling bonds but also decrease their concentration due to structural relaxation.

The dependence of the forbidden gap value on the component ratio in the films of the a–$Si_{1-x}Ge_x$:H system is shown in Fig. 5.16. One can see that changes in the silicon–germanium ratio lead to a linear variation in the material gap value in the limits from 1.1 to 1.8 eV.

Films of the a–$Si_{1-x}C_x$:H system are prepared by the glow discharge decomposition of a mixture of silane and methane (CH_4) or silane and ethylene (C_2H_4). The dependence of the material forbidden gap value on the carbon content (Fig. 5.17) has more complicated form in comparison with materials of the a–$Si_{1-x}Ge_x$:H system and changes under variations in film preparation conditions. The reason lies in the possibility of carbon atoms to have various electron state hybridizations and to form different structural units.

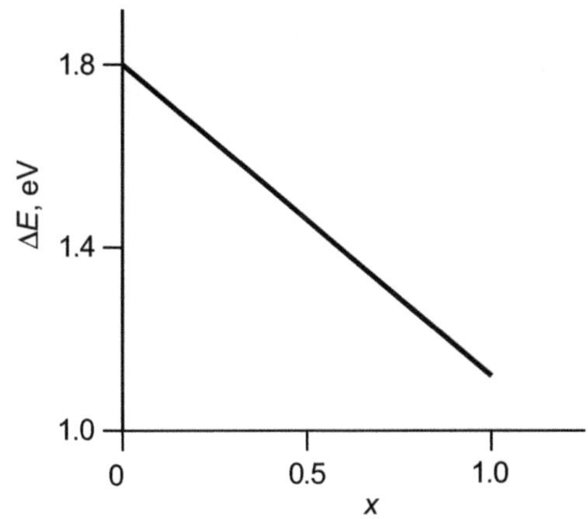

Figure 5.16 Dependence of the forbidden gap value on component relation in the films of the a–$Si_{1-x}Ge_x$:H system.

It is necessary to note that the increase in the number of components leads to a rise in the defect concentration in the film. Therefore, in comparison with a-Si:H, the density of localized states in the a–$Si_{1-x}C_x$:H and a–$Si_{1-x}Ge_x$:H systems is higher. Thus, the change in the material's gap value leads to degradation in the material's optoelectronic quality.

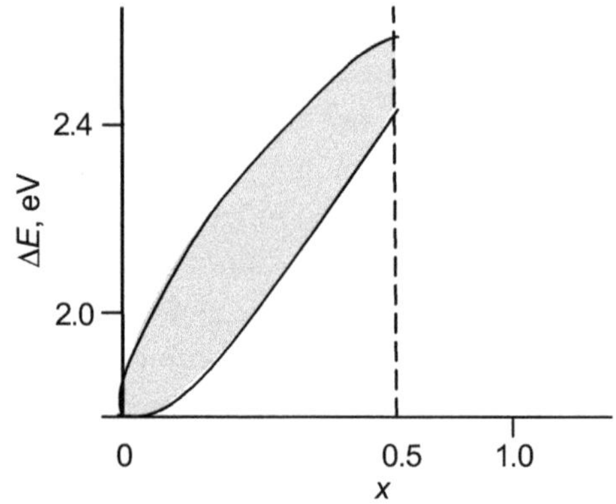

Figure 5.17 Dependence of the forbidden gap value on component relation in the films of the a–$Si_{1-x}C_x$:H system. Reprinted from Madan and Ovshinsky (1980), with permission from Elsevier.

5.4 Preparation of Hydrogenated Amorphous and Microcrystalline Silicon Films by Chemical Vapor Deposition Methods

The thermal decomposition (pyrolysis) of silicon-contained gases (SiH_4, $SiCl_4$) is widely used in the production of single-crystal silicon devices for the preparation of epitaxy layer. It would be convenient to apply this well-known technology to the preparation of hydrogenated amorphous silicon films. A thermal chemical vapor deposition (CVD) reactor consists of a quarts tube located in a furnace. A silicon-contained gas (silane, for example) is thermally decomposed in the tube, and silicon atoms deposit at the substrate. The film growth rate depends exponentially on the temperature. In the case of the epitaxy process, a temperature range of 900–1000°C is usually used. However, these values are not applicable to amorphous silicon film growth because the crystallization temperature of amorphous silicon lies in the limits of 670–700°C. Therefore, a temperature between 550 and 650°C is usually used in the latter case. It ensures amorphous silicon film growth rate of about 100 Å/min. This value is acceptable for some applications. But there is one more problem. A relatively high temperature of deposition leads to sharply decreasing concentration of hydrogen in a–Si:H films and to growing localized state density as a result.

Various methods have been used to overcome this problem (Kaplan, 1984). Additional film treatment in hydrogen plasma (post-hydrogenation) is one of the methods. It increases hydrogen concentration in the films, but material quality does not reach the level of films prepared by the glow discharge decomposition method. The other way is the photo-CVD method, which employs combined thermal and ultraviolet radiation treatment to decompose silane. This allows to decrease the decomposition process temperature and to raise hydrogen concentration in the films. The same result can be reached by the substitution of silane with disilane (see Section 5.2). However, significantly more complex production equipment is needed in both cases.

Another modification of the thermal CVD method is catalytic chemical vapor deposition (CatCVD) or hot wire chemical vapor deposition (HWCVD) (Matsumura et al., 2003). The gas mixture passes through tungsten or tantalum filaments heated to 1600–1800°C (Fig. 5.18; Sunayama et al., 2003) and silane decomposition takes place without heating the whole chamber space. The substrate temperature is kept in the limits of 250–350°C, which ensures sufficient hydrogen concentration in the growing film and low density of localized states in the material.

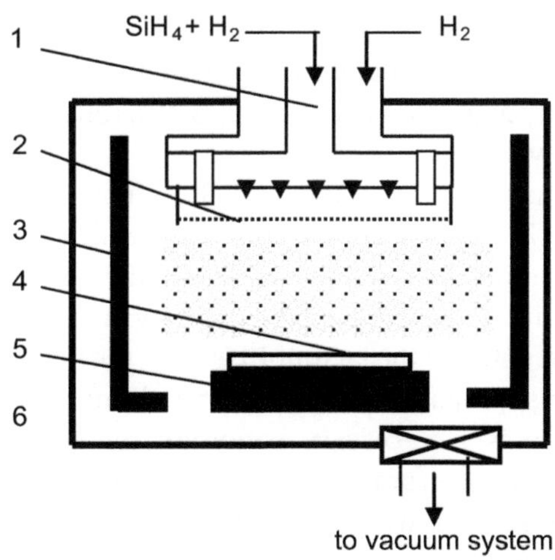

Figure 5.18 Schematic illustration of a hot wire chemical vapor deposition reactor (1 is gas-feeding system, 2 are hot wires, 3 are hot walls, 4 is substrate, 5 is pedestal, and 6 is gas-feeding system). Reprinted from Sunayama et al. (2003), with permission from Elsevier.

The following advantages are attributed to this method (Schroeder, 2003): requirement of much simpler setup for deposition, higher useable deposition rates, and higher stability of HWCVD a-Si:H films. High-quality hydrogenated amorphous and microcrystalline silicon films have been prepared by this technique. Characteristics of thin-film transistors and solar cells on the base of these films are not inferior to the devices on the base of films prepared by the glow discharge decomposition method (Schropp, 2009).

5.5 Preparation of Hydrogenated Amorphous Silicon Films by Sputtering Method

Hydrogenated amorphous silicon films are prepared by silicon target sputtering in an argon–hydrogen atmosphere. In the case of low-resistivity silicon target method, DC sputtering may be used (Cherfi et al., 2001), but the use of high-resistance target is required in the radio frequency sputtering method. Hydrogen atoms interact with the sputtered silicon particles and are incorporated in the growing film (reactive sputtering method). Hydrogen concentration depends on the hydrogen partial pressure (Fig. 5.19), on the discharge power, and the substrate temperature. Increasing the concentration of hydrogen with the growth of partial pressure leads to an increase in the conductivity activation energy (Fig. 5.19), reflecting a rise in the forbidden gap value.

The localized state density in a-Si:H films prepared by the sputtering method is higher in comparison with films prepared by the silane decomposition method. However, the former provides the possibility of film preparation with resistivity of up to 10^{13} Ωcm (Fig. 5.20). That is greater by the factor of 10^2 than the maximal resistivity values of films prepared by the silane decomposition method. High resistivity is a necessary condition for some applications (electrophotographic drums, for example; see Chapter 6).

There are several ways of material doping in the case of sputtering method (Thompson, 1984). First, phosphine or diborane may be added to the argon–hydrogen atmosphere in the chamber. Second, sputtering of the doped silicon target leads to doping of the growing film. Film conductivity changes by 5–7 orders of magnitude depending on the target conductivity (Fig. 5.20). Third, the thermal evaporation of phosphorus- or boron-containing compounds is carried out during the film preparation process.

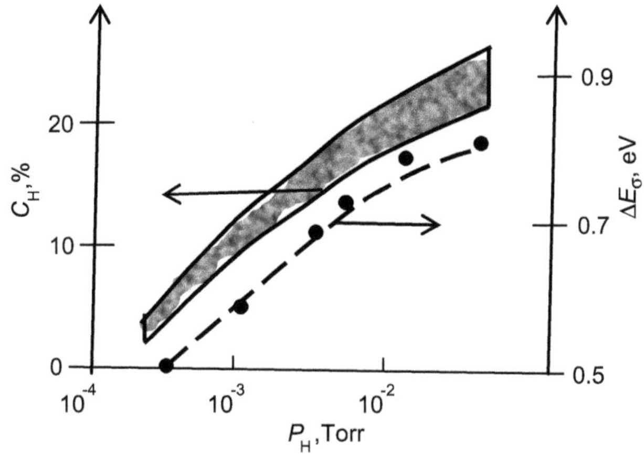

Figure 5.19 Hydrogen concentration (C_H) and conductivity activation energy value (ΔE_σ) of a–Si:H dependences on hydrogen partial pressure (ranges of values correspond to different substrate temperature; substrate temperature for ΔE_σ cover is 250°C).

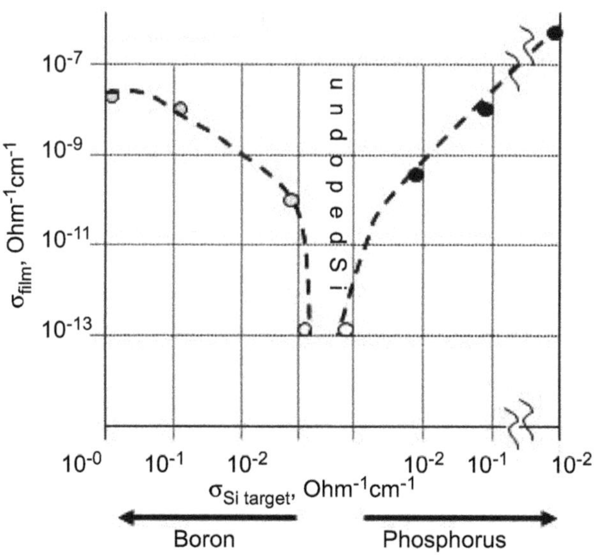

Figure 5.20 Dependence of a–Si:H film conductivity on silicon target.

Summing up, one may see that there are quite a few methods of preparation of disordered semiconductor films. These methods have various merits from the point of view of prepared film and device quality as well as complexity and cost of manufacturing. The specialist proficiency consists in the choice of a method that ensures appropriate device quality at reasonable manufacturing cost.

Chapter 6

Optical Information Storage and Transmission Devices

Disordered semiconductors are widely used as optical information storage and transmission devices as well as high-resolution photo- and electronic resists (Fig. 6.1).

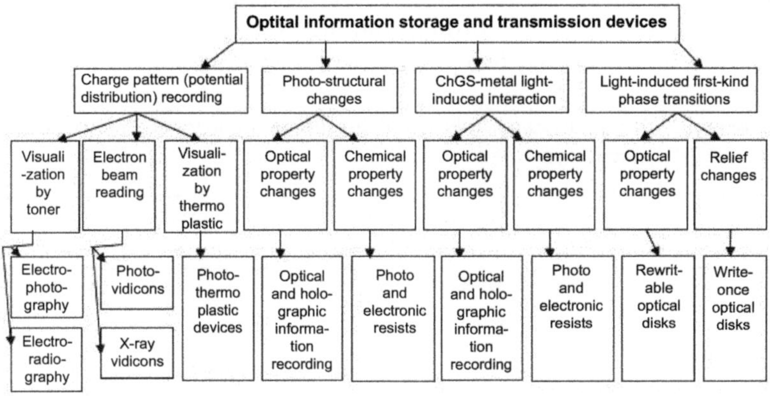

Figure 6.1 Optical information storage and transmission on the base of disordered semiconductors (ChGS is chalcogenide glassy semiconductors).

There are four different groups of optical information recording methods. All methods are based on creating the image on the surface of a semiconductor layer. The difference in the methods is in how

Disordered Semiconductors: Physics and Applications (2nd Edition)
Anatoly Popov

ISBN 978-981-4774-37-6 (Hardcover), 978-1-315-10042-5 (eBook)
www.panstanford.com

this image is recorded. The first group includes methods based on representation by electrical charge distribution on the surface of the semiconductor layer. The second group includes methods based on representation of the image by inducing structural changes in the non-crystalline phase under light exposure, leading to local changes in the material's optical and chemical properties. The third group includes methods based on representation of the image by inducing interaction between chalcogenide glassy semiconductor and metal under light exposure, leading to local changes in the material's optical and chemical properties, similar to group two described above. Last, but not the least, the fourth group includes methods of representation using light-induced first-kind phase transitions. These include local crystallization of non-crystalline phase or local amorphization of crystalline phase or local material evaporation under light irradiation. While all these methods and optical recording devices based on disordered semiconductors are characterized by reusability, real-time recording, and high resolution, they are also marked with relatively low photosensitivity (Table 6.1).

Table 6.1 Sensitivity and resolution of disordered semiconductor devices and human eye

Methods	First group	Second group	Third group	Fourth group	Human eye
Sensitivity (cm^2/J)	10^5–10^7	10^{-2}–10^1	10^2–10^3	10^{-1}–10^2	10^{10}
Resolution (mm^{-1})	10–500	10^3	$(5-10)\times10^3$	$\geq10^3$	10^1

6.1 Devices Based on Charge Pattern Recording

The first group (Fig. 6.1) includes devices with a recording medium and a photosensitive non-crystalline semiconductor layer (photoreceptor) on conductive substrate. The layer surface is charged to some initial potential via corona discharge or by a charging roller (in the case of electrographic or photo-thermoplastic devices, Fig. 6.2a) or via an electron beam (in the case of vidicons, Fig. 6.17).

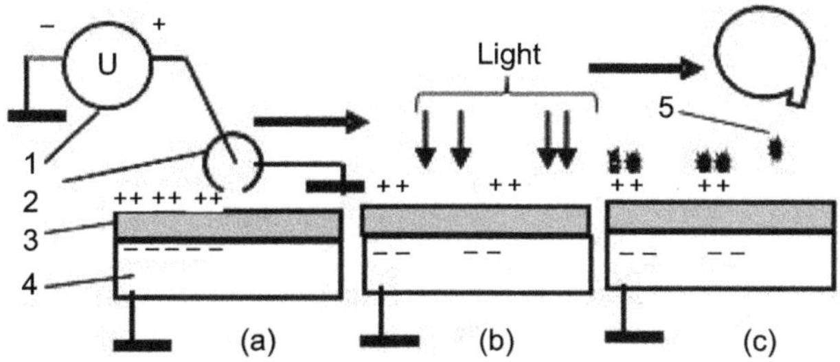

Figure 6.2 Creation and visualization of latent electrostatic image in electrophotography (see text): 1 is voltage source, 2 is korotron, 3 is semiconductor layer, 4 is conductive substrate, and 5 are charge toner particles.

Then an optical image is projected onto the charge surface. Surface potential is reduced in the regions exposed to light. Potential distribution that replicates the image pattern (latent electrostatic image) is formed due to semiconductor layer photoconductivity (Fig. 6.2b). Visualization of the latent electrostatic image or reading of potential distribution is realized by charged toner particles in the case of electrography (Fig. 6.2c), by heated thermoplastic film in photo-thermoplastic devices (Fig. 6.16), or by electron beam scanning in vidicons (Fig. 6.17).

The main function of the semiconductor layer is to convert optical image to electrostatic image. Therefore, the semiconductor layer has to have high value of photoconductivity in the visible spectrum to create sufficient differences in the surface potential between light and dark areas during exposure time. At the same time, the semiconductor layer has to have high resistance to maintain differences in the surface potential during the processing time. The time of exposure and reading of potential distribution is equal to TV frame time (0.04 s) in vidicons. The semiconductor layer resistivity has to be no less than 10^{11} Ω·cm to maintain potential distribution during this time. Even higher resistivity of semiconductor layers is necessary in the case of electrophotography and photo-thermoplastic devices.

The requirements mentioned above contradict one another. Semiconductor conductivity (and resistivity) is determined by concentration and mobility of charge carriers:

$$\sigma = 1/\rho = qn\mu_n + qp\mu_p \qquad (6.1)$$

where q is electron charge, n and p are electron and hole concentrations, and μ_n and μ_p are electron and hole mobility. The minimum concentration of charge carriers in intrinsic semiconductor is determined by the forbidden gap value:

$$n_i = p_i = A \exp[-\Delta E/(kT)] \tag{6.2}$$

where ΔE is the forbidden gap value, k is Boltzmann's constant, and T is temperature. Thus, it is desirable to use wide-band semiconductors to increase layer resistance.

On the other hand, the maximum photoconductivity of the material takes place when photon energy is equal to the semiconductor gap value (see Section 3.4). One usually records the image using visible light. Since the visible spectrum of electromagnetic radiation lies in the wavelength diapason from 0.40 to 0.75 μm, the semiconductor gap value has to be equal to 3.1–1.7 eV correspondingly ($\lambda = 1.234/\Delta E$) to ensure high photoconductivity. However, traditional crystalline semiconductors have insufficient resistivity at the given gap values to maintain potential distribution. The use of disordered semiconductors instead of single crystalline ones solves this contradiction because the former have resistivity 4–5 orders of magnitude higher at the same gap values. The reason for this lies in considerably lower mobility of charge carriers in the case of disordered semiconductors (see Section 3.2).

6.1.1 Electrophotographic Process

The first electrophotographic copy was created in 1938 by C. Carlson (Carlson, 1942). Twenty-one years later, in 1959, the first electrophotographic copying machine (model 914) was manufactured, which gained wide popularity. Since then the electrophotographic process became the most widespread method of copying. Today, about 4 billion copies are made daily in the world. Besides that, the electrophotographic process is used actively in other fields: printers, fax machines, etc. More than 5 million copiers and more than 15 million printers are produced annually.

A schematic illustration of the most widespread electrophotographic rotary copier is shown in Fig. 6.3 (Popov, 1997).

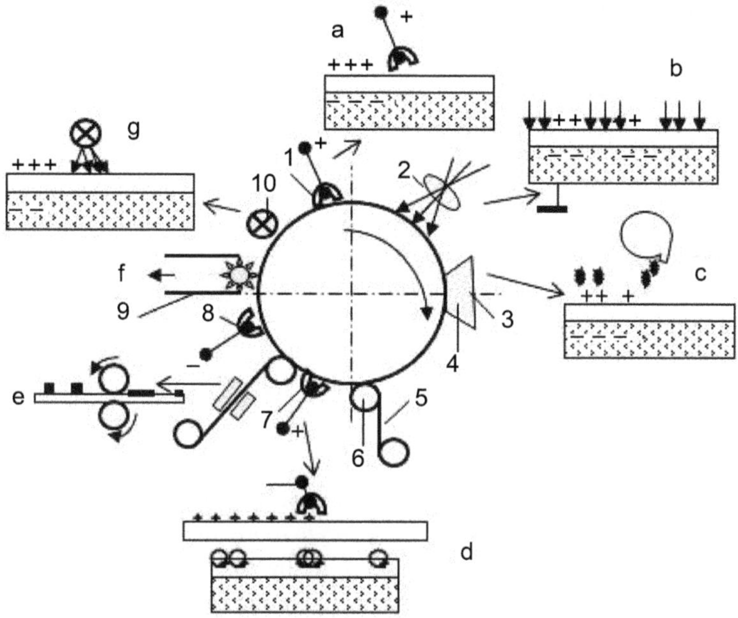

Figure 6.3 Schematic illustration of electrophotographic rotary copier (see text).

The photosensitive electrophotographic drum consists of high-resistivity disordered semiconductor layer on a cylindrical conductive substrate. The drum rotates around its own axis at a constant speed. All process steps are performed on the circumference of the drum. An electrical charge is sprayed on the surface of the semiconductor layer via corona discharge (or by charging roller) 1. A double-charge layer is formed by surface charges and screening charges in the conductive substrate as a result (stage *a*). A part of image pattern in the form of a narrow strip is projected onto the charged drum surface through the optical system 2. Due to the coordinated movement of the copied pattern and the drum, the projection of the former transfers completely on the surface of the semiconductor layer. At the illuminated regions, surface potential is reduced by the virtue of semiconductor photoconductivity. Discharge current flows normally to the surface, and the surface potential distribution or latent electrostatic image replicates the image pattern (stage *b*).

Image development is realized in chamber 3. A mixture of toner particles and cattier beads 4 spreads over the drum surface. Toner particles are charged by triboelectrical effect and are attracted to dark surface areas by Coulomb interaction. A visible image corresponding to the latent electrostatic image is formed as a result (stage *c*). The developed image transfers on paper 5 (or another carrier) through pressing it to the semiconductor layer surface by pinch rollers 6 and charging it via corona discharge 7. Charged toner particles displace on the paper under electric field action (stage *d*). Print fixing is carried out by heat, heat and pressure, or laser treatment (stage *e*). The electrophotographic copy is completed. However, it is necessary to prepare the surface of the semiconductor layer for the next cycle, that is, to clean the surface of residual toner particles and electrical charges. To facilitate cleaning, an electrical charge of opposite sign is sprayed on the surface via corona discharge 8. Residual toner particles are mechanically removed 9 (stage *f*). Finally, light irradiation 10 is carried out to delete residual electrical charges by photoconductivity (stage *g*). The drum is ready for the next cycle.

Printers differ from copiers only by the method of latent electrostatic image creation (stage *b*). The latent image is formed line by line by a selective exposure laser beam (laser printer) or a line of light-emitting diodes (LED printer). All the other stages remain unchanged.

The main performance targets of electrophotographic machines are copying rate, optical density of the copy, drum durability, original and copy sizes, the possibility of scale variation, weight-size, and power consumption characteristics. Most of them are determined by properties of the semiconductor material used in the electrophotographic drum.

The two major and contradictory requirements for these materials are high photoconductivity in the visible spectrum and high resistivity in the dark. This contradiction has been resolved through the use of disordered semiconductors (see above). Other requirements for the semiconductor material for electrophotography include: crystallization resistance, mechanical strength and stability of properties under influence of high electric fields, ionic currents, and toner particles. The current trend to reduce the size of electrophotographic machines has resulted in the deterioration of heat removal. As a result, the requirements for temperature stability of semiconductor materials are increasing.

Electrophotographic photoreceptors are produced on rigid (cylindrical or planar) or flexible substrates. In all cases, it includes several layers that perform different functions, namely (Fig. 6.4):

- The layer of carrier generation (2) where absorbed photons produce electron–hole pairs
- Transportation layer (3) through which photo-generated carriers move toward the substrate under electrical field action
- The layer creating a blocking contact between semiconductor and conductive substrate (4), which prevent carrier injection from the substrate, protective layer (1) on the photoreceptor surface

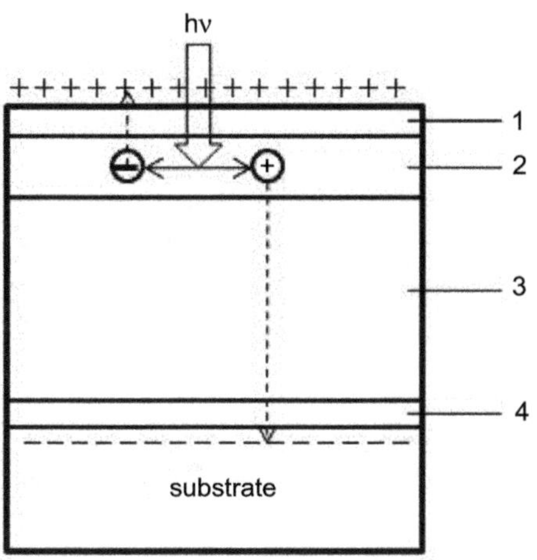

Figure 6.4 Schematic illustration of photoreceptor (see text).

It should be noted that the presence of all these layers is not necessary. In some cases, several functions are combined in a single layer.

The dark and light discharge characteristics of ideal and real photoreceptors are shown in Fig. 6.5. There is no initial surface voltage decay in the dark in the case of an ideal photoreceptor. With light turned on, the discharge rate is directly proportional to the light intensity and surface voltage decreases down to zero.

In practice, the situation differs significantly. Dark voltage decay is present because of the finite value of the photoreceptor's resistance. The dark discharge rate is characterized by the time of initial surface voltage reduction by half ($t_{½}$) or by the slope of the discharge curve at $t = 0$. The light discharge rate depends on surface potential, photon energy and is not directly proportional to light intensity. The voltage across the photoreceptor does not drop to zero after exposure due to the trapping of charge carriers on the deep localized states in the gap. Residual potential (R_{res}) remains even after the photoreceptor's complete light discharge.

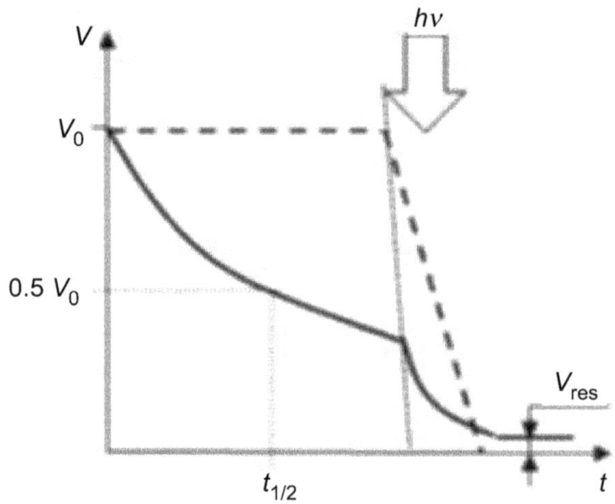

Figure 6.5 Dark and light discharge characteristics of ideal and real photoreceptors (V_0 is initial surface voltage and V_{res} is residual voltage).

Electrophotographic sensitivity (S) can be characterized as the light discharge rate (dV/dt) or as the exposure needed to reduce the initial voltage to a predetermined value. The latter is more convenient in practical application. However, it is necessary to specify the relative value of initial voltage reduction in this case: $[V_0 - \Delta V]/V_0$, where V_0 is the initial voltage and ΔV is the reduction in voltage as a result of exposure (contrast potential). Therefore, the sensitivity subscript 0.5 ($S_{0.5}$) means that measurements were carried out at the exposure, which reduces surface voltage by half and $S_{0.1}$ means reduction in surface voltage by 90%. Energy units of sensitivity (square meters per joule) are used if the spectral characteristics of photoreceptors are analyzed. In this case, sensitivity is the inverse value of radiation energy incident upon the surface unit. However, if the spectral composition of light is constant, the sensitivity is determined as the inverse value of light exposure required for a given surface voltage reduction ($lux^{-1}s^{-1}$).

6.1.2 Semiconductor Materials for Photoreceptors

Based on the requirements for the electrophotographic process mentioned in the previous section, five groups of semiconductor materials found application as photoreceptors in electrophotography:

1. Polycrystalline semiconductor (zinc oxide) in the dielectric medium (this was the first material to be used, but it is no longer used)
2. Glassy selenium and selenium alloys
3. Two- and multi-component chalcogenide glassy semiconductors
4. Organic semiconductors
5. Hydrogenated amorphous silicon and its alloys

Let us consider them starting with the second one (as polycrystalline zinc oxide semiconductors are no longer used).

Selenium photoreceptors. For a long time, selenium photoreceptors were the most widespread, mainly due to the match of their main characteristics to the requirements of the electrophotographic process. Chapter 5 describes the selenium drum manufacturing process and technology. Selenium photoreceptors are usually produced on duralumin substrate. The substrate is anodized to create a blocking contact. The photoreceptor itself is a multilayer system that includes the following parts: a blocking oxide film that is formed on the surface of the conductive substrate; a crystalline (trigonal) selenium layer that adjoins the oxide film; and a non-crystalline selenium transportation layer. The next layer is a carrier-generation layer, which is characterized by somewhat ordered non-crystalline selenium atomic structure created with the help of special deposition conditions. A protective layer on the photoreceptor's surface is formed by treatment in gas mixtures or by thin dielectric film deposition (organic films or diamond-like amorphous carbon films).

The spectral dependence of selenium photoreceptor sensitivity is shown in Fig. 6.6. One can see that there are two maxima of photosensitivity. The first maximum at the wave range 430–470 nm is defined by carrier generation in non-crystalline selenium. The second maximum at the wave range 720–750 nm is defined by carrier generation in crystalline selenium layer adjoined to the substrate. This peak appears because of weak absorption of long-wave photons in non-crystalline selenium. However, the intensity of this peak is by an order of magnitude lower compared to the intensity of the first peak.

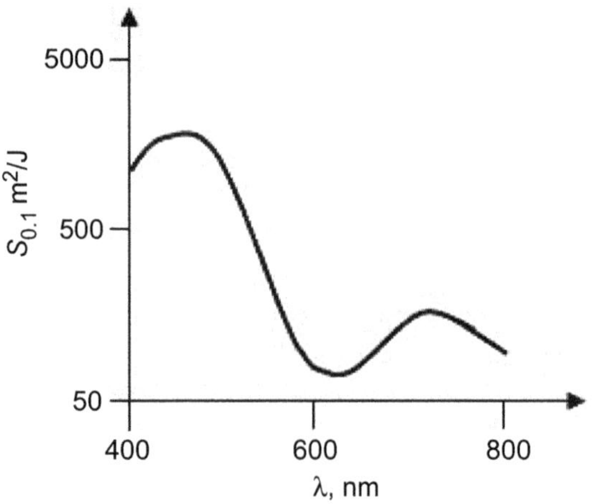

Figure 6.6 Spectral dependence of selenium photoreceptor sensitivity.

The presented spectral dependence of photosensitivity provides good-quality copies in the case of plain monochrome printing. However, this spectral characteristic is not suitable for the realization of color printing. In light-emitting monitors, true color images are formed by using a combination of three colors: red, green, and blue (RGB system). In the case of printers and copier machines, the eye perceives reflection of light from the image. In this case, higher quality is provided by the CMY (cyan, magenta, and yellow) system and extended CMYK system, where the key (black) color is added (Dvorkovich and Dvorkovich, 2012). As such, the photoreceptor should be sensitive over almost the entire spectral range of visible light in order to obtain a true color image with either of these systems. It is certainly possible to control some characteristics of photoreceptors (crystallization resistance and mechanical strength, for example) by means of structural modification at the level of medium-range order (see Chapter 4). On the other hand, this method does not ensure broadening of or shift in the photosensitivity spectral range because the latter depends on the forbidden gap value. That is why the photosensitivity spectra are modified by the addition of an admixture to selenium.

Admixtures of arsenic, antimony, germanium, bismuth, and tellurium shift the selenium photoreceptor sensitivity to the long-wave spectrum range. At the same time, these admixtures provoke deterioration of other characteristics: decrease in carrier drift mobility and in the absolute value of photosensitivity, increase in dark discharge rate and residual potential value, and change in crystallization activation energy. Bismuth and tellurium admixtures offer the best application-tailored compromise (Popov, 1997).

Photoreceptors of the $Se_{100-x}Bi_x$ system are usually prepared by the thermal co-evaporation (see Chapter 5) of selenium and bismuth. X-ray and electron diffraction analysis do not display crystalline phase in $Se_{100-x}Bi_x$ films up to $x = 30$ (Bowman and Schottmiller, 1968; Schottmiller, Bowman, and Wood, 1968). Bismuth admixture broadens the valence band and reduces the selenium forbidden gap value due to the formation of selenium–bismuth bonds, which have lower energy in comparison to selenium–selenium bonds (Popov, Mikhalev, Karalynets, and Vasillyeva, 1991). Reduction in gap value leads to decreasing resistivity of the material. $Se_{100-x}Bi_x$ film resistivity as a function of bismuth concentration is shown in Fig. 6.7a. Thirty percent of bismuth decreases the material resistivity by 11 orders of magnitude (Schottmiller, Bowman, and Wood, 1968). The most considerable resistivity changes take place at low bismuth concentration.

The spectral characteristics of photoreceptor sensitivity change as well. $Se_{100-x}Bi_x$ photosensitivity maximum as a function of bismuth concentration is shown in Fig. 6.7b. One can see that the photosensitivity maximum shifts to the long-wave range with bismuth concentration increasing up to 5 at.%. Subsequent increase in bismuth concentration has weak influence on the photosensitivity maximum position.

Thus, from the point of view of photosensitivity maximum shift, it is not really useful to increase bismuth concentration to more than 5 at.%. The material resistivity falls down to 10^{10} Ω·cm at bismuth concentration of 5 at.%. This is too low a value to maintain differences in surface potential during processing time. Hence bismuth concentration from 0.5 to 2 at.% is used in photoreceptors of the $Se_{100-x}Bi_x$ system. Such systems have spectral dependence of photosensitivity shifted to the long-wave part of the spectrum up to the red light region (Fig. 6.8a).

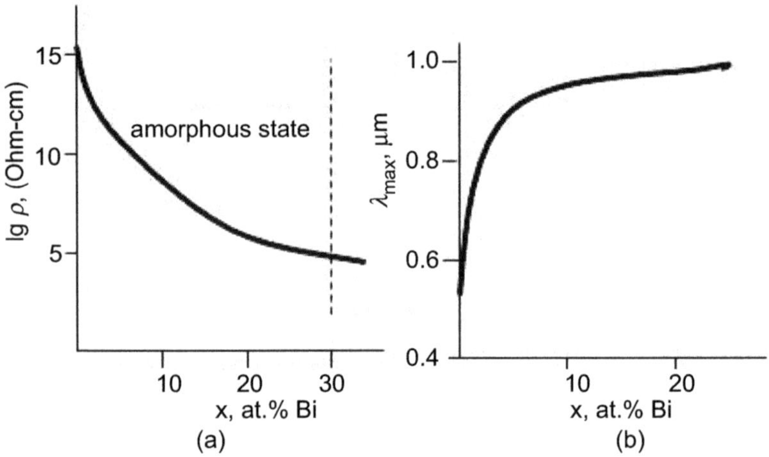

Figure 6.7 Concentration dependences of $Se_{100-x}Bi_x$ film resistivity (a) and wavelength corresponded to photosensitivity maximum (b).

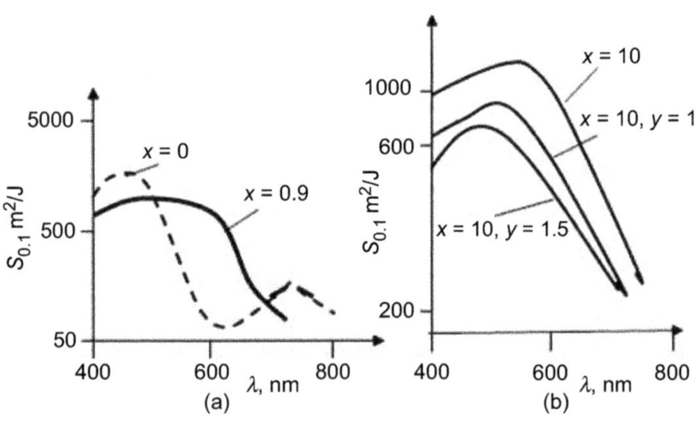

Figure 6.8 Spectral dependences of $Se_{100-x}Bi_x$ (a) and $Se_{100-x-y}Te_xP_y$ (b) photoreceptor sensitivity.

It should be noted that bismuth admixture sharply decreases the drift mobility of holes (from 0.16 $cm^2/V{\cdot}s$ in pure non-crystalline selenium to 10^{-3} $cm^2/V{\cdot}s$) and increases the residual potential value.

In general, a bismuth admixture to selenium leads to:

- Shift in photoreceptor sensitivity to the long-wave spectrum region
- Increase in dark discharge rate
- Increase in the residual potential value

In contrast to bismuth, an admixture of tellurium to selenium has no influence on the carrier drift mobility. This leads to broadening of the photosensitivity spectral dependence toward long waves up to 800 nm at 10 at.% of tellurium. At the same time, it leads to an inadmissible rise in the photoreceptor dark discharge rate due to decrease in resistivity.

There are two ways to address this problem (Popov, 1997). The first is limiting the tellurium concentration at the level of 1–3 at.%. The second involves simultaneous admixture of tellurium and other elements that reduce the dark discharge rate. For example, a phosphorus admixture to $Se_{100-x}Te_x$ layers leads to a decrease in the dark discharge rate by an order of magnitude. However, at the same time, it decreases photosensitivity by 30–40% as well (Fig. 6.8b).

The last but not the least of the shortcomings of $Se_{100-x}Te_x$ photoreceptors are lower crystallization activation energy and, as a result, lower durability.

Arsenicum triselenide photoreceptors. Glassy arsenicum triselenide has a layered polymeric structural network, consisting of pyramidal blocks $AsSe_{3/2}$ (see Chapter 2). This type of structure facilitates the through passage of current carriers and provides high discharge efficiency with light exposure. The photosensitivity of As_2Se_3 photoreceptors under exposure of incandescent lamp light considerably exceeds selenium photoreceptor photosensitivity, mostly due to the broadening of the spectral sensitivity dependence toward the long waves (Fig. 6.9a). BiI_3 additive to arsenicum triselenide (2–6 molecular %) shifts the peak of the spectral dependence of photosensitivity to the long-wave region (Table 6.2) (Popov, 1997). This photosensitivity increase and the peak shift are accompanied by a rise in the dark discharge rate and a decrease in the initial surface potential. From the manufacturing point of view, arsenic is not among the favorites due to its high toxicity and relatively high cost.

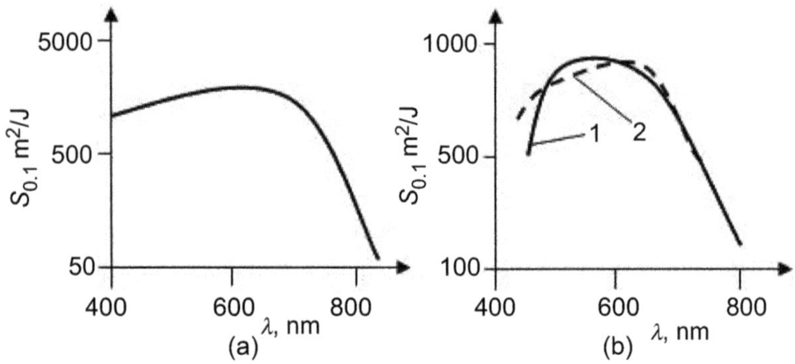

Figure 6.9 Spectral dependences of As_2Se_3 (a) and a–Si:H (b) photoreceptor sensitivity in the case of negative (1) and positive (2) charge.

Table 6.2 $(As_2Se_3)_{100-x}(BiI_3)_x$ photoreceptor sensitivity (m^2/J) in infrared range

Wavelength (nm)					
Composition	700	800	850	900	950
As_2Se_3	150	80	40	10	—
$(As_2Se_3)_{98}(BiI_3)_2$	260	160	110	70	30
$(As_2Se_3)_{96}(BiI_3)_4$	300	190	140	95	40
$(As_2Se_3)_{94}(BiI_3)_6$	240	130	80	40	10

Photoreceptors on the basis of hydrogenated amorphous silicon first went on sale in 1984. These photoreceptors have the following advantages (Kawamura et al., 1983; Shirai and Kajita, 1993):

- High photosensitivity throughout the visible spectrum (Fig. 6.9b) and, in the case of multilayer photoreceptors, in the infrared spectrum as well
- High drum durability (≥3 × 10^6 copies) because of high material microhardness (1500–2000 kg/mm²)
- Feasibility of structures where the voltage of any polarity could be used for initial surface charging (Fig. 6.9b)
- High-temperature stability, allowing the realization of compact machine construction

However, there are at least two technological problems in the manufacturing of a–Si:H photoreceptors. First, it is necessary to increase the resistivity of hydrogenated amorphous silicon layers above 10^{12} Ω·cm, the threshold value for electrophotography. Commonly, the highest value of resistivity of hydrogenated amorphous silicon films prepared by the glow discharge decomposition method (with a small amount of boron) does not exceed 10^{11} Ω·cm (see Chapter 4). Increasing the resistivity may be achieved by simultaneous boron and oxygen (or nitrogen or carbon) doping of a–Si:H. In all these cases, it is possible to achieve resistivity values not less than 10^{12} Ω·cm. This allows to create high-resistance photoreceptors with the structure shown in Fig. 6.10a. However, the addition of oxygen, nitrogen, or carbon leads to the growth of localized state concentration, which are the centers of free charge carrier capture. It leads to an increase in the likelihood of ghost images and in residual potential and a number of other problems. A structure where the carrier-generation layer is formed of a–Si:H <B> film and the carrier-transport layer is a high-resistance film containing additives of O, N, or C has been proposed to overcome these difficulties. However, the surface recombination of charge carriers increases when the carrier-generation layer is the outer layer. This is why an upper high-resistance layer is added to the structure (Fig. 6.10b).

Another approach to increasing the resistance of a–Si:H photoreceptors is to create a multilayer structure where the generation layer (a–Si:H <B>) is enclosed between two layers of high resistance, blocking the injection of charge carriers (Fig. 6.10c). A good quality of electrophotographic process is also achieved in this case. The disadvantage of this structure is the strong dependence of photoreceptor characteristics on the polarity of the applied voltage.

Last, the third approach to preparing a–Si:H films with a resistivity of 10^{13} Ω·cm is to use the method of silicon target sputtering in an atmosphere of argon and hydrogen (see Section 5.5).

The second problem is the need to substantially increase the film growth rate in order to provide the requisite surface potential without excessively increasing the field strength of the layer with the targeted layer thickness of 20 μm and above. In the glow discharge decomposition method, this problem is addressed either by increasing silane concentration (up to 40%) in the gas mixture, or by substituting silane for polysilanes (Si_2H_6 or Si_3H_8). In the former case, the growth rate increases to 5 μm/h. In the latter case, the growth rate increases up to 20 μm/h (see Chapter 5). Increasing the film growth rate when the sputtering method is used may be achieved by using magnetron sputtering.

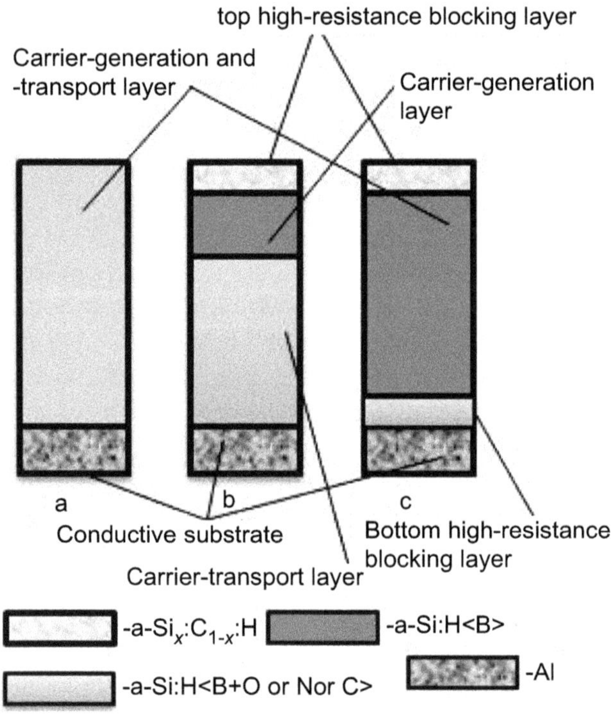

Figure 6.10 Schematic cross sections of a–Si:H photoreceptors (see text).

Organic photoreceptors. Organic semiconductors are widely used in the production of electrophotographic drums. Modern organic electrophotographic drums constitute a multilayer system (Fig. 6.11).

If surface alignment is required, then the primer coat is applied to the substrate. After the primer coat, a high-resistance blocking layer is formed. This layer prevents the injection of charge carriers into the semiconductor from the substrate and also provides the necessary adhesion to the substrate. Next is the carrier-generation layer, where light absorption occurs and electron–hole pairs are generated. The generation layer is a heterogeneous system composed of photosensitive microparticles dispersed in the polymer. Energy barriers at the boundaries of microparticles contribute to the separation of excitons, formed by the absorption of light quanta (Frenkel excitons, see Section 3.2.5), which leads to an increase in quantum efficiency. The spectral characteristics of organic drums are very diverse and depend on the type of sensitizing additives. Layers based on phthalocyanine pigments (e.g., titanyl phthalocyanine or metal-free phthalocyanine of the γ-form) and some other are used to operate in the red and infrared regions of the spectrum. For photosensitivity at shorter wavelengths, layers based on bisazo pigments are used (Fig. 6.12). Then the carrier-transport layer, through which charge carriers are transported to the surface, is formed. The transport layer is a dielectric polymer matrix (polycarbonates or polymethyl methacrylate), with embedded molecules of material that provide the transfer of the charge carriers. Aromatic amines of hydrazones are used in the latter (Fig. 6.13) (Uralova, Vannikov, and Churkin, 2005).

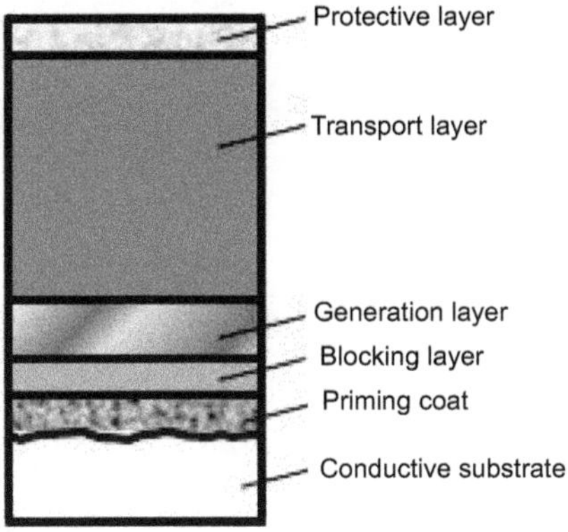

Figure 6.11 Schematic cross sections of organic photoreceptor.

The mobility of charge carriers in such a system is determined by the concentration of the molecules in which they move. The minimum concentration should provide the average distance between molecules, which does not exceed 1.0–1.5 nm. The upper concentration limit is determined by the beginning of the association of molecules in agglomerates.

All these layers are prepared from solutions by dipping or other similar methods. These methods are simple to implement and provide low-cost products, which is the main advantage of organic drums. However, organic materials have low resistance to mechanical stress and, consequently, low drum durability. It can be improved by protective coating from an amorphous diamond-like carbon, diamond-like silicon–carbon film (see Section 2.5.4), or some other materials. However, all these films are prepared by vacuum methods, which substantially increases the cost of the product.

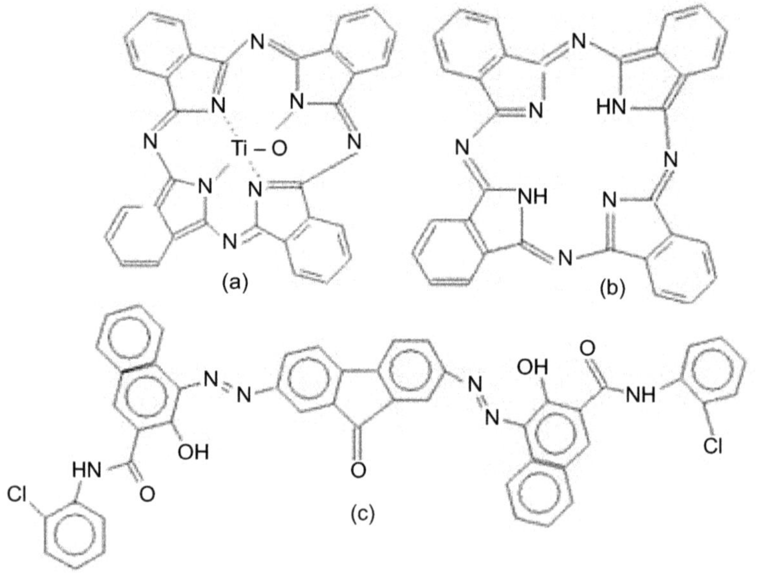

Figure 6.12 Structure of the phthalocyanine pigment molecules: titanyl phthalocyanine (a), metal-free phthalocyanine of the γ-form (b), and bisazo pigment (c).

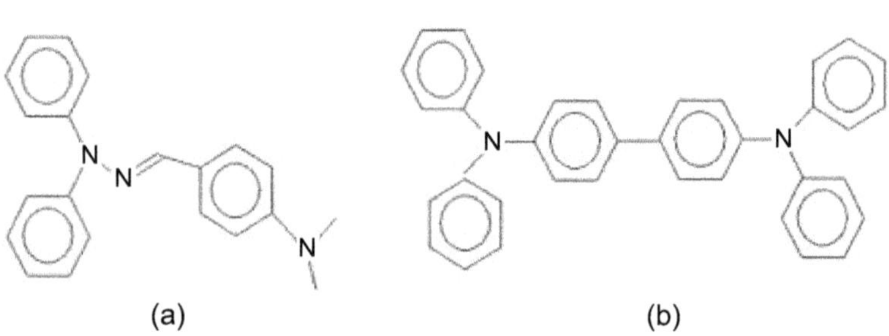

Figure 6.13 Structure of the transport molecules: hydrazone (a) and polyarylamine (b).

The organic photoreceptor sensitivity usually does not exceed 100 m^2/J and drum durability is (5–20) × 10^3 copies. These values are an order of magnitude worse than in inorganic photoreceptors. Nevertheless, organic photoreceptors successfully withstand competition through ease of manufacturing and, consequently, low cost.

6.1.3 Multilayers and Variband Photoreceptors

As was mentioned before, the photoreceptor contains several functionally different layers. Therefore, tuning the material characteristics to the function of the layer can improve the parameters of the photoreceptor.

In the simplest case, a photoreceptor may contain two layers of semiconductor materials with different gap values. A wide-gap semiconductor is placed near the free surface, and a narrow-gap semiconductor is located underneath. In this case, the short-wave part of light is absorbed in the wide-gap semiconductor, and lower energy photons pass through this layer and are absorbed in the narrow-gap semiconductor. As a result, the spectral dependence of photoreceptor sensitivity is expanded. Photoreceptors with a double carrier-generation layer, a–$Si_{1-x}N_x$–a–Si:H, Se–$Se_{1-x}Bi_x$, and Se–$Se_{1-x}Te_x$, have been created on this basis. The distribution of chemical composition across the thickness of the photoreceptor Se–$Se_{1-x}Te_x$ and its photosensitivity spectral dependence are shown in Fig. 6.14 (Popov, 1997). As one can see from the figure, the carrier-generation region consists of a selenium layer with 0.5 μm thickness and a $Se_{65}Te_{35}$ layer with a thickness of 0.3 μm. The transportation layer is a selenium layer 50 μm thick. The spectral dependence of the photoreceptor sensitivity is the superposition of characteristics of the materials used in the generation layer (Fig. 6.14b).

It is obvious that the integral sensitivity of a photoreceptor should increase with the number of layers with different gap values in the carrier-generation region. A natural extension of this process was the creation of variband photoreceptors, in which the gap value gradually decreases from the surface to the depth of the layer. The dashed line in Fig. 6.14a shows tellurium concentration changing in this case. Such variband photoreceptors have high value of sensitivity in the broad spectral range (from 400 to 800 nm) while maintaining an acceptable level of other electrophotographic characteristics (initial surface potential and dark discharge rate).

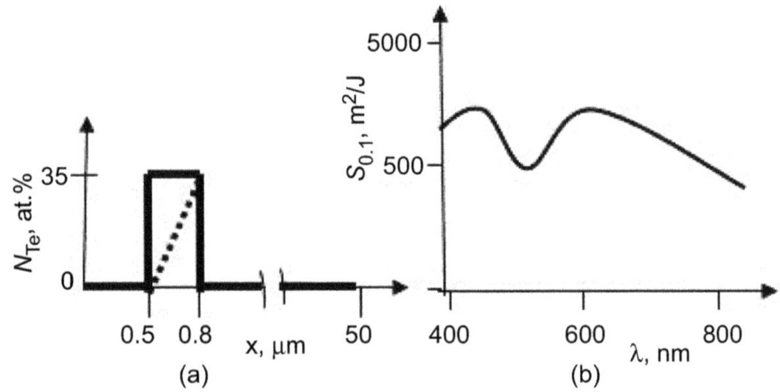

Figure 6.14 Distribution of chemical composition across the thickness of the Se–$Se_{1-x}Te_x$ photoreceptor (a), and its photosensitivity spectral dependence (b).

6.1.4 Electroradiographic Process

In contrast to electrophotography, an electroradiography (see Fig. 6.1) image is created by X-rays, and not visible light radiation. This allows to use the method for diagnostics in medicine and in industrial nondestructive quality control. Thicker semiconductor layers (more than 100 μm) and additional lead layers on the substrate to facilitate X-ray absorption are characteristic for these applications.

The key parameters of X-ray receptors in the case of medical application are size, spatial resolution, image frequency, and signal-to-noise ratio. For example, conventional thorax radiographs are 43 × 35 cm^2 in size. Mammograms need to display microcalcifications of some 100 μm in detail. Fluoroscopic images are taken at a rate of 30 images per second, or even faster. Moreover, all receptors need to deliver optimum image quality at the lowest tolerable dose (Hoheisel and Batz, 2001). On the basis of these requirements, X-ray receptors on amorphous selenium and hydrogenated amorphous silicon have been developed.

Commercial electroradiographic systems based on amorphous selenium layer were introduced at the end of the last century.

They produced high-quality mammographic images and enjoyed a special attribute called edge enhancement, which amplified the contrast at object edges and thereby facilitated diagnosis. However, an electroradiographic equipment was cumbersome, which made it unattractive. One approach to modernize electroradiography involved the introduction of an array of electrometers to read the electrostatic image. Later, scanned electrometer readout was used. It was based on electronic readout technique, which enables the digitization of X-ray image. The photoconductor in this system was a 500 μm thick amorphous selenium layer coated on aluminum drum, which was rotated in the imaging system during charging, exposure, and readout. Much progress was also made in using a laser scanning technique.

While the aforementioned techniques presented significant improvements in electroradiography, the fundamental principle remained unchanged, that is, the photoreceptor surface was first charged and then selectively discharged by the incident X-rays passing through the object to form a latent electrostatic image. Then the charge distribution was read out.

The situation changed fundamentally due to the development and broad commercialization of active matrix arrays based on hydrogenated amorphous silicon thin-film transistors (see Section 8.2). Initially, these matrix arrays were created to control liquid-crystal displays. Their application in electroradiography allowed to create digital flat-panel X-ray image detector (Kasap et al., 2009). Figure 6.15 shows a schematic view of one pixel of X-ray detector on the basis of an amorphous selenium layer on a glass substrate with an active matrix array.

The photoreceptor includes a charge carrier-generation layer of several hundred microns in thickness and two blocking layers required to reduce dark current. A top electrode is deposited on the photoreceptor surface to apply a bias voltage. Each pixel has a bottom (charge collection) electrode (which is part of storage capacitance C_{px}) and thin-film transistor (TFT). When a signal is applied to the gate of the transistor, the charge stored at the pixel capacitance flows to a data line and hence to an output amplifier and then into a computer as a pixel of the image.

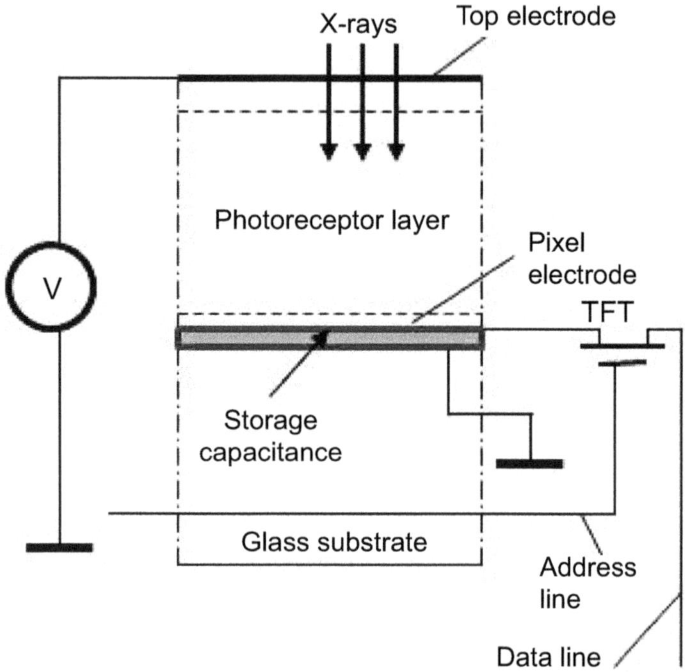

Figure 6.15 Schematic diagram of X-ray detector pixel.

Thus, generation of charge carriers due to the absorption of X-ray radiation in the selenium layer and a reading of this signal through the TFT occur in each pixel. Popov et al. (1998a, 1998b) suggested a different X-ray receptor consisting of an array of thin-film diodes (TFDs) based on a–Si_xN_{1-x}. TFD is both an electrical signal generator (due to current increase because of the absorption of X-ray radiation) and an element for reading the signal. This significantly simplifies the design of the receptor, but it complicates the readout circuit.

These detectors are called direct conversion as opposed to indirect conversion detectors in which X-rays are first converted to light by a suitable scintillating phosphor and the emitted light is detected by a photodiode integrated into each pixel.

6.1.5 Photo-thermoplastic Process

As mentioned above, visualization of the latent electrostatic image or reading of potential distribution in photo-thermoplastic devices is realized by heated thermoplastic film. Thermoplastic is a polymer organic material that turns to liquid (or flexible) state when heated up to ~100°C and becomes solid when cooled. Photo-thermoplastic films are prepared on the lavsan (dacron) transparent substrate. The semi-transparent bottom electrode is made of tin oxide, ITO (In_2O_3 + SnO_2), or metal. There are two types of photo-thermoplastic devices. In the first type (one-layer structure), the photosensitive organic polymer film (e.g., polyvinylcarbazole) is formed on the transparent electrode (Fig. 6.16a).

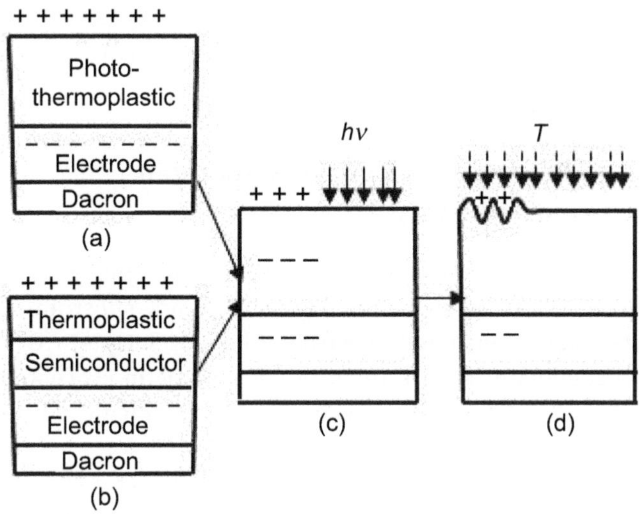

Figure 6.16 Stages of photo-thermoplastic process: charging in the case of one-layer (a) and two-layer (b) structures, exposition (c) and development (d).

In the second type (two-layer structure), the photosensitive layer is a high-resistance glassy chalcogenide film ($As_4S_3Se_3$, for example), and a layer of thermoplastic organic polymer (polyethylene, polimetilametakril, or styrene) is deposited on it (Fig. 6.16b). Electrical charging (Fig. 6.16a,b) and light exposition (Fig. 6.16c) stages are analogous to the corresponding stages of the electro-photographic process. The electrical field of the latent electrostatic image creates an electrostatic pressure, which is proportional to the square of the surface charge density. Thus, the electrostatic pressure distribution replicates the image pattern. Under the influence of high temperature, thermoplastic turns to the flexible state and is deformed under the influence of electrostatic pressure (Fig. 6.16d). As a result, the relief corresponding to the latent image appears, which is fixed by the cooling thermoplastic. This relief alters the phase of the readout beam in the process of recorded information reading.

The separation of functions in the two-layer structures between the semiconductor layer, responsible for the formation of the potential relief, and the thermoplastic layer providing its visualization, enables the most complete implementation of both materials' capabilities. Re-heating of the photo-thermoplastic film without charging its surface leads to erasing of the information due to the effect of surface tension. Copies of the recorded information is made by the hot pressing method.

6.1.6 Vidicons on the Basis of Disordered Semiconductors

Vidicon is a television camera pickup tube. A schematic illustration of vidicon design is shown in Fig. 6.17. Glass substrate 2 with deposited semi-transparent electrode (SnO_2 or $SnO_2 + In_2O_3$) 3 and the semiconductor layer 4 are at the bottom of an evacuated glass bulb 8. The gun 6 forms the electron beam 5, which can be moved along the layer surface with the help of deflecting plates 7. The image is projected on the semiconductor layer through an optical system 1, glass substrate, and semi-transparent electrode. The surface on the opposite side of the semiconductor layer is scanned by the electron beam. The current in the electric circuit—electron beam, semiconductor layer, resistor R, and voltage source E—depends on the semiconductor layer resistance. The resistance of illuminated areas decreases due to photoconductivity, and the electrical current in the circuit and the voltage drop across the resistor increase. The video signal from the resistor R is fed to an amplifier.

Vidicons based on disordered materials usually make use of As_2Se_3, As–Se–Te system, or a-Si:H (Hirai and Maruyama, 1982). Vidicon sensitivity is determined by the ratio of resistances between lit and unlit layer areas, and hence semiconductor photoconductivity (see Eq. 3.53). In the case of semiconductor layer thickness of about 1 μm and applied voltage 50 V, the electric field in the semiconductor is about 10^5 V/cm. This corresponds to quantum efficiency close to 1.

However, it was mentioned in Section 3.4 that quantum efficiency in selenium layers exceeds 1 if the electric field is more than 10^6 V/cm, due to the effect of avalanche multiplication (Fig. 3.34). Moreover, quantum efficiency increases with semiconductor layer thickness at a constant electric field value.

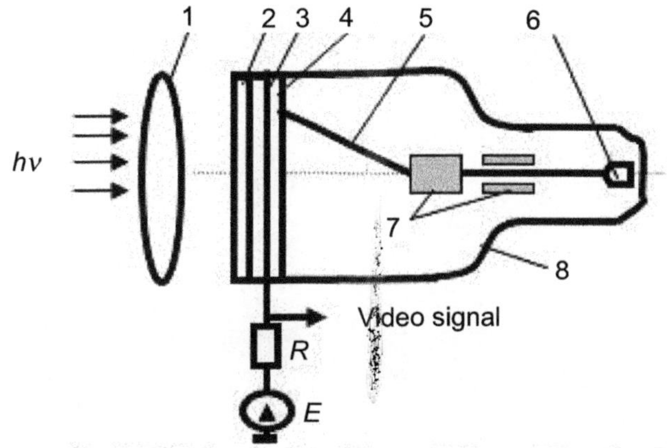

Figure 6.17 Schematic illustration of vidicon design (see text).

High-sensitive and super high-sensitive vidicon camera tubes named HARP (high-gain avalanche rushing amorphous photoconductor) tubes were developed on this basis (Kurashige et al., 1993; Park and Tanioka, 2003). Quantum efficiency and sensitivity of a tube increase by tens and hundreds of times with an increase in the selenium layer thickness from 1 to 35 micron. However, thicker layers require higher target voltage, for example, 240 V for 2 μm thickness, 460 V for 4 μm, and so on. Aerial photographs taken at night with the conventional and HARP vidicons are shown in Fig. 6.18 (Kurashige, unpublished).

The high sensitivity of vidicons opens up new possibilities of their application. For example, let us consider the problem of detecting objects under water at a great depth. The spectral dependence of sea water optical transmission is shown in Fig. 6.19. The maximum transparency is near 450 nm, which corresponds to the region of maximum photosensitivity of selenium layers. However, the radiation intensity decreases by orders of magnitude at a depth of 100–150 m, even in the area of maximum transparency. On the other hand, the tenfold increase in the HARP tube sensitivity gives the ability to detect an object at such a depth, with the same success as conventional vidicon detects it on the water surface (Maruyama, 1992).

Figure 6.18 Airscape taken at night with the CCD camera (top) and the Super-HARP camera (bottom) (Kurashige, unpublished).

Initially, HARP tubes had short operating lifetime due to the low crystallization activation energy of the selenium layer and the acceleration of crystallization process in high electric field. As a rule, crystallization process begins on either the outside or the inside (adjacent to the substrate) surface of the film. A multilayer structure has been used to increase the activation energy while maintaining high quantum efficiency. Blocking contacts on both surfaces of the selenium film are prepared from materials with higher activation energy: for example, selenium doped with arsenic and arsenicum triselenide or antimony trisulfide. In this case, the carrier avalanche multiplication occurs only in the pure selenium (or selenium doped with tellurium) layer, while the areas in contact with the substrate and the electron beam have higher crystallization activation energy due to the presence of threefold coordinated arsenic atoms.

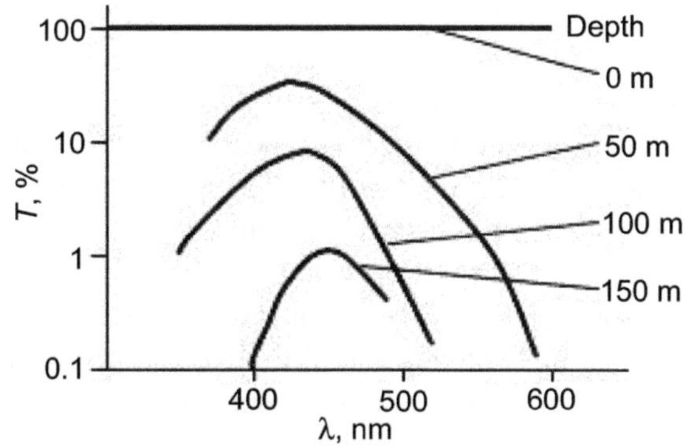

Figure 6.19 Spectral dependence of sea water optical transmission.

Soon after practical implementation of such structures, it was discovered that they do not prevent crystallization during operation at lower temperatures. It was found that operating the device below room temperature leads to the appearance of image defects caused by crystallization of the semiconductor layer. This fact is quite surprising, because decreasing temperature should lead to a decrease in the rate of activation processes in accordance with the Arrhenius equation (1.6). The reasons for accelerating crystallization during device operation at low temperatures have been explained (Resnik et al., 2006) on the basis of photo-structural changes in glassy semiconductors. This offers an opportunity to find a solution to the operating lifetime problem mentioned earlier (see Section 6.2).

6.2 Devices Based on Photo-induced Transformations in Chalcogenide Glasses

A number of effects observed in chalcogenide glassy semiconductors are specific only to this class of materials and are not observed in crystals of the same chemical composition. These effects include reversible and irreversible photo-induced changes in optical, chemical, and mechanical properties, photo-induced dissolution of metals, photo-crystallization, photo-compression, and photo-expansion. Photo-induced changes in the properties of chalcogenide glassy semiconductors may be caused by effects associated with the change in the defect state and by effects affecting the basic structural network of the material (photo-induced structural changes). Photo-induced changes in the charge state of defects arise as a result of capture of non-equilibrium carriers on localized states in the mobility gap. As a result, the charge state of defects is changed, which leads to changes in the optical absorption coefficient, photoluminescence, ESR signal, photoconductivity, and other properties. The irradiation of glassy As_2Se_3 at 6 K leads to an increased absorption at about mid-gap energies (0.9–1.6 eV) and to the appearance of an electron spin resonance signal. The induced absorption is stable at low temperature, but it disappears on thermal annealing. The study of the short-range order of atomic arrangement in glassy selenium by the in situ EXAFS method (see Section 2.3) at 30 K showed a reversible increase in the first coordination number by about 5% in the photoexcited amorphous selenium film. After the cessation of illumination, the local changes in coordination disappear (Kolobov and Tominaga, 2012). Investigations of radiation effect on the state of the defects are important in terms of understanding the physics of disordered materials. However, the instability of arising states makes these effects unattractive from a practical point of view. In this respect, the effects affecting the main structural network of chalcogenide glassy semiconductors are much more interesting.

6.2.1 Photo-structural Changes in Glassy Semiconductors

Photo-induced structural transformation is based on the following process. Photon absorption leads to electronic excitation, which, in turn, causes changes in the atomic structure of the material. These changes may be both reversible and irreversible. The changes in atomic structure lead to reversible and irreversible transformation of optical, chemical, mechanical, and others properties of the material.

Photo-induced transformation of optical properties (absorption coefficient, reflection factor, and refractive index) under the effect of photons with energy corresponding to the fundamental absorption edge can be observed in various chalcogenide glassy semiconductors: two-component glasses of As–S, As–Se, Ge–S, Ge–Se systems, three-component systems (As–S–I, As–S–Se, As–Se–Ge, As–Se–Cu), and others. Most of the research is done on thin films (thickness less than 1 μm), but similar effects are observed in bulk samples (Zakery and Elliott, 2003).

Typical changes in the optical transmission spectra of chalcogenide films under the influence of light radiation and annealing are shown in Fig. 6.20.

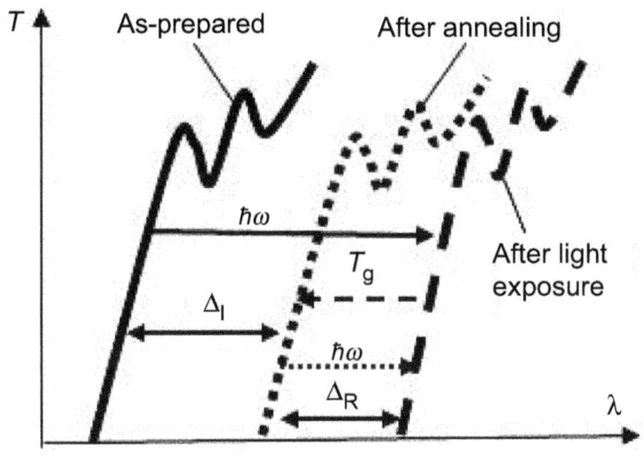

Figure 6.20 Typical changes in the optical transmission spectral dependence of chalcogenide film under the influence of light radiation ($\hbar\omega$) and annealing (T_g) (Δ_R is a reversible change and Δ_I is an irreversible change).

Irradiation of the as-prepared film by photons with energies corresponding to the fundamental absorption edge of the material shifts the optical transmission spectrum to the long-wave region. It corresponds to photo-darkening of the film. Annealing at the temperature near the glass transition value (T_g) leads to a curve shift in the opposite direction to some intermediate position. The subsequent cycles of irradiation and annealing are accompanied by reversible shifts in the curve between the last two positions. Thus, photo-induced changes in the optical properties of chalcogenide glassy films have both a reversible part (irradiation of annealed film) and an irreversible part (first irradiation of the as-prepared film). The values of these changes depend on the film's chemical composition, methods and conditions of film preparation, photon energy, temperature, and other factors.

The dependence of the reversible part of arsenicum trisulfide transmission spectrum shift on irradiation photon energy is shown in Fig. 6.21 (Tanaka, 1980).

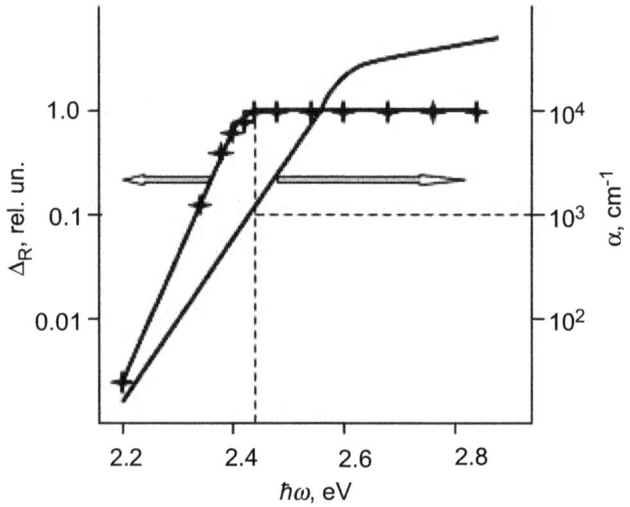

Figure 6.21 Dependence of the reversible part transmission spectrum shift (Δ_R) on irradiation photon energy and spectral dependence of absorption coefficient (α) of As_2S_3 film. Reprinted from Tanaka (1980), with permission from Elsevier.

Data are normalized to the number of absorbed photons. The figure shows the spectral dependence of the optical absorption coefficient of the material as well. It was mentioned in Section 3.3 that the value of the optical gap may be determined as the photon energy corresponding to the optical absorption coefficient α = 10^3 cm^{-1}. One can see from the figure that the maximum shift in the optical transmission spectrum is observed at a photon energy approximately equal to the gap value (2.45 eV). The spectrum shift remains unchanged with increasing photon energy. In contrast, the decrease in the photon energy leads to a sharp reduction in the effect value by orders of magnitude.

Table 6.3 shows the photo-induced changes in the optical properties of thermally evaporated films from As–Se systems with various component ratios (Gurevich et al., 1976). The films were exposed to radiation with a wavelength of 633 nm at a temperature of 300 K. As one can see from the table, significant changes in optical properties are observed only in the case when the material is irradiated at a temperature that is significantly lower than the glass transition temperature. The values of reversible and irreversible photo-induced property changes grow with increasing arsenic content. Tanaka (1983) analyzed the dependence of the reversible shift in the optical transmission spectrum on irradiation temperature, normalized with T_g for different materials (Fig. 6.22).

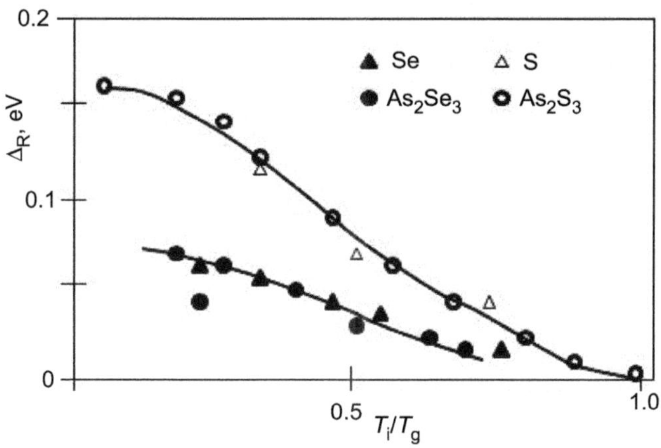

Figure 6.22 Dependence of reversible part transmission spectrum shift on normalized temperature during irradiation. Reprinted from Tanaka (1983), with permission from Elsevier.

It can be seen that the magnitude of the photo-induced changes in the optical properties decreases to zero when the irradiation temperature approaches T_g. Hence, it becomes clear why photo-darkening effect is not observed in glassy selenium films at room temperature.

Table 6.3 Photo-induced changes in optical properties of films from the systems As–Se (λ = 633 nm, T = 300 K)

	Irreversible changes			Reversible changes		
Composition	Δ_I, **eV**	**Contrast**	**Refractive index changes**		$T_{annealing}$, **K**	T_g, **K**
Se	–	–	–	–	–	~305
As_2Se_7	–	1.5	≤ 0.02	–	370–400	389 ± 2
As_2Se_5	0.025	2.0	≤ 0.02	–	408–418	408 ± 2
As_2Se_3	0.06	8.0	0.13	0.02	463–473	464 ± 2
AsSe	0.145	80–100	0.21	0.13	463–468	455 ± 2
As_3Se_2	0.19	180–200	0.30	0.24	423–438	414 ± 2

Δ_1 is irreversible shift of optical transmission spectrum.

At the same time, as demonstrated by investigations of the kinetics of this process (Fig. 6.23; Gurevich et al., 1976), the material's sensitivity to photo-induced changes decreases with increasing arsenic content, ceteris paribus.

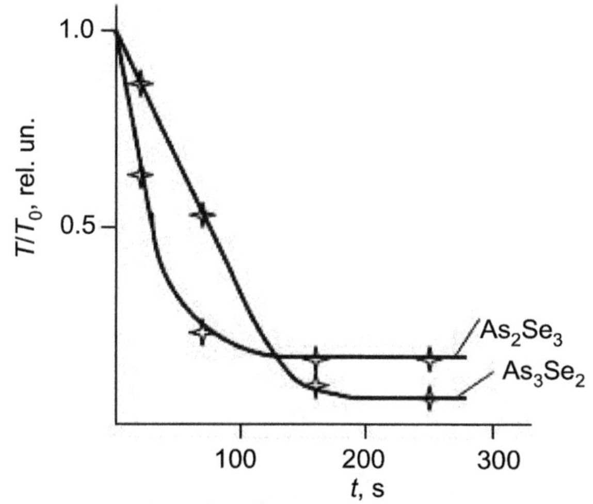

Figure 6.23 Dependence of film optical transmission on the time of irradiation (λ = 633 nm, T = 300 K).

The sample temperature during irradiation determines not only the magnitude of photo-induced changes in the optical properties but also their sign. Under certain conditions, irradiation may cause an increase in the material's optical transmission (photo-bleaching). Figure 6.24 shows the reversible part of As_3Se_2 film optical transmission changes under the radiation of helium–neon laser at different temperatures (Averynov et al., 1980; Kolomiets, 1981). Transmission changes from point *A* to point *B* correspond to the photo-darkening of the film irradiated at room temperature. Heating of the irradiated film leads to a slight decrease in transmission, first due to the temperature dependence of gap value (section *BC*). A sharp increase in transmission occurs when the temperature reaches the annealing temperature value (section *CD*). Section *DA* corresponds to transmission changes in the annealed film with temperature decreasing.

If irradiated at room temperature, the film is heated to temperature T_1, and when irradiated again, the transmission increases from point E to point F. In other words, photo-bleaching of the film takes place. On the other hand, irradiation at the same temperature of annealed film (point G) reduces the optical transmission to the same value, characterized by point F. Finally, if irradiated at room temperature, the film is heated; while being irradiated, its transmittance will change along the curve *BLFD*. Thus, the final value of the film optical transmission is determined only by the temperature of the last exposure and does not depend on the sequence of irradiation and annealing processes.

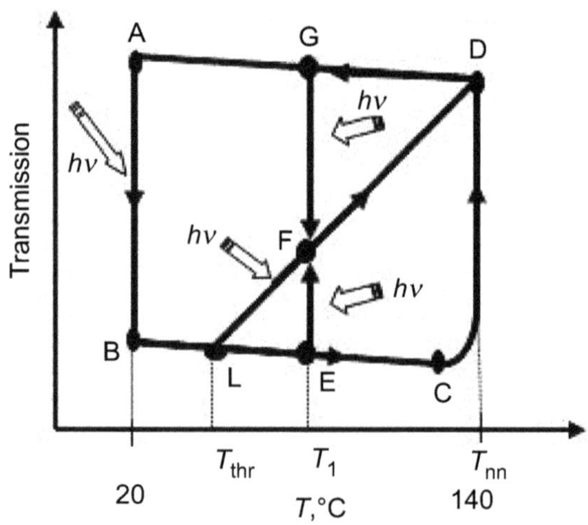

Figure 6.24 Reversible changes in the optical transmission in As_3Se_2 film at various temperatures.

Photo-induced anisotropy of optical and other properties induced by linearly polarized light is also observed in glassy chalcogenide films (Zakery and Elliott, 2003; Kolobov and Tominaga, 2012).

Along with the optical properties, photo-induced changes in the electrical and photoelectrical properties, microhardness, glass transition temperature, and the rate of material etching are observed in glassy chalcogenide films. These changes have reversible and irreversible components as well. The changes in the microhardness, and glass transition temperature of As_3Se_2 films under the influence of light exposure and annealing are shown in Fig. 6.25 (Gurevich et al., 1976b). Characteristic values increase under irradiation and decrease with subsequent annealing. The relative changes in microhardness reach 40– 50%. The density of the films varies similarly. The increase in the density of the material leads to a decrease in film thickness (photo-contraction effect).

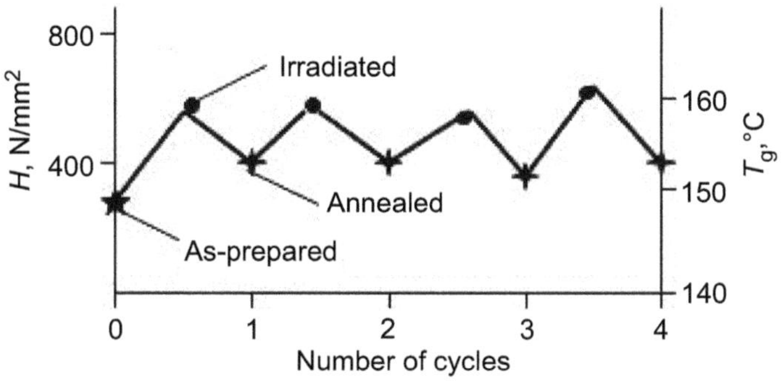

Figure 6.25 Changes in the microhardness and glass transition temperature in As_3Se_2 film under irradiation and annealing.

The effect of photo-stimulated changes on the etching rate of chalcogenide glasses in certain solvents is very interesting from the application point of view. Alkaline enchants dissolve the irradiated areas several times faster than non-irradiated areas (Fig. 6.26a) (Kolomiets, Lyubin, and Shilo, 1978). Therefore, a positive image is obtained after the exposed film is processed in an alkaline etchant. However, if the same film is processed in potassium dichromate ($K_2Cr_2O_7$), the effect will be opposite. The irradiated areas of the film are practically insoluble, and the non-irradiated areas dissolve quickly and a negative image is obtained as a result. Thus, both positive and negative image can be obtained on the same material, depending on the type of solvent.

Photo-induced changes in the etching rate of chalcogenide glasses accelerate with increasing arsenic concentration. The ratio of the dissolution rates of irradiated and non-irradiated areas of As–S films increases by 12 times with arsenic concentration increasing from 25 to 45 at.% (Fig. 6.26b) (Petkov and Fedorov, 1982).

Analysis of experimental results shows that photo-induced property changes described above are observed in materials that meet at least two conditions: chalcogen atoms have to be in the material composition and the material must be in non-crystalline state.

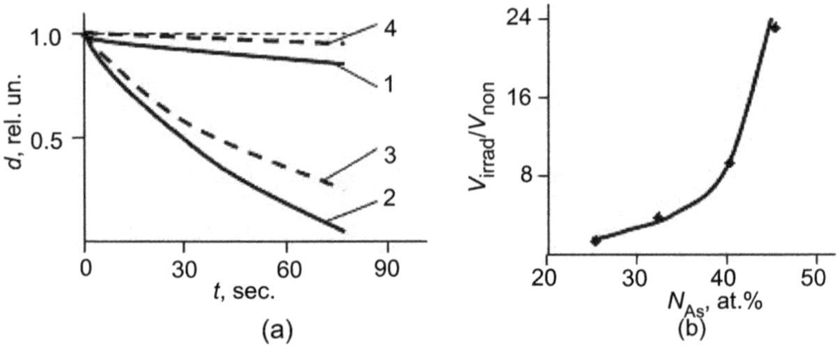

Figure 6.26 Thickness of non-irradiated (curves 1 and 3) and irradiated (curves 2 and 4) areas of AsSe film dependencies on etching time in NaOH (curves 1 and 2) and $K_2Cr_2O_7$ (curves 3 and 4) (a) and dependence of dissolution rates ratio of irradiated (V_{irrad}) and non-irradiated (V_{non}) areas of As–S films on arsenic concentration (b).

Initially, the photo-induced changes in chalcogenide glass properties were associated with photo-chemical reactions, accompanied by the formation of individual component clusters of the material. However, diffraction and other studies have established that these changes were caused by changes in the atomic structure. As an example, Fig. 6.27 shows the diffracted X-ray intensity curves of As_2S_3 films after annealing at 170°C and exposure for 10 h to light at 290 and 77 K (Tanaka, 1975). As shown in the figure, the intensity curve after exposure to light at 290 K has a lower maximum at $2\Theta \approx 17°$ and a higher minimum at around 20° compared with those before exposure, and the first diffuse peak shifts to a higher angle. Such tendency is enhanced in the case of exposure at 77 K. Transition from one curve to another is reversible with good reproducibility. All this confirms the structural nature of the photo-induced changes in properties.

Changes in chalcogenide glass properties occur not only under light irradiation but also under the action of electrons and X-rays. These effects have reversible and irreversible components and are associated with structural changes in the material. Figure 6.28 shows the blackening curves of glassy $As_{65}Se_{35}$ film electron diffraction patterns obtained after different periods of electron exposure. As seen from the figure, changes in the diffraction pattern occur in the area from 1.5 to 4.5 $Å^{-1}$ during the first 2 min of sample irradiation. The intensity of the second and, especially, the third peaks decreases, and their mutual rapprochement takes place. At the same time, there are no significant changes in the height and position of the first peak at 1.1–1.2 $Å^{-1}$. After approximately 2 min (at 50 kV accelerating voltage), the diffraction pattern is stabilized, and further exposure of the sample to electrons does not lead to changes in it.

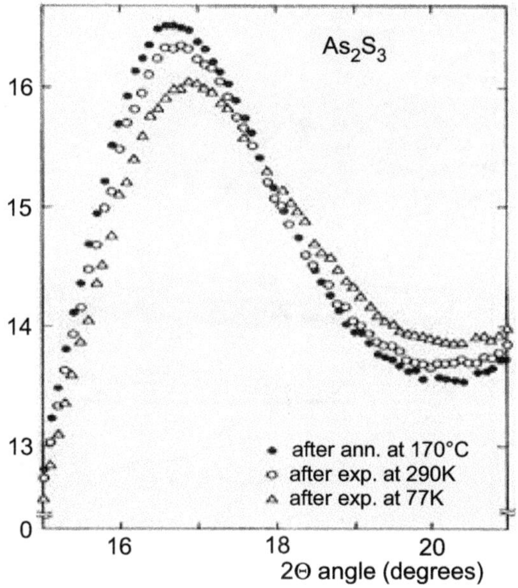

Figure 6.27 X-ray intensity curves of As_2S_3 amorphous films after annealing and illumination. Reprinted from Tanaka (1975), with permission from AIP Publishing LLC.

These results are a direct evidence of structural changes occurring in the material under electron radiation as well. The observed broadening of the peaks in the diffraction pattern indicates increase in structural disorder under electron radiation. It should be noted that it is related to changes in the short-range order of atomic arrangement, since the peak at S = 1.1–1.2 $Å^{-1}$, which characterizes the medium-range order, remains practically unchanged.

A number of phenomenological models have been created to explain the photo-induced changes in the properties of chalcogenide glassy semiconductors (Kolobov and Shimakawa, 1996; Kolobov and Tominaga, 2012). These models are based on the presence of atoms in the structural network that can be in two different metastable states. The appearance of metastable states may be due to the photo-induced formation of valence-alternation pairs on the atoms of chalcogen (neutral C_1^0–C_3^0 or charged C_1^-–C_3^+; see Fig. 3.11) (Kolobov and Tominaga, 2012), or a photo-induced transformation of pyramidal units As_2S_3 into a fivefold coordinated As site having four As–S bonds and one As–As bond (Fig. 6.29) (Uchino, Clary, and Elliott, 2000).

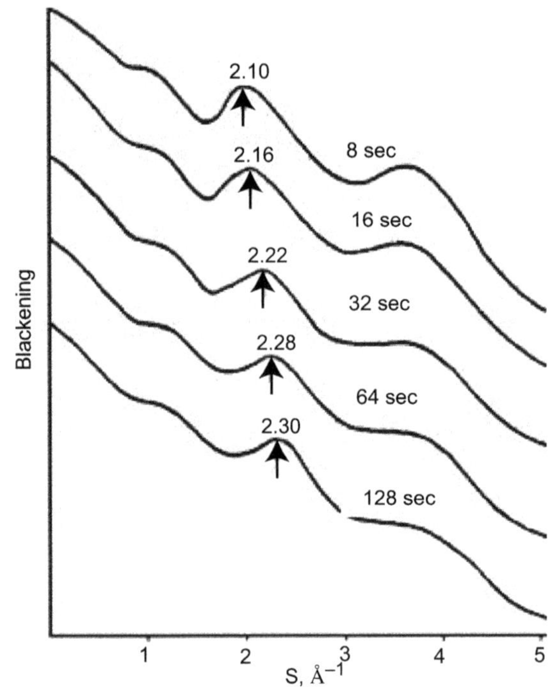

Figure 6.28 Blackening curves of glassy $As_{65}Se_{35}$ film electron diffraction patterns obtained after different periods of electron exposure (accelerating voltage is 50 kV).

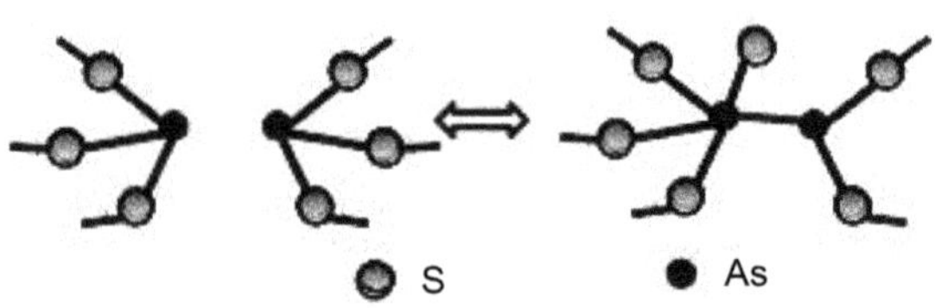

Figure 6.29 Schematic representation of photo-induced transformation in As_2S_3. Reprinted from Uchino et al. (2000), with permission from American Physical Society.

The state with lower energy is the ground, but a state with higher energy is excited. Optical and thermal transitions between these two states take place. Energy versus configuration dependencies have a parabolic form. It is assumed that the parabolas, corresponding to two states, have significantly different steepness (Fig. 6.30) (Kolomiets, 1981).

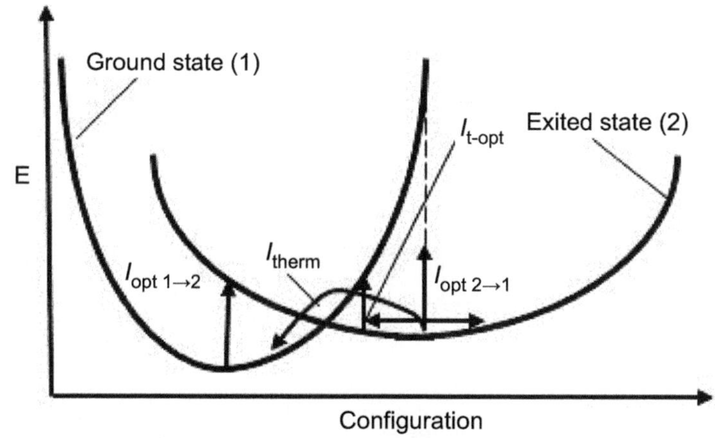

Figure 6.30 Illustration of photo-induced changes on the basis of configuration diagram (I_{opt} is optical transition, I_{term} is thermal transition, and $I_{t\text{-}opt}$ is thermo-optical transition). Reprinted from Kolomiets (1981), with permission from EDP Sciences.

As seen from Fig. 6.30, the energy required for the optical transition from ground state 1 to excited state 2 (direct transition $I_{opt1\rightarrow 2}$ by the selection rule) is significantly less than the energy of optical transition from state 2 to state 1 ($I_{opt2\rightarrow 1}$). As a result, the first process is dominating under irradiation, and most of bistable sites are transferred to state 2. This leads to photo-darkening and to changes in other properties described above. Recurrence of atoms to state 1 is possible by thermal transition I_{therm} during annealing at temperatures close to glass transition temperature. However, there is another possibility of transferring bistable sites from state 2 to state 1. Heating of the sample reduces the energy required for the optical transition from state 2 to state 1, due to increasing thermal vibrations of atoms. As a result, the bistable site is transferred to state 1 by means of thermo-optical transition $I_{t\text{-}opt}$. This explains the effect of photo-bleaching under material irradiation at heightened temperature (transition from point E to point F in Fig. 6.24).

The effect of acceleration of semiconductor layer crystallization in the case of vidicon operation at low temperatures (see Section 6.1.6) can be explained based on the above model. Let us consider the configuration diagram of glassy and crystalline states of the semiconductor (Fig. 6.31).

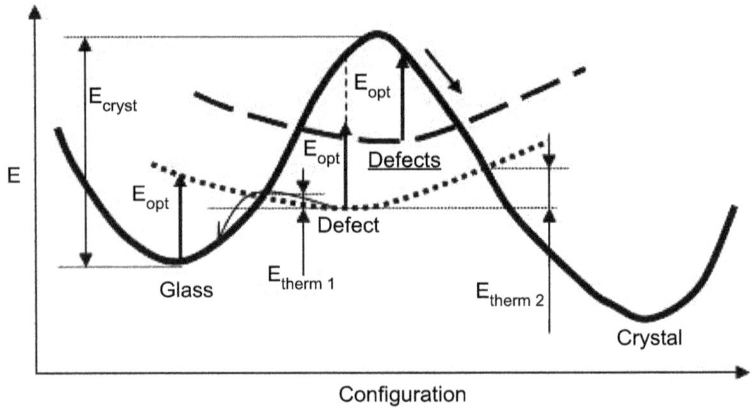

Figure 6.31 Configuration diagram illustrating the effect of light-induced selenium layer crystallization at low temperature (see text).

The left parabola in the figure corresponds to the glassy state, and the right parabola corresponds to the crystalline state. The activation energy of crystallization is equal to barrier height E_{cryst}. Optical transitions E_{opt} occur under irradiation, and excited (defect) states corresponding to the dotted parabola are formed. Reverse optical transition to the ground glassy state is not possible because of different steepness of the parabolas. The energy required for thermal transition to the crystalline state is equal to E_{therm1}, and the energy required for thermal annealing to the glassy state is equal to $E_{therm2}<E_{therm1}$. However, there is not enough energy for any of these transitions at low temperatures and photo-induced excited states are accumulated in the material. This leads to energy increase and a shift in the parabola corresponding to the excited state (the dashed curve in the figure). At a certain moment, the photon energy becomes sufficient for optical transitions from the excited states. But, as can be seen from the figure, these transitions lead to material crystallization.

The above configuration diagram not only explains the crystallization acceleration in the case of vidicon operation at low temperatures, but also shows the way to solve the problem. Elevating the temperature of the selenium vidicon target to 35°C during illumination acts to instantaneously anneal the photo-induced excited defect states and eliminates the crystallization (Resnik et al., 2006).

The common advantage of devices based on photo-induced transformation is high resolution (at least 10^3 lines per millimeter), and the general drawback is low photosensitivity (10^{-2}–10^1 cm^2/J). Other advantages of these media are the possibility of multiple recording and high diffraction efficiency in hologram recording, due to significant changes in the refractive index. Photo-induced changes in the etching rate make it possible to obtain relief images. It also provides high diffraction efficiency in holographic recording and the possibility of material application as submicron photoresists in microelectronics. The existence of electron-stimulated and X-ray-stimulated changes in chemical properties makes it possible to use these materials in electronic lithography and X-ray lithography.

Low propagation losses and photo-induced effects make chalcogenide glasses suitable candidates for the fabrication of devices such as gratings and waveguides (Zakery and Elliott, 2003). Transmissive diffraction gratings for use at infrared wavelengths have been fabricated in chalcogenide glasses using the photo-darkening effect. These gratings can be used as efficient beam combiners, couplers and have significant application in monochromators, laser-tuning devices, shapers, and coupling of optical fibers. Planar channel waveguides with very low losses (less than 0.3 dB/cm) have been fabricated in different compositions of chalcogenide glasses using the photo-darkening effect in thermally evaporated, sputtered, and pulsed-laser deposited films.

6.2.2 Photo-induced Metal Dissolution in Chalcogenide Glasses

It has been shown (Kostyshin et al., 1966) that the diffusion of some metals (Ag, Cu) in chalcogenide glasses As–S, As–Se, and some other systems is accelerated by orders of magnitude under illumination by light of greater than band edge energy. This process readily occurs at room temperature at illumination level as low as 10^{-4} J/cm^2. Until now the nature of this effect has only been explained at a qualitative level. The photons pass through a semi-transparent silver film of about 30 nm thick and are absorbed in chalcogenide film (Fig. 6.32). As a result, electron–hole pairs are generated in the chalcogenide film. Since the mobility of electrons in these materials is significantly less than the hole mobility, the electrons are captured by traps and charged C_1^- centers are formed. The holes diffuse from the generation region and recombine with the electrons of silver atoms. The diffusion of positively charged silver ions is accelerated by the Coulomb interaction with the negatively charged C_1^- centers.

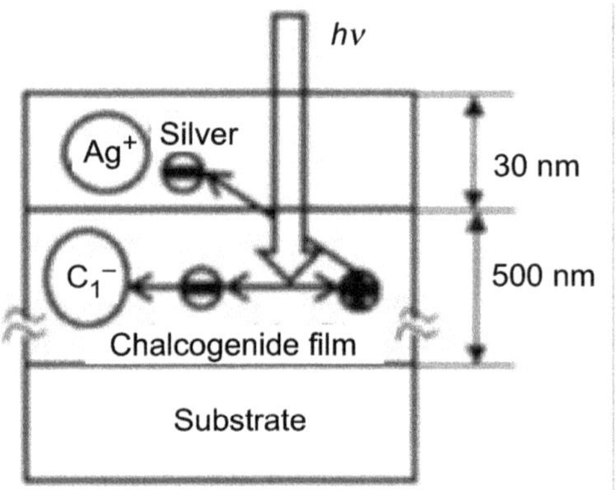

Figure 6.32 Schematic illustration of photo-doping process (see text).

The above model assumes that the diffusion front progresses through film at the same rate. However, studies of silver photodiffusion dynamics in Ge_xS_{1-x} by time-resolved neutron reflectivity measurements (Sakaguchi et al., 2014, 2016) have found that the effect has more complex nature. Two silver diffusion processes (fast and slow) were observed. At the first stage of fast diffusion, an Ag-rich layer is formed and the layer of silver on the surface of the film is exhausted. At the second stage, the diffusion of silver from the Ag-rich layer into the rest of the film occurs with a substantially slower rate. The position of the interface between the Ag-rich and Ar-poor reaction layers remains almost fixed. The rates of diffusion processes depend to a large degree on the ratio of components in the Ge–S system, which is associated with a change in structural flexibility.

Photo-doping leads to changes in optical and chemical properties. It has been shown that chalcogenide films of Ge–S and Ge–Se systems photo-doped with silver are extremely resistant to etchants used in semiconductor manufacturing. This, together with high resolution (see Table 6.1), allows to create photoresists and electronresists for high and super high resolution (Masters and Goldberg, 1981).

The photolithographic process based on photo-induced silver dissolution is as follows. A chalcogenide film of the Ge–Se system with thickness less than 1 μm and a thin silver film (30–50 nm thick) are deposited sequentially on the surface of an oxidized silicon wafer. This structure is exposed. Silver on the exposed areas diffuses into the chalcogenide film. Silver remaining on the unexposed areas is removed in acid solution. Then wet or plasma etching of the chalcogenide film and SiO_2 is carried out. The dependencies of the plasma etching depth of $Ge_{25}Se_{75}$ photoresist on etching time in CF_4 are shown in Fig. 6.33 (Mizushima and Yoshikawa, 1982).

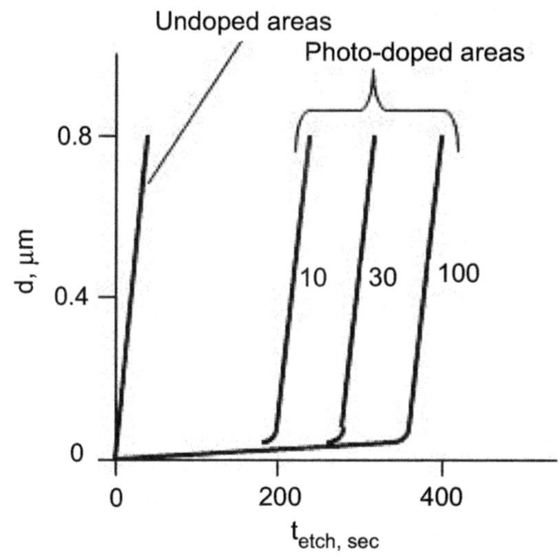

Figure 6.33 Dependences of the plasma etching depth of $Ge_{25}Se_{75}$ photoresist on etching time in CF_4 (numbers on the curves indicate the exposure time, seconds).

These results indicate high efficiency of the photolithographic process. Transmission electron microscopy shows that the spatial resolution for photo-doping effect is at least as fine as 10 Å.

The following advantages of these resists can be mentioned:

- High resolution
- High resistance to etchants, which allows to use thin resist films and, as a consequence, increase the resolution of the photolithographic process overall
- High hardness, which reduces the probability of defect formation in the manufacturing process
- High atomic mass of elements in the resist composition effectively protects wafer surface in the case of ion implantation doping

6.2.3 Photo-induced Phase Transitions

The ability to store information is an important aspect of our civilization's development. Among multiple methods of information storage, optical data storage (optical disks or compact disks) based on the first-kind phase transitions (evaporation, melting, crystallization, and amorphization) in chalcogenide films is one of the most successful. An optical disk is a device for recording and storage of optical information. Its principle is based on the creation of spots (pits) in its active layer, the properties of which differ from the rest of the layer. The presence or absence of a pit represents "0" or "1" in binary digital code, that is, one pit is equal to one bit of information.

Photo-induced evaporation leads to the formation of surface relief (pits) and to changes in optical reflection from the exposed areas. Optical data storage based on crystalline amorphous transitions uses significant property differences of crystalline and amorphous phases. At the same chemical composition, the band gap of the amorphous phase is 30–50% larger than the crystal band gap, and the dielectric constant is 2–2.5 times and resistivity is 10^4–10^5 times larger. The reversibility of the crystallization–amorphization process allows fabrication of rewritable memory (write-and-read-many-times memory). Phase transitions between amorphous and crystalline states originally have been applied to the creation of non-volatile memory switchers on the base of multi-component chalcogenide films (Ovshinsky, 1968). The difference in the electrical conductivity of amorphous and crystalline phases has been used in these devices (see Section 8.1.2). In contrast, the differences in the coefficients of reflection from the amorphous and crystalline phases are used in optical disks. Optical contrast due to recoded pits is determined as follows:

$$S = (I_{R1} - I_{R2}) / (I_{R1} + I_{R2}) \tag{6.3}$$

where I_{R1} and I_{R2} are light intensities reflected from the original surface and pit area, respectively.

Optical disks came to replace gramophone (phonographic) disks and magnetic tapes, which were analog storage media. The first commercial music digital compact disk and disk player were developed by Sony and Phillips alliance in 1982. A disk diameter of 120 mm (4.72 inches) provided 72 min of audio recording duration, which was determined by the requirement to hold Beethoven's 9th Symphony on a single disk (Kolobov and Tominaga, 2012).

The modern optical disk is a flat circular polymer substrate (usually polycarbonate) 1.2 mm thick and 120 mm in diameter with deposited recording and certain additional layers. The encoding pattern follows a continuous spiral path covering the entire disk surface and extending from the innermost track to the outermost track (Fig. 6.34). The distance between the centers of two neighboring tracks lies in the range of 1.6–0.32 μm.

Information is read in the following manner. The laser beam collimated in lens 1 passes through a beam splitter and is focused by lens 2 on a recording layer. The reflected beam passes through lens 2, beam splitter, and lens 3 before entering the photodetector (Fig. 6.34). If the pit is recorded at that point (left side of the figure), then the direction and intensity of the reflected beam changes, causing a change in photodetector signal. Thus, the laser beam reacts to the presence of a pit. The method to read a pit position is called pit-position readout method. In some cases, the data are actually addressed at the edges (front and tail) of the pits. This method is called edge-position readout and allows to increase the recording density by 50%. A distinctive feature of this method is the different length of pits (Fig. 6.35).

Optical disks can be classified into three large groups according to the method of information recording and reading:

- **Read only memory (ROM) disk**: Information from these disks can be read out any number of times. However, it is impossible to erase, rewrite, or make changes to recorded information.

- **Recordable (R) disk**: Information on this disk can be recorded once on a standard drive unit, but the recorded information cannot be changed (irreversible recording).
- **Erasable or rewritable (RW) disk**: These disks allow not only to record but also to rewrite information (at least up to 10^4 times).

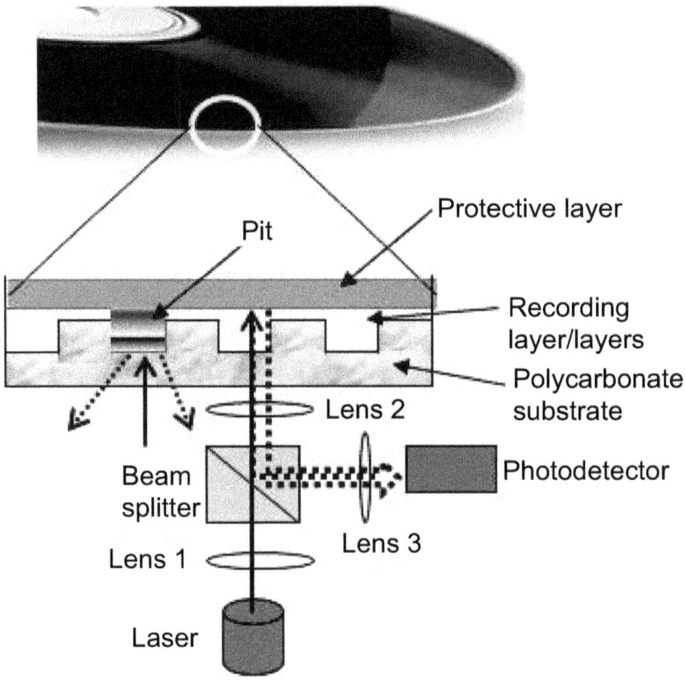

Figure 6.34 A schematic representation of compact disk structure and the reading information process.

Recording information on ROM disks is carried out during the manufacturing process by changing the surface relief by stamping. The so-called master disk needs to be produced first using the methods of photolithography and galvanic deposition of metal. The master disk has surface protrusions instead of pits. Then the master disk replicas are produced, and these replicas are used for hot forming polycarbonate disks with the required surface relief. The final stage is deposition onto the disk surface of the reflective (aluminum) and protective layers. An image of the ROM disk surface after stamping, which was obtained at the atomic force microscope, is shown in Fig. 6.35. The described technology is characterized by low cost for high-volume production.

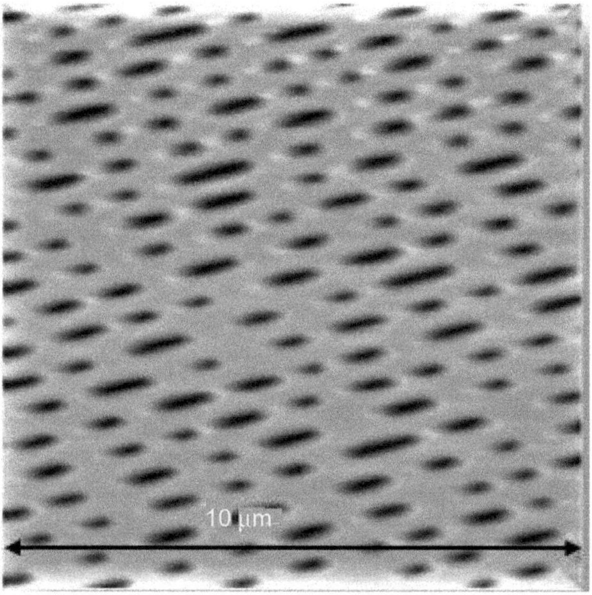

Figure 6.35 Image of ROM disk surface at the atomic force microscope.

Recordable or write-once-read-many-times memory (Moller et al., 2003) disks are produced by two technologies. The first of these is the creation of pits by the photo-induced evaporation of low-melting reflective layer (bismuth, selenium–tellurium system, and some other). In the second case, an organic dye layer and a reflective layer (aluminum) are successively deposited on the polycarbonate substrate. Pits are created by changing the optical transmission of the organic dye (cyanine dye, for example) under the effect of radiation. Recording is carried out by laser radiation with pulse duration of 50–500 picoseconds forming pits with the size of 0.5–0.05 microns. At present, the main type of recordable disks is disk based on organic dyes because of their lower cost.

The concept of erasable or rewritable (RW) disks can be described as follows (Fig. 6.36). A short, high-intensity laser pulse locally heats the crystalline phase of the phase-change material above the melting temperature (transition 1). The melted volume quenches into the amorphous phase after pulse finishing (transition 2) due to the high cooling rate (10^8–10^9 K/s). This amorphous dot has lower optical reflectivity in comparison with the surrounding crystalline phase and can be easily detected with a low-intensity laser beam. To erase the stored information, a laser pulse with lower intensity and longer duration locally heats the amorphous dot to a temperature slightly above the glass transition temperature. It leads to increasing mobility of atoms and crystallization of the amorphous dot (transition 3).

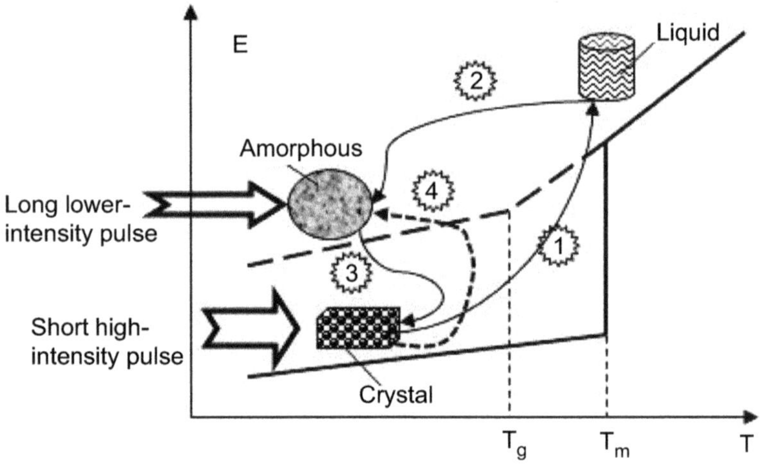

Figure 6.36 Schematic illustration of phase-change optical recording (see text).

The phase-change material in the optical disk should satisfy the following requirements (Wuttig and Yamada, 2007):

- High speed of phase transition induced by nanosecond laser pulse
- High optical contrast determined by considerable difference in the refractive index of amorphous and crystalline phases
- Long thermal stability of amorphous state to ensure data retention (at least several decades at room temperature)
- Large cycle number of amorphous–crystalline state transitions without material decomposition (more than 10^5 cycles)
- High chemical stability and water resistivity

The above requirements are rather strict and contradictory. For example, amorphous dots should be stable at room temperature for at least 10 years (10^8 s). On the other hand, fast crystallization within several tens of nanoseconds (10^{-8} s) must occur under the laser pulse at the temperature range between the glass transition and melting temperatures. It means that increasing temperature by a few hundred degrees has to change the phase-transition time by 16 orders of magnitude. Therefore, very few materials have the combination of properties mentioned above. As a result of research and development, the following systems and materials—Ge–Sb, Ge–Sb–Te, $Ge_2Sb_2Te_5$ named GST 225, Ag–In–Sb–Te named AIST—were found to be the best in terms of speed and stability.

It should be noted that fast and stable phase-change transitions in these materials are difficult to quantitatively describe by means of conventional melting, amorphization, and crystallization processes. Based on extended X-ray absorption fine structure spectroscopy (EXAFS, see Section 2.3) of $Ge_2Sb_2Te_5$, Kolobov et al. (2004) concluded that phase changes under laser pulse occur without material melting in a conventional sense. Germanium atoms occupy the octahedral position in the crystalline state and have stronger covalent and weaker bonds (Fig. 6.37a).

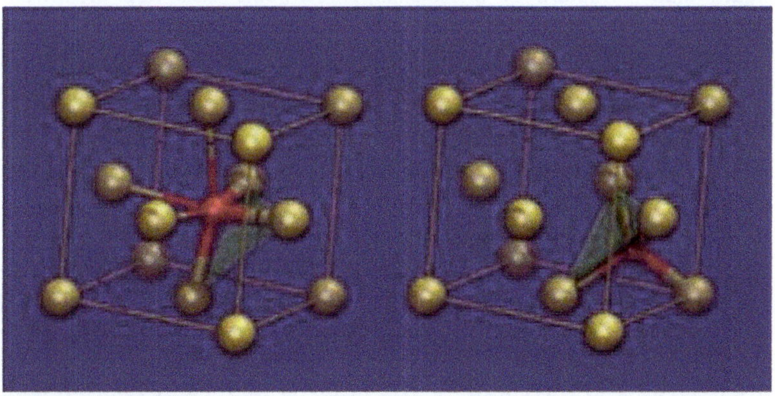

(a) (b)

Figure 6.37 Fragments of the local structure of GST225 around germanium atoms in the crystalline (a) and amorphous (b) states (stronger covalent bonds are shown as thicker lines, whereas weaker bonds are shown as thinner lines). Reprinted from Kolobov et al. (2004), with permission from Nature Publishing Group.

An intense laser pulse induces rupture of the weaker bonds due to electronic excitation, and germanium atom flips into tetrahedral position corresponding to the amorphous state (transition 4 in Fig. 6.36 called "umbrella-flip" structural transformation, Fig. 6.37b). Notice that three covalent bonds remain intact. This structural transformation involves a change in electron orbital hybridization and accounts for a large change in the optical properties of the material in phase transition.

A typical structure of a phase-change rewritable optical disk is shown in Fig. 6.38 (Kolobov and Tominaga, 2012).

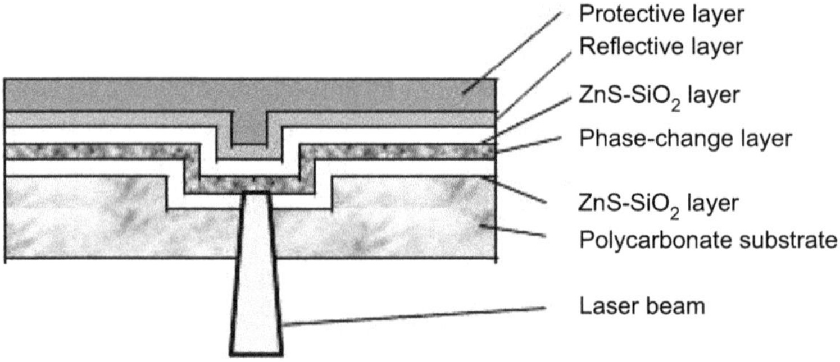

Figure 6.38 Structure of phase-change rewritable disk. Reprinted from Kolobov and Tominaga (2012), with permission from Springer.

The laser beam passes through polycarbonate substrate and the bottom $ZnS–SiO_2$ layer, which is used to reduce reflection from the interface between the substrate and the phase-change layer. In the phase-change layer, the laser beam provides recording/erasing or reading of the information. The top $ZnS–SiO_2$ and aluminum alloy reflective layers provide increased optical contrast between crystalline and amorphous areas. Also the aluminum alloy layer contributes to heat removal from the phase-change layer during the recording of information. The latter is important for reducing pit size and increasing the density of information recording (see below). Figure 6.39 shows the atomic force microscope image of the phase-change layer surface after information recording. Recorded pits that represent the areas of amorphous phase in the crystalline layer are clearly visible as lighter areas.

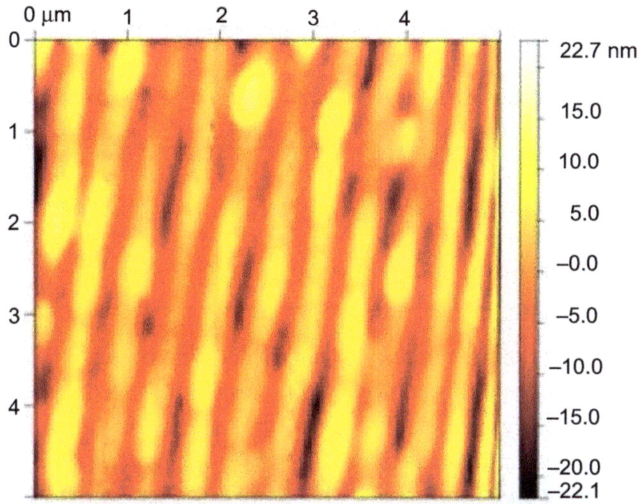

Figure 6.39 Atomic force microscope image of phase-change layer surface after information recording.

The data storage capacity of optical disks is directly related to the size of amorphous/crystalline dots. In its turn, the dot size depends on the wavelength of the laser used for recording and on the numerical aperture of the optical system. The spot size of the focused laser beam is determined by the equation

$$d = \lambda/(2NA) \tag{6.4}$$

where λ is the wavelength of light and NA is the numerical aperture of the lens. That is why decreasing the recording laser wavelength has become the main way to increase the overall density of information storage achievable in a single disk. However, the problem of good optical contrast has appeared at reduced wavelengths. The GST225 alloy, which showed good optical contrast in the infrared and red spectral range, is not ideal for shorter wavelengths. Optical property changes due to amorphous–crystalline state transitions in alloys of the Ge–Sb–Te system with different chemical composition are presented in Table 6.4 for a wavelength of 405 nm. As the table shows, the difference in the optical properties and, consequently, the optical contrast increase with increasing germanium concentration (Yamada, 2002). Alloys such as $Ge_8Sb_2Te_{11}$ or those with more germanium content possess higher optical contrast and are preferable for short-wave optical recording.

Table 6.4 Change in refractive index (n) and absorption coefficient (k) in Ge–Sb–Te alloys for a wavelength of 405 nm

Composition	Ge content (at. %)	$\Delta n = n_{cryst} - n_{amorph}$	$\Delta k = k_{cryst} - k_{amorph}$
$Ge_2Sb_2Te_5$	22	1.20	1.05
$Ge_4Sb_2Te_7$	31	1.26	1.09
$Ge_6Sb_2Te_9$	35	1.45	1.27
$Ge_8Sb_2Te_{11}$	38	1.48	1.35

Along with the decreasing wavelength of the recording laser, the development of optical memory went toward improving the optical and mechanical systems of disk drives and software algorithms. The progress of optical disks is shown in Table 6.5 (Wuttig and Yamada, 2007; Kolobov and Tominaga, 2012; Sony and Panasonic, 2015).

Table 6.5 Comparison of optical disk characteristics

Year	1997	2000	2003 (single)	2004 (double-sided)	2015 (double-sided with triple layer per side)
Recording capacity	650 MB	4.7 GB	23.3 GB	50 GB	300 GB
Laser wavelength	780 nm	650 nm	405 nm	405 nm	405 nm
Cross section of focused beam	0.866 μm	0.54 μm	0.238 μm	0.238 μm	0.238 μm
Track pitch	1.60 μm	0.615 μm	0.32 μm	0.32 μm	0.225 μm
Numerical aperture	0.45	0.60	0.85	0.85	0.85
Minimal pit length	0.833 μm	0.41 μm	0.149 μm	0.149 μm	0.0795 μm
Distance from disk surface to phase-change layer	1,2 mm	0.6 mm	0.1 mm	0.075 + 0.025 mm	

Reduction in the laser wavelength from infrared to red and after that from red to blue-violet, increase in the numerical aperture of the lens from 0.45 to 0.85, and increasing the number of recording layers ensured the rise in recording density by more than two orders of magnitude and increase in the data-transfer rate by three orders of magnitude.

As seen in the table, the largest storage capacity (300 GB) is achieved by Archival Disc, which was created by the alliance of Sony and Panasonic. The disk is designed for cloud service companies and archival services. The Archival Disc has a double-sided structure with three layers per side. The main feature of this disk is a technology to record and/or playback simultaneously on both sides, which gives double the transfer rate achieved with one-sided recording or playback. However, the companies are not positioning the disks as a medium for consumer storage.

Table 6.5 also shows that the pit size (minimal pit length) for all disks is less than the cross section of recording laser beam focused on the surface of recording layer [calculated using formula (6.4)]. Moreover, this difference was approximately 10% in the disks of the first generation, but in the latest generations of disks, these dimensions differ by 1.5–3 times. This effect is explained as follows. During information recording, the laser light is absorbed in an area the size of the laser beam cross section. However, due to the heat sink through the lateral and edge surfaces, the temperature reaches the value of the phase transition only in the central part of this area. Consequently, the phase transition also occurs only in this central part. Thus, increasing heat sink during the recording of information allows to increase the recording density at constant laser light wavelength and numerical aperture of the lens.

In conclusion, we note that at present research and development of optical disks is aimed at increasing recording capacity and recording/reading speed. This objective is achieved in one of two ways. The first way is to create multilayer recording disks with electronic selection of the layer on which information is recorded or read (Layer Selection Recordable optical disk) (Hirotsune, Mukoh, and Terao, 2006). An estimate shows that a disk with 20 layers will have a recording capacity of 1 terabyte. The second way is to create a holographic versatile disk, in which the information recording is carried out in a volume (three-dimensionally) in the form of interference fringes from two overlaid laser beams (reference beam and a signal beam). Thus, the information is stored in the three-dimensional matrix form throughout the thickness of the layer instead of pits (Watabe, 2010). Three-dimensional recording of information allows to increase significantly both the recording capacity and the recording/reading speed, since the information is transmitted a few bytes at a time, and not one bit after another, as in conventional disc drives.

Chapter 7

Photoelectric Devices Based on Disordered Semiconductors

7.1 Renewable Energy Resources

The development of civilization is inseparably linked with the increase in energy consumption. According to the International Energy Agency (IEA), the average energy consumption per capita increased by 10% between 1990 and 2008. At the same time, the world population has increased by 27%. Until now the growing needs of humanity were largely met through power generation from fossil (hydrocarbonaceous) fuels: oil, gas, and coal. Of course, one can argue about how long the hydrocarbon fuel reserves will last, but it is clear that it is a non-renewable natural resource, as its rate of consumption is incomparable with the time of its accumulation, which is millions of years. Furthermore, the burning of fuel for electric power production leads to several negative environmental effects, such as pollution of the atmosphere, the "greenhouse" effect. In this regard, it is necessary to look for alternative renewable energy sources. Renewable sources of energy, as a rule, include energy of solar radiation, wind, streams of water, biomass, thermal energy of the upper layers of the earth's crust, and oceans. Solar radiation delivers the most energy among all of these sources. There are three ways to use solar energy: heating and lighting of buildings, water heaters, and photovoltaic process, that is, conversion of sunlight

Disordered Semiconductors: Physics and Applications (2nd Edition)
Anatoly Popov

ISBN 978-981-4774-37-6 (Hardcover), 978-1-315-10042-5 (eBook)
www.panstanford.com

directly into electric current. The direct conversion of solar energy into electricity is very attractive for several reasons:

- Sunlight is free and inexhaustible.
- Sunlight is abundant (the power of solar radiation at the earth's surface is 10^{16} TW, which is four orders of magnitude greater than human beings' needs in energy production today).
- Photovoltaic systems burn no fuel and thus produce neither atmospheric emissions (e.g., greenhouse gases), nor hazardous waste to cause detrimental effects on the environment.
- Since photovoltaic systems have no moving parts, they are silent and have high reliability.

The disadvantages of solar energy are the low energy density, instability, and seasonality of solar radiation fluxes. However, the main problem is the cost of electricity generation.

The first solar cell based on single-crystal silicon was manufactured in 1954, which had an efficiency of about 6%. Significant progress in the development of solar cells based on single-crystal silicon was made in the 1960s fuelled by the development of space technology. At present, the efficiency of such photovoltaic cells is more than 20%. Solar cells based on single crystals of gallium arsenide were also created in the same period. Currently, we observe intensive growth both in the production of photovoltaic solar cells and in electricity generation at photovoltaic plants. According to the *BP Statistical Review* (2015), the installed capacity of photovoltaic power plants in the world has increased by two orders of magnitude over the past decade. The annual increase in electricity generation was between 35% and 90%.

The leaders in power generation through photovoltaic plants are Germany and China (NPD, 2013). It should be noted that 90% of all solar modules in Germany are located on the roofs of houses. A significant event happened in Germany on June 9, 2014. About 23.1 GWh or 50.6% of the consumed electricity in the country on that day was produced by photovoltaic plants (PR Newswire, 2014). Despite the fact that it was a day off, when energy consumption is minimal, this is an achievement of the solar industry.

Following the leaders, other countries are trying to encourage the development of renewable energy sources. A change in the rules for permitted development rights in the United Kingdom has increased the development threshold for rooftop solar panels on commercial buildings by a factor 20. Under the new rules, installations up to 1 MW no longer require full planning permission for development (Appleyard, 2015a). A recently passed legislation in France has mandated that the roofs of all new buildings in commercial areas must be so-called green roofs, which are at least partially covered either in solar modules or with living plants (Appleyard, 2015b). Russia is going to construct 178 renewable energy facilities with a total capacity of 146 MW within the coming 5 to 6 years at the Russian Far East and close to the Artic Circle. These regions have low population density and weak network connection and get power from 500 diesel power plants (Vorotnicov, 2015).

The vast majority of photovoltaic solar cells are manufactured from single-crystal silicon. However, the production cost of solar cells based on single crystals is rather high. And more than half of this cost is the cost of manufacturing of ingots and wafers (Fig. 7.1). Furthermore, since the sizes of single crystals are limited, module assembly cost is also quite high. Thus, possibilities to reduce the cost of photovoltaic modules based on single-crystal semiconductors are very limited. This leads to the high cost of electricity produced. Cost reduction by a factor of 10 is necessary for solar cells to be competitive with the traditional sources of electricity.

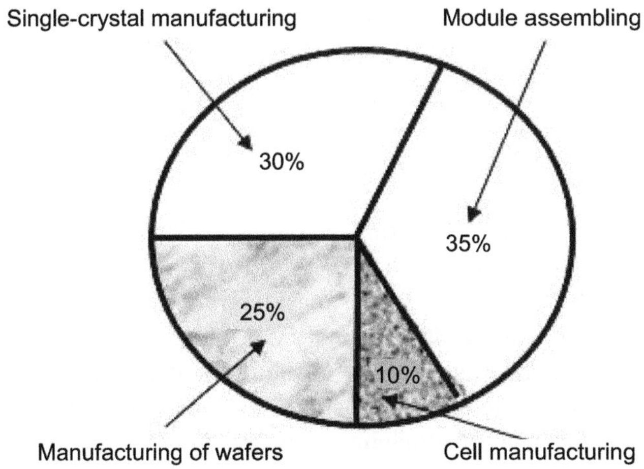

Figure 7.1 Price structure of the photovoltaic module based on single-crystal silicon.

The solution lies with the development and manufacture of thin-film solar cells based on polycrystalline or amorphous semiconductor material. While such cells have lower efficiency, the significantly lower production costs and the possibility of manufacturing elements with a large surface area make it possible to reduce the cost of produced electricity.

Thus, today there are two trends in photovoltaic solar cell development:

- Solar cells based on single crystals characterized by high efficiency and high cost
- Solar cells based on thin films with non-crystalline or even organic structure characterized by lower efficiency and significantly lower cost

Devices based on hydrogenated amorphous silicon, its alloys, and microcrystalline silicon represent a significant share of the world's production of thin-film solar cells (see Section 7.3). Organic semiconductors, as well as hybrid systems (organic–inorganic semiconductors), including organometallic compounds with a perovskite structure (Rhee et al., 2013; Yeh and Yeh, 2013), are relatively new but a rapidly developing class of materials for thin-film solar cells. Solar cells based on these materials are considered in Section 7.5.

It should be noted that the third trend in the construction of photovoltaic silicon solar cells has been actively developed in recent years. It is called silicon heterojunction technology (SHJT) or heterojunction with intrinsic thin-layer (HIT) technology. SHJT cells are created on the basis of heterojunction between single-crystal and hydrogenated amorphous silicon. This technology was suggested 25 years ago by Sanyo (Tanaka et al., 1992). The reason for the renewed interest in this technology is largely due to a sharp decline in polycrystalline silicon prices. Its price in 2006–08 was in the range of 200–450 US dollars per kilogram, and in 2011–13, the price decreased below 30 US dollars per kilogram (Terukov, 2016). Accordingly, there was a decline in the price of silicon ingots made from polycrystalline material. Using SHJT technology, it is possible to sharply increase cell efficiency with a reasonable increase in their cost (see Section 7.4).

7.2 Conversion of Solar Energy into Electricity

The principle of photovoltaic cells is based on the internal photoelectric effect (see Section 3.4). Irradiation with light in the fundamental absorption region of the semiconductor structure with a p–n junction leads to the generation of electron–hole pairs. The concentration of electron–hole pairs in the generation region increases, so that they diffuse to the p–n junction. The electric field of the p–n junction separates electrons and holes (Fig. 7.2a). As a result, the open-circuit voltage (U_{oc}) across the terminals of photovoltaic cell appears if the external electric current equals zero. The magnitude of this voltage is determined by the p–n junction contact potential difference (barrier height). If the external circuit is closed (load resistance is equal to zero), the current in the circuit will be equal to the short-circuit current (I_{sc}), determined by the internal resistance of the cell. In the case of finite values of load resistance, current and voltage will have some intermediate values. The current–voltage characteristic of photovoltaic cells is shown in Fig. 7.2b.

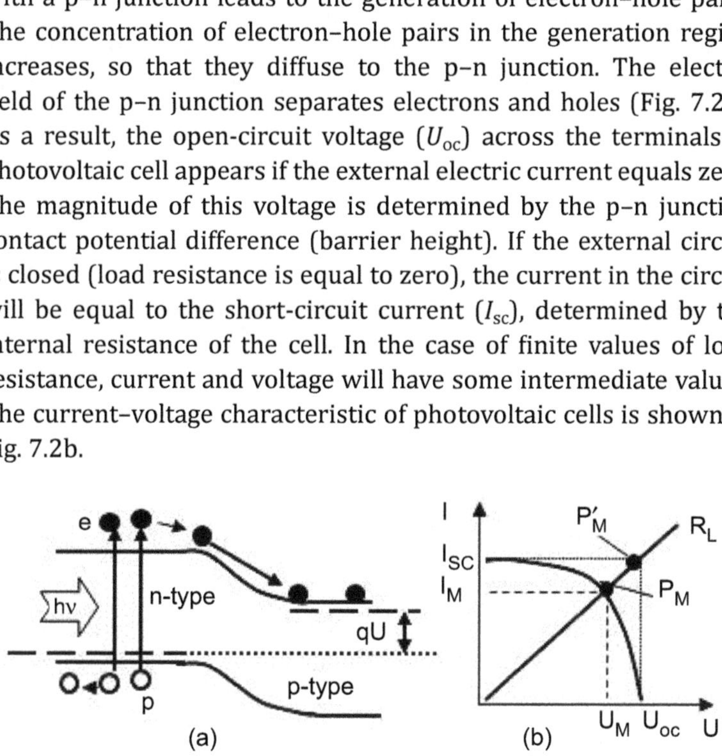

Figure 7.2 Schematic illustration of photovoltaic cell principle (a) and current–voltage characteristic (b).

The efficiency of a photovoltaic cell can be written as follows:

$$\eta = I_M U_M / P_i = (I_{SC}\, U_{OC}\, FF) / P_i \qquad (7.1)$$

where I_M and U_M are current and voltage of photovoltaic cells at the point of maximum power, FF is the fill factor, and P_i is the power of solar flux incident on the cell surface under standard test conditions (temperature 25°C and an irradiance of 100 mW/cm^2 with an air mass 1.5 (AM1.5) spectrum). AM1.5 corresponds to the spectrum of sunlight incident on a clear day with the sun at an angle of 48° above the horizon. The fill factor is determined from the expression (7.1) as follows:

$$FF = (I_M U_M)/(I_{SC} U_{OC}) \quad (7.2)$$

and is the ratio of the maximum output power of photovoltaic cell (point P_M in Fig. 7.2b) to the power of the cell with an idealized rectangular current–voltage characteristic (point P'_M in Fig. 7.2b).

The efficiency of a photovoltaic cell depends on its ability to use all the absorbed solar energy without loss. There are four main channels of losses in photovoltaic cells (Fig. 7.3):

1. Absence of absorption of photons with energies less than the band gap (channel 1)
2. Thermalization (loss of energy) of those charge carriers, which have arisen upon absorption of photons with energies significantly higher than the band gap (channel 2)
3. Losses due to voltage drop on the cell's contacts (channel 3)
4. Losses due to the recombination of charge carriers in the semiconductor (channel 4)

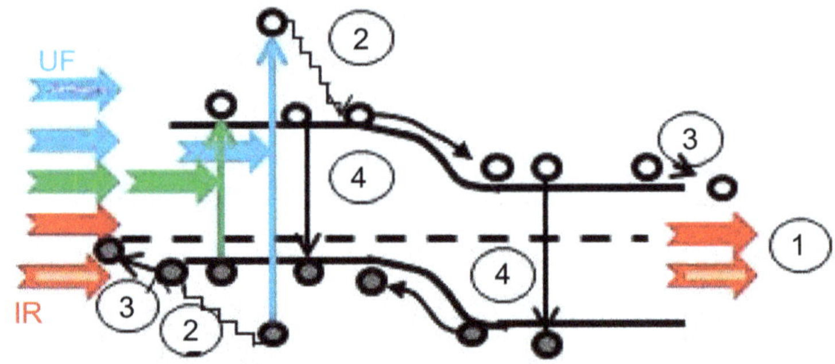

Figure 7.3 Energy losses in photovoltaic solar cells: 1 is the absence of absorption at $h\nu < \Delta E$, 2 is the loss of thermalization, 3 is losses on the contacts, 4 is recombination losses.

Consequently, conditions for the efficient conversion of solar energy into electricity in photovoltaic cells may be formulated as follows:

- A significant part of solar radiation has to be absorbed in the active region of the cell

- Photo-generated electrons and holes must be effectively collected on the contact electrodes
- Potential barrier of p–n junction should be large enough, since it determines the value of the cell's open-circuit voltage
- Internal resistance of the cell, which includes the resistances of semiconductor layers, contacts, etc., should be small enough, since it determines the value of short-circuit current.

The efficiency of solar radiation absorption is determined by the absorption coefficient of semiconductor material in the range of photon energies from 1.0 to 3.5 eV and by the thickness of the semiconductor. At the same time, the thickness of the semiconductor layer affects the collection efficiency of photo-generated electrons and holes, and should not exceed the diffusion length of carriers. Otherwise, the charge carriers recombine before their separation by the electric field of the p–n junction. In the case of single-crystal semiconductors, the above requirements are consistent with a large diffusion length of carriers. In disordered semiconductors, due to the lack of long-range order in the arrangement of atoms and high density of localized states, the diffusion length of carriers is small. And despite the fact that the optical absorption of sunlight in the hydrogenated amorphous silicon is 20 times higher than in single-crystal silicon, these requirements are in conflict and require special solutions for implementation.

7.3 Photovoltaic Cells Based on Hydrogenated Amorphous Silicon

For separation and subsequent collection of photo-generated charge carriers in the semiconductor layer, an internal electrical field has to be created. In the cells based on single-crystal semiconductors, this field is created by the space charge of the p–n junction.

The width of the p–n junction space charge in non-crystalline semiconductors is small because of the high density of localized states. Consequently, separation of charge carriers occurs only in a small part of the semiconductor layer. This situation, together with a small diffusion length of charge carriers, leads to low collection efficiency of carriers. To improve this, either Schottky barriers or p^+–i–n^+ structures are generally used in the photovoltaic cells based on hydrogenated amorphous silicon. Photovoltaic cells using Schottky barrier have a simpler structure (n^+–i–Me structure). The metal, forming Schottky barrier, should have a high work function of electrons (Pt, Rh, or Pd) since the undoped silicon layer has weak n-type conductivity (see Section 4.1). High work function of electrons from the metal causes the formation of positive space charge in the semiconductor layer.

However, the higher efficiency of photovoltaic cells has been achieved by using p^+–i–n^+ structures (Fig. 7.4a). There are two junctions (p^+–i and n^+–i) in this structure. The concentration of charged centers in i-type layer is relatively low due to the absence of doping. Therefore, in the case of film with 0.4–0.8 μm thickness, the space charge region extends to the full thickness of i-type layer (Fig. 7.4b). Consequently, electric field exists in the whole i-type layer (Fig. 7.4c), and the charge carriers photo-generated in any part of i-layer are separated by an electric field that will ensure their efficient collection. Electrons and holes photo-generated in the n^+- and p^+-layers do not contribute significantly to the photocurrent of the cell because of the low lifetime of charge carriers in heavily doped a–Si:H layers. Therefore, thickness of the n^+- and p^+-layers typically ranges from 8 to 20 nm.

The next question, most important from the standpoint of efficiency, is from which side (through the p^+- or n^+-layer) sunlight be brought to the cell's active region. Since the absorption of photons is more probable nearer to the illuminated surface, in the case of illumination from the p^+-layer (as shown in Fig. 7.4c), the path for electron transition in i-layer is longer than the path for hole transition. This is the so-called "direct design" of photovoltaic cell. If the cell is illuminated from the n^+-layer side (reversed construction of photovoltaic cell), holes pass along the longer path in i-layer.

The diffusion length of electrons is longer than the diffusion length of holes in hydrogenated amorphous silicon. From this point of view, the direct design of photovoltaic cells is preferable because it ensures better collection of photo-generated carriers at the contacts. However, it is necessary to consider one more factor when choosing the design of photovoltaic cells. Any doping increases the optical absorption coefficient of a–Si:H. Moreover, doping with boron increases the optical absorption coefficient of a–Si:H to a much larger degree than doping with phosphorus (Fig. 7.5).

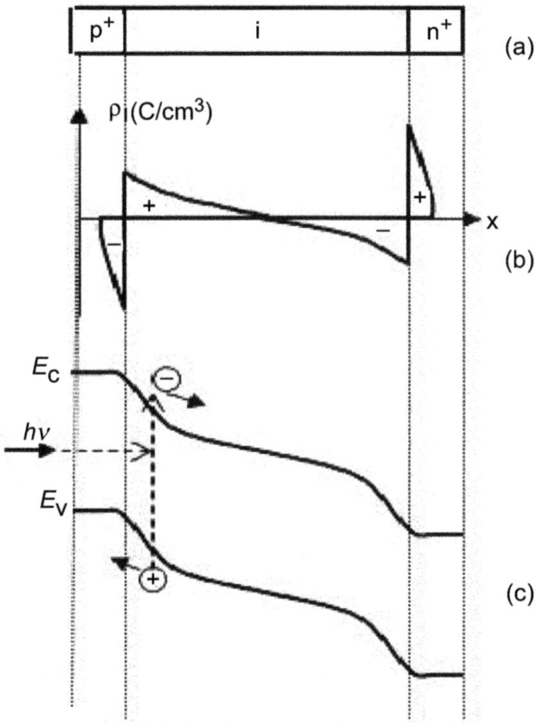

Figure 7.4 Schematic illustration of p$^+$–i–n$^+$ structure (a), distribution of electrical charge (b), and energy band diagram (c).

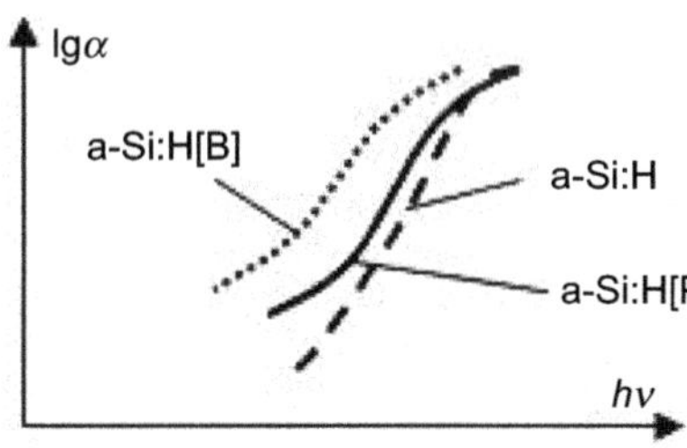

Figure 7.5 Schematic illustration of doping influence on spectral dependence of a–Si:H optical absorption coefficient.

Thus, losses caused by photon absorption in a heavily doped layer in the case of direct design (illumination through boron-doped layer) are higher in comparison with the reversed design of the cell. Based on the above factors, designers of a–Si:H photovoltaic cells use both direct and reversed designs of the cells (Carlson, 1984). Examples of these designs are shown in Fig. 7.6.

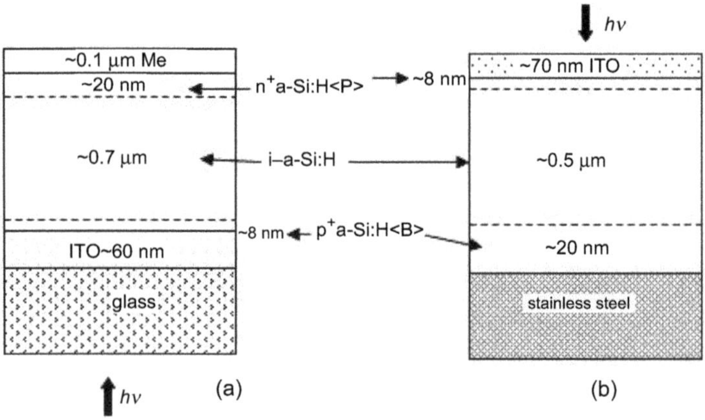

Figure 7.6 Direct (a) and reversed (b) constructions of a–Si:H photovoltaic cells (ITO is indium tin oxide).

In the case of a cell with direct design, it is possible to decrease the absorption losses in heavily doped p^+-layer by replacing p^+a–Si:H with p^+a–SiC:H. The latter has a larger band gap and is transparent to photons with energy equal to the width of a–Si:H gap. Properties of hydrogenated amorphous silicon device quality films are given in Table 7.1 (Sherchenkov and Budagyn, 2007).

Table 7.1 Properties of hydrogenated amorphous silicon device quality films

Property	**i-a-Si:H**	**p-a-Si$_{1-x}$C$_x$:H[B]**	**n-a-Si:H[P]**
Gap value (eV)	1.8	2.0	1.8
Refractive index	4.3		
Absorption coefficient at λ = 600 nm (cm^{-1})	4×10^4	10^4	4×10^4
Conductivity (Ohm^{-1}cm^{-1})	10^{-10}	10^{-5}	10^{-3}
Photoconductivity (Ohm^{-1}cm^{-1})	10^{-5}		
Conductivity activation energy (eV)	0.8	0.5	0.3
Hydrogen content (at.%)	8–12	8–12	8–12
Density of defect states (eV^{-1}cm^{-3})	10^{15}		

The above designs ensure *efficient collection of photo-generated carriers on the contact electrodes.* Let us consider implementation of other conditions for efficient conversion of solar energy into electricity in a photovoltaic cell.

7.3.1 Condition for the Absorption of a Significant Part of Solar Radiation in the Active Region of the Cell

As mentioned above, the thickness of i-layer in a photovoltaic cell (0.4–0.8 μm) is determined by the requirement of the electric field existence throughout the volume of the layer. However, i-layer of this thickness does not ensure absorption of a considerable part of solar radiation, which has a broad spectral range from 0.3 to 1.0 μm. Only radiation with photon energy approximately equal to the semiconductor band gap is converted effectively into electricity (Fig. 7.7a). Photons with lower energy are not absorbed by semiconductor material. Photons with higher energy, as a rule, are absorbed in the heavily doped layer. Even in the case of high-energy photon absorption in the active i-layer, a significant portion of its energy is spent on thermalization of generated electron and hole and is converted into heat (Fig. 7.3).

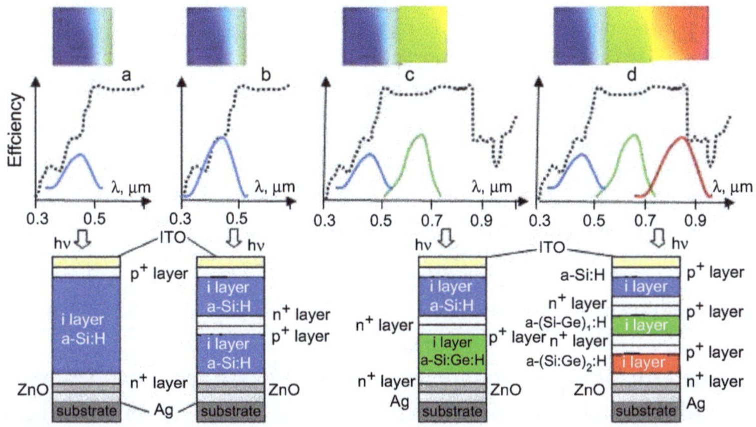

Figure 7.7 Schematic illustration of the spectral dependence of the efficiency (top) and structure (bottom) of multiple-junction solar cells: single-junction (a), double junction with the same gap (b), double junction with dual gap (c), and triple junction (d) (The dashed line is the spectrum of sunlight on the earth's surface).

The solution to this problem was implemented on the basis of cascade or multiple-junction solar cells. In the case of two a–Si:H cells, formed one on top of another as a sandwich, instead of one cell, the radiation that is not absorbed in the first cell may be absorbed in the second. This increases the efficiency of sunlight absorption (Fig. 7.7b). However, efficiency increases still occur only in a narrow region of the spectrum at photon energies close to the value of the a–Si:H band gap. It is necessary to use semiconductors with various values of the band gap for the expansion of the spectral characteristics of the cell's optical absorption. In the case of a double-junction solar cell, the upper element should have a wider and the lower element should have a narrower band gap (Fig. 7.7c). In the case of triple-junction solar cells, the band gap of elements should decrease from the illuminated surface to the opaque electrode (Fig. 7.7d). This ensures efficiency of a photovoltaic cell in a wider part of the solar radiation spectrum.

Solid solutions of silicon with other elements of the periodic table's fourth group (germanium, carbon, nitrogen) and microcrystalline silicon are used to create various band gap materials based on hydrogenated amorphous silicon (see Sections 5.2 and 5.3). This makes it possible to obtain semiconductor layers with band gap value from 1 to 3 eV. The structures of single and multiple-junction photovoltaic cells are listed in Table 7.2.

Table 7.2 Structures of single and multiple-junction photovoltaic cells

Single junction	Double junction Top/Bottom	Triple junction Top/Middle/Bottom
a–Si:H	a–Si:H/a–Si:H	a–Si:H/a–Si:H/μc–Si:H
—	a–Si:H/a–Si_xGe_{1-x}:H	a–Si:H/a–Si_xGe_{1-x}:H/μc–Si:H
—	a–Si:H/μc–Si:H	a–Si:H/a–Si_xGe_{1-x}:H/a–Si_yGe_{1-y}:H

The efficiency dependencies of solar cells on band gap values of the middle and bottom elements are shown in Fig. 7.8. The band gap of the top element is equal to 1.95 eV (Yukimoto, 1983). The figure shows that the maximum efficiency is achieved with the following gap values: E_{g1} = 1.95 eV, E_{g2} = 1.45 eV, and E_{g3} = 1.0 eV. It should be noted that a necessary condition for the maximum efficiency of multiple-junction photovoltaic cell is the equality of short-circuit currents of all its elements (assuming the equality of their fill factors). Since the band gap and resistivity of i-layers are different, layer thicknesses should be equal d_{i1} = 0.2 μm, d_{i2} = 0.5 μm, and d_{i3} = 1.0 μm to fulfill this condition.

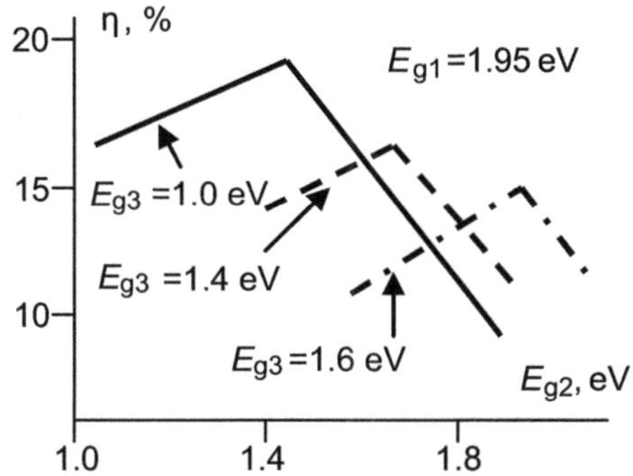

Figure 7.8 The solar cell efficiency dependencies on band gap values of the middle and bottom elements.

Another way to reduce optical losses in photovoltaic cells is to use reflective layers or the so-called light trapping. The idea of the method is that the special textured reflective layers are placed at the front and back surfaces of the cell. The light-trapping structure of a double-junction solar cell that provides multiple internal reflections is shown in Fig. 7.9. This leads to the increase in the optical path up to 50 times and a sharp growth in light absorption.

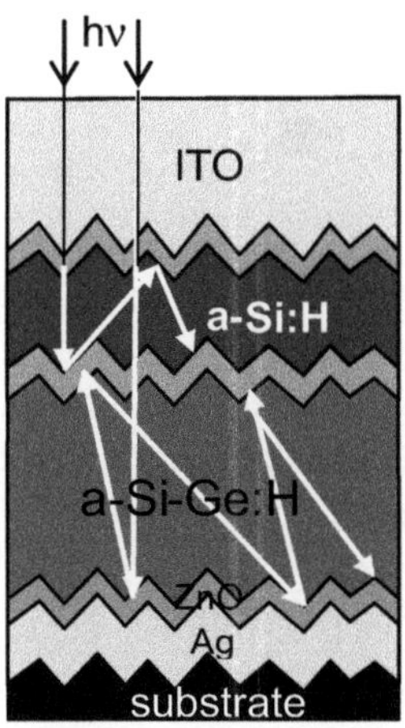

Figure 7.9 Schematic illustration of a solar cell with light-trapping structure.

7.3.2 Condition for Small Value of Internal Photovoltaic Cell Resistance

As shown above, highly doped n^+ and p^+ layers of a–Si:H or μc–Si:H form contacts with electrodes in the photovoltaic cells. Therefore, the internal resistance of the cell should be small. This is confirmed in the case of small-area cells. However, increasing the photovoltaic cell's area leads to a sharp rise in internal resistance and a corresponding decrease in short-circuit current and efficiency. The reason for this effect, called the size effect, becomes clear from the results of a simple experiment (Kuwano and Ohnishi, 1983). An a–Si:H photovoltaic cell 1×1cm in size was made on a glass substrate of 10×10 cm size coated with a transparent indium tin oxide (ITO) electrode. Metal contacts to the transparent electrode were formed at the corners of the substrate (Fig. 7.10a). The length of the current path (l) through the transparent ITO electrode was varied by attaching the cell through various contacts (1–4). The resistance of ITO electrode varies from 10 to 30 Ω/□. Therefore, increasing the length of the current path causes a significant increase in the internal resistance of the cell and reduces its efficiency (Fig. 7.10b).

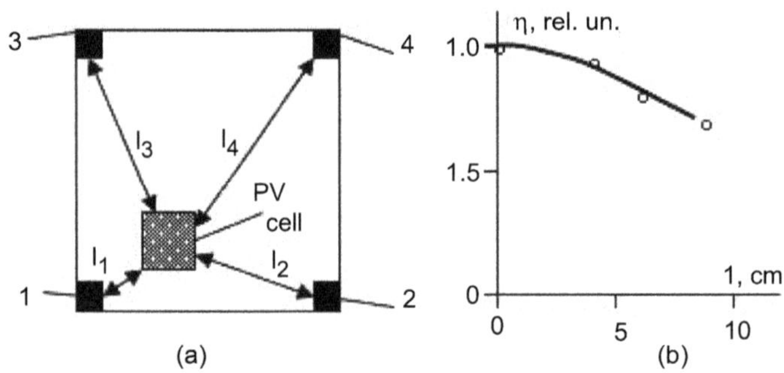

Figure 7.10 Illustration of the photovoltaic cell size effect (see text).

The influence of size effect on photoelectric characteristics can be reduced by reducing the length of the current path through the transparent electrode. To achieve this, the integral design of photovoltaic modules was developed (Fig. 7.11). The production process for these modules is as follows. Rectangular transparent ITO electrodes (4) are deposited on the substrate (5). Spacing between the electrodes is equal to *d*. Photovoltaic cells (2) are formed after that. The final step is the formation of metal contacts to the ITO electrodes (3) and the cell upper electrodes (1). Moreover, the upper electrode of the first cell is connected with the ITO electrode of the second cell along all cell length. The upper electrode of the second cell is connected with the ITO electrode of the third cell and so on. Thus, all the cells in the photovoltaic module are connected in series. As one can see, the length of the current path through transparent electrodes is sharply decreased in this case and hence power losses are reduced as well. The replacement of one cell to 15 cells formed in the same area reduces the power losses in the transparent electrode by approximately 100 times. However, the active module area decreases with increasing number of cells due to the rise in the total cell interconnection area. Therefore, optimization of the number of cells in the module is required.

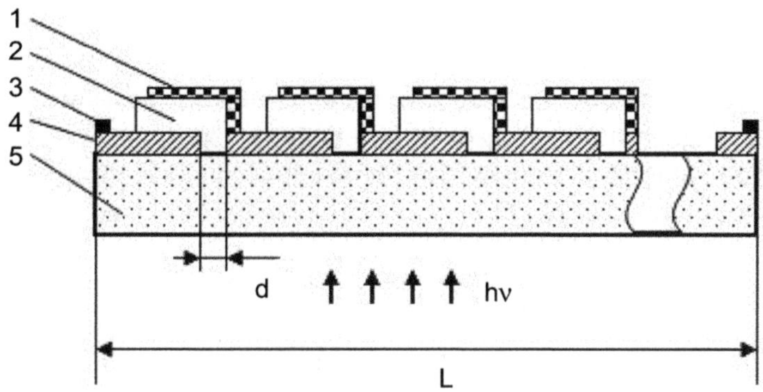

Figure 7.11 Design of integrated photovoltaic module (1 is upper metal electrode, 2 is photovoltaic cell, 3 is metal contact, 4 is transparent ITO electrode, and 5 is glassy substrate).

The efficiency of the integrated module is given by

$$\eta_M = \eta_0\{1 - [(N + 1)d/L]\}[1 - (P_{los}/P_\Sigma)] \qquad (7.3)$$

where η_0 is the efficiency of the small-area cell in which the power losses in the transparent electrode are neglected, N is the number of cells in the module, d is the spacing between the electrodes, L is the module width, P_{los} is the power losses in the transparent electrode, and P_Σ is the total module power.

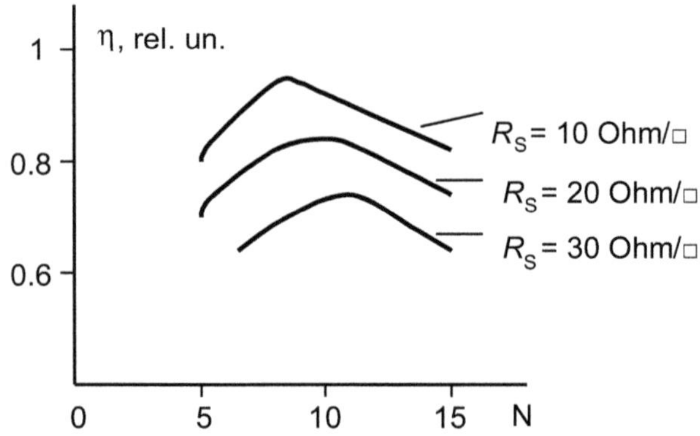

Figure 7.12 Dependence of module efficiency on the number of cells (R_S is the surface resistance of the transparent electrode).

The results of module efficiency calculation in the case of different values of the transparent electrode's surface resistance are shown in Fig. 7.12. The module size was 10 × 10 cm, and the spacing between the electrodes is $d = 1$ mm. As seen from the figure, the optimal number of elements in the module increases with the increasing resistance of the transparent electrode.

The integral design of the photovoltaic module ensures the implementation of *condition for high value of open-circuit voltage* as well. Open-circuit voltages of all cells are summed up due to the serial connection of cells in the integral module. As a result, the output voltage varies from 70 to 140 V at 1 m module width.

The guaranteed service life of photovoltaic modules based on hydrogenated amorphous silicon is assumed to be 20 years. The factors limiting the service life can be divided into two groups. The first group includes the design and technological factors, such as the reliability of element hermetical sealing, the choice of electrode material to minimize its diffusion in the semiconductor, and others. The influence of these factors can be reduced by improving the design and manufacturing process. The second group includes factors that depend on the fundamental properties of the semiconductor material.

The Staebler–Wronski effect in hydrogenated amorphous silicon belongs to the latter (Staebler and Wronski, 1977). The effect is to reduce both the dark- and photoconductivity of hydrogenated amorphous silicon under the long-term effect of illumination. The effect is reversible in nature. Annealing at 150–200°C for an hour leads to a complete renewal of the dark- and photoconductivity. It was found that the conductivity changes occur due to the formation of new metastable localized states in the gap of the semiconductor under the influence of illumination. The dependence of the localized states density in the gap of a–Si:H on the illumination time (Sherchenkov and Budagyn, 2007) is shown in Fig. 7.13a.

The nature of these metastable states is not completely clear. Usually their appearance is associated with the breaking of the weak bonds between silicon atoms due to the recombination of photo-generated carriers, the interaction of the dangling bond with hydrogen atoms to form an Si–H bond, and localized state Si_3^0 (Fig. 7.13b). At increased temperatures, the effusion of a hydrogen atom and restoring of the Si–Si bond take place (Fig 7.13c).

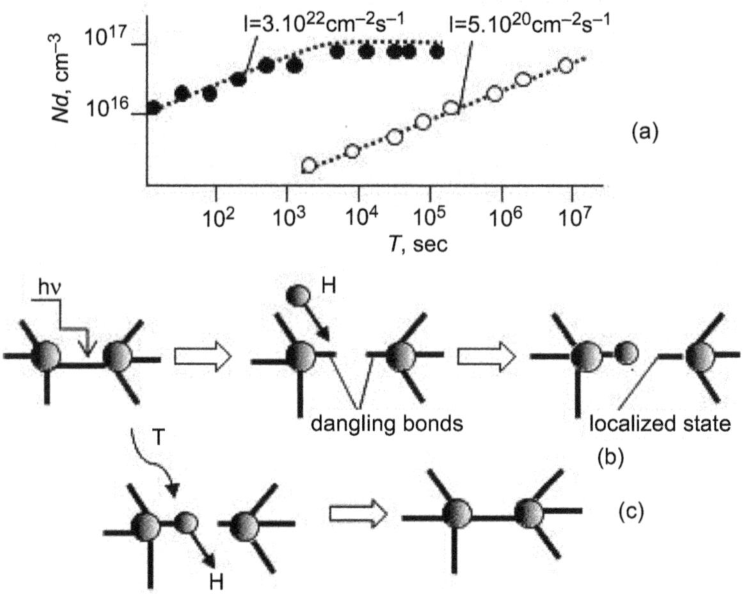

Figure 7.13 The dependence of the localized states density in the gap of a–Si:H on the illumination time at various intensities of illumination (a) and schematic illustration of appearance under illumination (b) and annealing (c) of localized states.

The Staebler–Wronski effect is observed in photovoltaic cells based on hydrogenated amorphous silicon as well. As a result, efficiency is reduced by 10–20% from its initial value during the first year of operation. After that the efficiency is stabilized and only slightly changes during the subsequent operation. At present, the stabilized efficiency of commercially produced photovoltaic modules based on hydrogenated amorphous silicon is 10%.

Examples of amorphous silicon ground-mounted network photovoltaic power plant (5 MW) (a) and autonomous photovoltaic power plant (100 kW) (b) both in Siberia, Russia, are shown in Fig. 7.14.

Figure 7.14 Amorphous silicon ground-mounted network photovoltaic power plant (5 MW) (a) and autonomous photovoltaic power plant (100 kW) (b) both in Siberia, Russia. Image courtesy: E. I. Terukov.

7.4 Photovoltaic Cells Based on Silicon Heterojunction Technology

As noted in Section 7.1, the creation of photovoltaic silicon solar cells on the basis of SHJT has generated renewed interest in recent years. Let us consider the features of photovoltaic cells manufactured by this technology. It is known (see Fig. 7.1) that the manufacture of crystalline ingot and wafer accounts for most of the cost of traditional photovoltaic modules based on single-crystal silicon. Therefore, reducing the silicon wafer's thickness would seem a natural way to reduce its cost. However, this approach is faced with several significant problems.

The first is that a decrease in the thickness of the active region of the photovoltaic cell leads to an increase in the role of surface recombination of photo-generated carriers. This reduces the efficiency of the collection of charge carriers on the cell contacts and reduces the open-circuit voltage. The second reason is that reduction in the absorption of photons in the active region leads to a lower short-circuit current. Both reasons lead to reduced efficiency of the photovoltaic cell. Moreover, fabrication of photovoltaic cells based on single-crystal silicon includes the step of diffusion being done at a temperature above 700°C. Such heat treatment at small thicknesses leads to wafer warping by thermal stress.

SHJT is an approach that puts together technologies used in the production of traditional solar cells based on single-crystal silicon and thin-film solar cells based on hydrogenated amorphous silicon. It combines the advantages of the above-mentioned technologies and is free from many of their drawbacks. The temperature of process operations does not exceed 200°C, thus eliminating the thermal deformation of wafers at small thicknesses. The SHJT solar cell structure is shown in Fig. 7.15a. The single-crystal silicon wafer of n-type conductivity with a textured surface is a substrate on both sides of which thin (5–10 nm), undoped hydrogenated amorphous silicon layers are deposited. These layers are deposited for the purpose of a single-crystal surface passivation to reduce the surface recombination rate of photo-generated charge carriers (Garcial et al., 2016). Due to the passivation of dangling bonds on the surface of the silicon wafer, the surface recombination rate becomes substantially lower than the bulk recombination rate. This reduction in thickness of the wafer does not decrease the effectiveness of the charge carrier collection at the contacts. It has been shown (Mishima et al., 2011) that the open-circuit voltage of the cell with the passivated surfaces is not decreased, and even increased slightly with decreasing plate thickness from 170 to 80 microns. The selection of silicon wafer n-type conductivity is also associated with the problem of surface passivation. It is known (see Section 4.1) that the undoped hydrogenated amorphous silicon has weak n-type conductivity. Hence, it provides better surface passivation of crystal n-type conductivity. Additionally, the lifetime of charge carriers in the crystal silicon n-type conductivity is higher than that in p-type crystals.

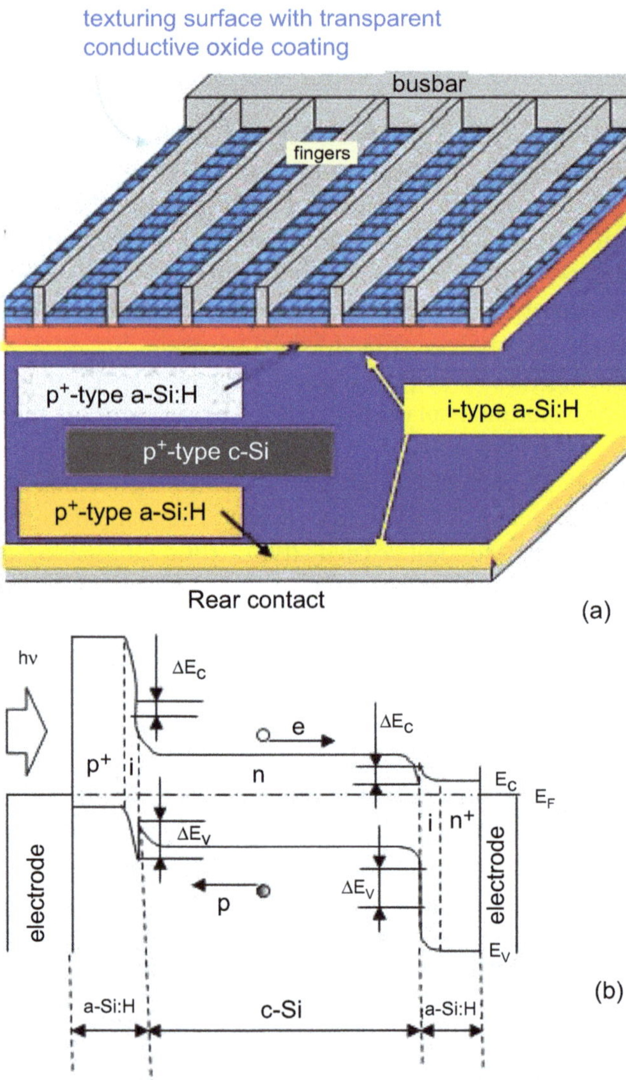

Figure 7.15 The structure (a) and energy band diagram (b) of SHJT solar cell.

After passivation, a layer of a–Si:H p-type conductivity is deposited on one side of a silicon wafer, and a layer of a–Si:H n-type conductivity is deposited on the other side. Thereafter, electrodes and contacts are deposited. Thus, the cell has the following structure: Me - transition conductive electrode (TCE) – p + a–Si:H – ia–Si:H – nc–Si – ia–Si:H – n + a–Si:H – Reflection Coating – Me. Its energy diagram is shown in Fig. 7.15b.

The decrease in the absorption of photons by reducing the thickness of the active region can be compensated by actions for light management, namely: wafer surface texturing, formation of reflective layers inside device, and antireflection coating on the surface. In addition, a larger band gap of hydrogenated amorphous silicon compared with the crystalline one reduces light losses in the upper doped layer of photovoltaic cells.

The use of indicated approaches has allowed to create SHJT solar cells (Fujishima et al., 2010; Mishima et al., 2011; Descoeudres et al., 2013) with the thickness of the silicon wafer less than 100 microns and with efficiency of more than 22%. The output power temperature coefficient for these elements is 1.5 times less than in the cells based on single-crystal silicon and corresponds to the values for the cells based on a–Si:H. On the other hand, unlike the cells based on a–Si:H, the Staebler–Wronski effect (see Section 7.3) is not observed in SHJT cells; therefore, there is no degradation of the parameters under exposure to light. Thus, the cost of SHJT module production is lower than traditional modules based on single-crystal silicon by 10–20%. Table 7.3 shows an assessment of the three types of silicon photovoltaic module characteristics. It is made on the basis of the above-mentioned articles and Information Reference (Directory for Fuel and Energy Sector, 2013).

Table 7.3 Comparison of the three types of silicon photovoltaic modules

PV module	Single crystal	SHJT	a–Si:H
Maximum temperature at the device fabrication (°C)	≥700	≤200	≤200
Light degradation	No	No	Yes
Commercial modules efficiency	17–19	20–22	10–12
Output power temperature coefficient (%/K)	0.40–0.45	0.30	0.30

7.5 Organic and Hybrid Photovoltaic Cells

It was noted in Sections 2.5.5 and 3.1.5 that electrons located on the p-orbitals of conjugated organic molecules form relatively weak π-bonds (Fig. 2.20). As a result, bonding and anti-bonding π-orbitals are formed. The difference in energy between the highest occupied molecular orbital (HOMO), which is the bonding π-orbital, and the lowest unoccupied molecular orbital (LUMO), which is an anti-bonding π-orbital (forbidden gap in Fig. 3.19), corresponds to the energy of the light quanta. This is the reason for the presence of internal photoelectric effect and leads to the possibility of creating solar cells based on these materials.

The search for organic materials suitable for photovoltaic applications dates back to the 1950s. Photovoltaic effect in magnesium phthalocyanine was reported by Kears and Calvin in 1958 (Yeh and Yeh, 2013). Since then the topic has caught the attention of researchers and has been actively investigated due to the low cost, light weight, and elasticity of polymer materials. Additionally, organic semiconductors have very high absorption coefficients, which allow very thin films to be used, while still absorbing a sufficient portion of the solar spectrum. As a result, the power conversion efficiency of organic solar cells has improved rapidly from less than 1% to more than 12% in the past 20 years (Fig. 7.16; Zhou, Yang, and You, 2012; Scharber and Sariciftci, 2013). Among the disadvantages of organic solar cells are their rather low stability and strength, as well as susceptibility to oxygen and water.

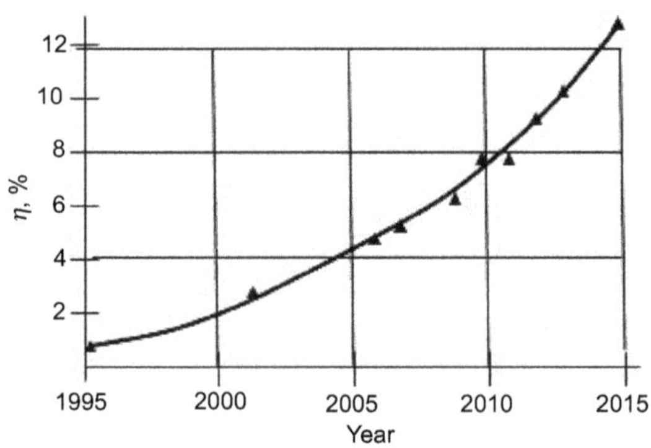

Figure 7.16 Development of the power conversion efficiency of organic solar cells.

When creating solar cells based on organic semiconductors, it is necessary to take into account a number of essential differences in comparison with classic inorganic semiconductors.

The various types of electrical conductivity in inorganic semiconductors can be created by doping of material with the appropriate impurity that creates donor or acceptor levels in the band gap. In organic semiconductors, conductivity type depends on the ratio of electron work function from organic material and from material of the electrode. If the electron work function from semiconductor (φ_{sem}) is more than the electron work function from electrode (φ_{el}), electrons are transferred from the electrode to the conduction band (LUMO π-orbitals) of the semiconductor, creating an n-conductivity type (in this case, the semiconductor is called an acceptor because it accepts electrons). If the electron work function from the semiconductor is lower than the electron work function from the electrode ($\varphi_{sem} < \varphi_{el}$), the electrons leave from HOMO π-orbitals of the semiconductor, which leads to an increase in the hole concentration and p-type conductivity (in this case, the semiconductor is called a donor, as it gives up electrons). Typical materials used for the creation of regions with p-type conductivity (i.e., donor of electrons) are conjugated polymers, and those used to create regions with n-type conductivity (i.e., acceptor of electrons) are fullerene derivatives.

Another feature of organic semiconductors is that photo-generation of electron–hole pairs in these materials leads to the formation of Frenkel excitons with high binding energy (see Section 3.2.5). Therefore, it requires potential barriers for separation of excitons. These potential barriers are created at the boundaries between the donor and acceptor materials. The energy band diagram of organic donor–acceptor system that illustrates operation of the cell is shown in Fig. 7.17.

Photoelectric effect includes the following steps in this case. As a result of photon absorption, Frenkel exciton arises in the material (step 1). The formed exciton diffuses to the heterojunction between the donor and acceptor materials (step 2). Due to the potential barrier of the heterojunction, exciton dissociation into free charge carriers takes place (step 3). Transport of the free charge carriers to the respective electrodes then takes place (step 4).

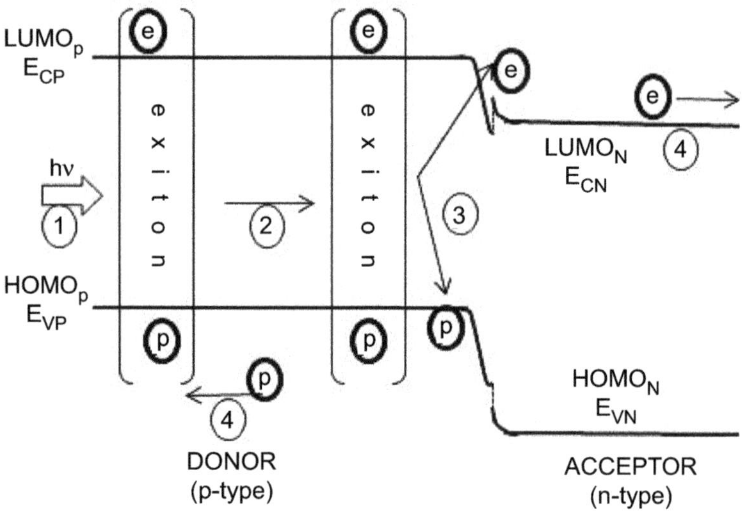

Figure 7.17 Energy band diagram of organic donor–acceptor system and sequence of photocurrent generation process: step 1 is generation of exciton by photon absorption; step 2 is diffusion of exciton to heterojunction; step 3 is dissociation of exciton; and step 4 is the transport of free charge carries to electrodes.

An organic-based photovoltaic cell structure is shown in Fig. 7.18a. A layer of poly(3,4-ethylenedioxythiophene)-poly(styrenesulfonate) (PEDOT: PSS) is deposited on glass or dacron substrate coated with ITO, which acts as a transparent electrode. The function of this layer is to improve the electrical contact between the ITO and the organic polymer layer and to adjust energy levels (Zhou, Yang, and You, 2012). After that, the acceptor (n-type) and donor (p-type) organic semiconductor layers and a top metal electrode are deposited.

As mentioned in Section 7.3, double-junction and triple-junction solar cells (Fig. 7.7) are used in thin-film solar cells based on hydrogenated amorphous silicon in order to reduce losses and improve the efficiency of conversion of solar radiation. The same approach is used in the case of organic solar cells. To improve efficiency, multiple-junction solar cells are created in which there are several heterojunctions between semiconductors with different band gap (Fig. 718b).

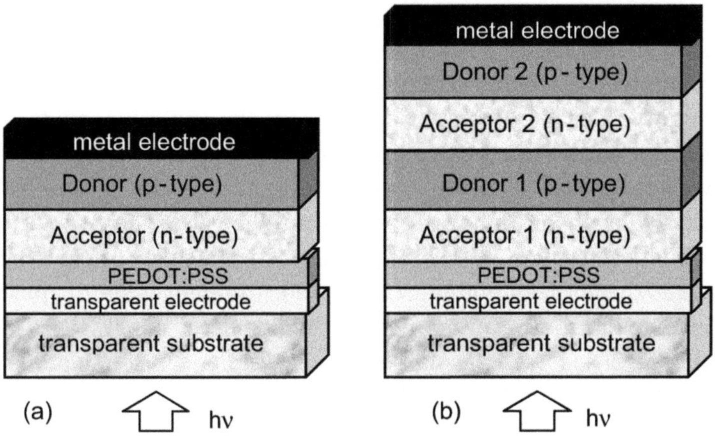

Figure 7.18 Schematic illustration of the structure of single heterojunction (a) and double heterojunction (b) organic solar cell.

In the cases shown in Fig. 7.7 (for inorganic materials) and Fig. 7.18 (for organic materials), cell efficiency increases due to the increasing amount of consecutively arranged planar p–n junctions. However, organic materials represent another opportunity to improve the efficiency of solar cells by creating bulk heterojunction. In this case, the organic n-type and p-type semiconductors are blended and mutually penetrate into each other, creating a bulk heterojunction (Fig. 7.19a). Such construction provides a significant decrease in the distance to the heterojunction boundary from any point of semiconductor layers. This reduces the diffusion length of excitons and the probability of their recombination and increases the collection efficiency of free charge carriers. The next step is to create cells with ordered heterojunction (OHJ) (Fig. 7.19b). This structure provides the conditions in which the distance from any point of a photon absorption in the active layer to the boundary of the heterojunction is less than exciton diffusion length. This leads to dissociation of the majority of excitons into free carriers (Yeh and Yeh, 2013).

Comparison of solar cells fabricated on the basis of inorganic and organic materials (Sections 7.3–7.5) shows that each class of devices has its own advantages and disadvantages. This is why creation of hybrid organic–inorganic solar cells, which would have the advantages of the aforementioned devices, but would be free of their disadvantages, is a promising direction of solar converters development. Research using various materials, such as lead and cadmium chalcogenide, silicon nanoparticles, titanium oxide and zinc oxide, and others has been carried out (Wright and Uddin, 2012). However, the use of metal-organic compounds with a perovskite structure has been the most successful. The classic perovskite has the chemical formula $CaTiO_3$. The material in which the calcium ion is replaced by methylammonium ion and the titanium oxide is replaced by lead halide: methylammonium led-iodide perovskite $CH_3NH_3PbI_3$ (or Br_3 or Cl_3) is used in solar cells. It is a direct-gap semiconductor with a band gap of 1.5–1.6 eV, which ensures effective absorption of solar radiation. In addition, $CH_3NH_3PbI_3$ films have high values of mobility, lifetime, and diffusion length of charge carriers due to the low density of localized states. Due to this, the efficiency of solar cells based on metal-organic perovskite increased five times from 4% to 20% in the period from 2009 to 2015 (Brenner et al., 2016) and continues to grow

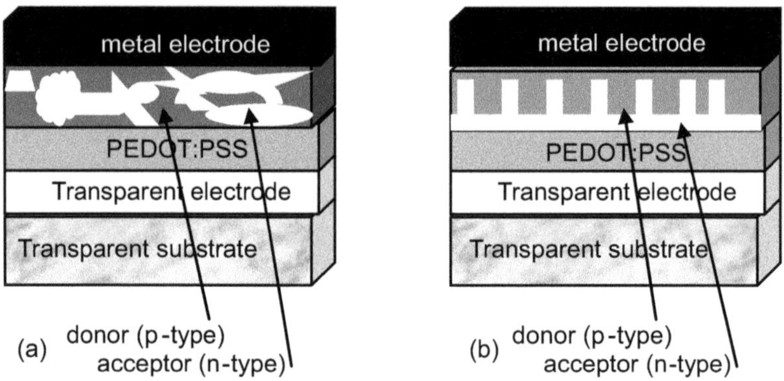

Figure 7.19 Schematic illustration of the structure of bulk heterojunction (a) and ordered heterojunction (b) organic solar cell.

Chapter 8

Electronic Devices Based on Disordered Semiconductors

8.1 Switching and Memory Devices on the Basis of Chalcogenide Alloys

8.1.1 Threshold Switching Effect in Chalcogenide Glassy Semiconductors

Chalcogenide glassy semiconductors demonstrate the threshold switching effect or rapid transition from the high-resistivity state to the high-conductivity state and back. The first discovery of switching in chalcogenide glasses was independently made by S. Ovshinsky and B. Kolomiets in 1963. Despite numerous studies, a unified theory of the switching effect has not yet been created. There is a number of coexisting hypotheses and models to explain this effect.

A typical chalcogenide glass threshold switching device is about 1 μm thick film sandwiched between two metal electrodes (Fig. 8.1a). The device has a symmetrical voltage–current characteristic with negative differential resistance (Fig. 8.1b).

The initial state of the device is a high-resistance state. At low applied voltages, the conductivity varies linearly, and the resistance is about 10^5 Ω. When the electric field strength reaches about 10^4 V/cm, the linear law is replaced by the nonlinear (exponential) law,

Disordered Semiconductors: Physics and Applications (2nd Edition)
Anatoly Popov

ISBN 978-981-4774-37-6 (Hardcover), 978-1-315-10042-5 (eBook)
www.panstanford.com

which corresponds to the space charge limited current. Switching the device in low-resistivity state occurs when a voltage rises to the threshold value (U_{th}). The threshold voltage is linearly dependent on the film thickness (Fig. 8.2). It conforms to consistent voltage distribution across the film and constant value of threshold electric field (approximately 10^5 V/cm). The device resistance decreases by five to six orders of magnitude. Dynamic resistance in low-resistance state is typically about 1 Ω. Reducing the current in the low-resistance state to a critical value (I_h) causes the device to switch to its original high-resistance state.

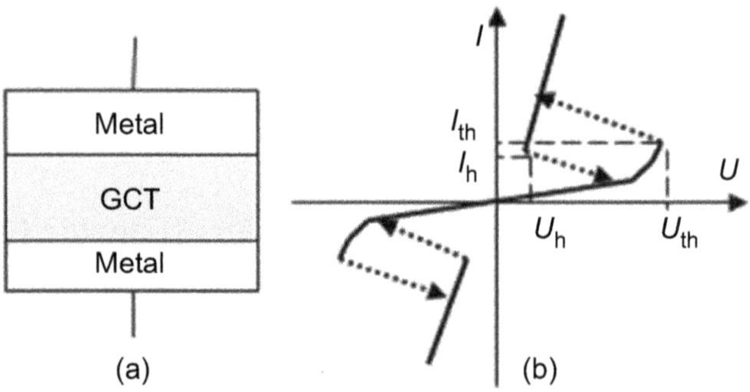

Figure 8.1 Structure (a) and voltage–current characteristic (b) of threshold switcher on the basis of chalcogenide glassy semiconductor.

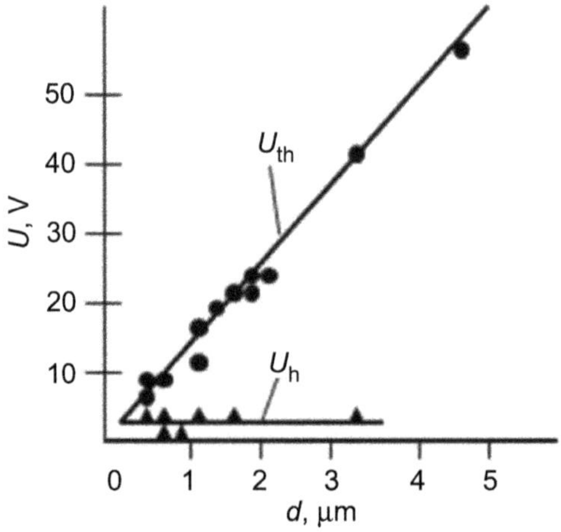

Figure 8.2 Dependences of threshold voltage (U_{th}) and holding voltage (U_h) on chalcogenide glassy film thickness.

The current is proportional to the area of electrodes in the high-resistance state, but in the low-resistance state, it does not depend on the area of electrodes. This means that a current channel (filament) with small cross section and with high current density is formed in the low-resistance state. Since the current flows through the small cross-section channel in the low-resistance state, the maximum current of the switch, as a rule, does not exceed several mA.

Holding the voltage of low-resistance state (U_h) does not depend on the thickness of the semiconductor film (Fig. 8.2). This indicates that most part of the voltage applied to the device drops at the contacts, but not in the bulk of the semiconductor. As one can see from Fig. 8.2, there is a critical thickness of the semiconductor film (approximately 0.2 μm) at which the threshold voltage and the holding voltage are equal to each other. Consequently, when film thickness is smaller than this value, switching is impossible in this type of devices. It should be added that the threshold voltage decreases with increasing temperature and increasing pressure, but does not depend on the illumination. Under illumination, only the threshold current increases at a constant value of the threshold voltage.

An important characteristic of switches is their switching speed. The time response of the threshold switch to a voltage pulse is shown in Fig. 8.3. The time of switching from high-resistance in the low-resistance state (switch-on time) consists of two components: delay time (t_d) and time of the switching process (t_s). The delay time is the time that elapses from the applying of a voltage pulse equal or greater to the threshold value and before a sharp increase in current through the device. The delay time depends strongly on the applied voltage, varying from 10^{-6}–10^{-7} s with $U = U_{th}$ to about 10^{-9} s with voltage increasing. The time of the switching process does not exceed 10^{-9}–10^{-10} s.

The renewal time of the high-resistance state (switch-off time t_{sb}) is defined as the minimum time that has elapsed since the end of the pulse that switched the device into the low-resistance state, after which a pulse with amplitude of 10% less than the threshold voltage does not switch the device into the low-resistance state. This time is usually about 10^{-6} s and increases with the thickness of the chalcogenide glassy film.

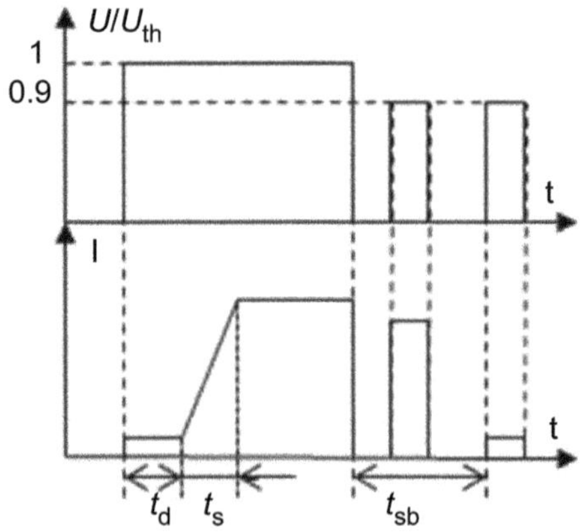

Figure 8.3 The time response of the threshold switch to a voltage pulse (see text).

The mechanism of threshold switching in chalcogenide glassy films is not fully understood. The existing hypotheses can be divided into three groups:

1. Thermal hypotheses, based on the consideration of thermal processes in the material
2. Hypotheses based on the consideration of electronic processes
3. Mixed electron–thermal hypotheses

Historically, *thermal hypotheses* based on the theory of thermal breakdown were developed first. According to these hypotheses, the current flowing through the sample raises the temperature of the material, which leads to an increase in electrical conductivity and current value and to the heating of the sample. Insufficient heat removal causes an avalanche process that leads to a loss of electrical strength of the sample. At first approximation, the process described above is consistent with the experimental dependences, observed at the threshold switching in the chalcogenide glassy semiconductors. Channel (filament) with high current density is formed in both cases. At the qualitative level, the theory of thermal breakdown explains the presence of switching delay time (time required for pre-heating of the sample), the presence of high-resistance renewal time (time required for cooling of the sample), and the dependence of delay time on the voltage (higher voltage causes higher current and more rapid heating). However, quantitative analysis showed that the thermal hypotheses cannot explain a lot of experimental dependences.

Temperature dependences of the threshold voltage for thick (300 μm) and thin (2 μm) samples are shown in Fig. 8.4a. In the case of a thick sample, the threshold voltage decreases sharply with increasing temperature. This is both qualitatively and quantitatively consistent with the thermal breakdown theory. However, in the case of a thin sample, the aforementioned dependence is much weaker and cannot be explained from the standpoint of the thermal hypothesis.

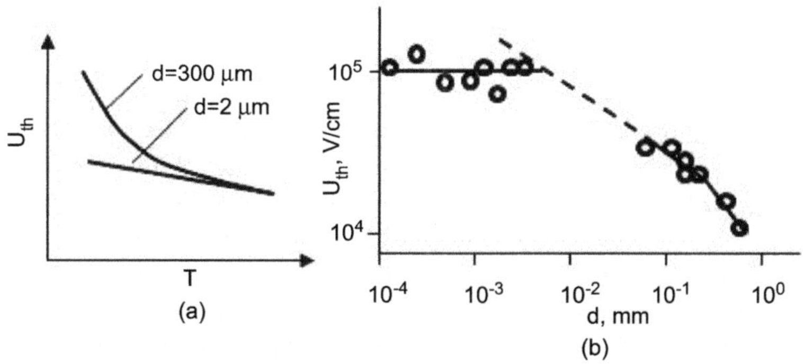

Figure 8.4 Schematic illustration of threshold voltage temperature dependences for samples with different thickness (a) and dependence of threshold electric field on sample thickness (b).

The dependence of the electric field at which switching occurs (threshold field) on the thickness of the sample is shown in Fig. 8.4b (Kolomiets et al., 1969). The threshold field decreases with increasing thickness in the case of samples with a thickness of 100 μm and more. In terms of thermal breakdown, this fact is well explained by heat transfer deterioration. However, the threshold field does not depend on the thickness of a sample of less than 10 μm thick. Moreover, it was found that the threshold field in thin samples depends on the electrode material. Besides that, the influence of the electrode material is not due to its thermal conductivity, but due to the value of the metal–semiconductor work function. All these facts do not fit into the thermal switching hypothesis.

Thus, quantitative analysis allows to conclude that the model of reversible thermal breakdown is valid only for samples with semiconductor layer thickness of hundreds of microns or more.

Since film thickness of less than 1 micron is used in devices, it is necessary to consider other models of threshold switching.

Electronic hypotheses use various effects and processes to explain threshold switching. These are double injection of carriers from the electrodes, electronic phase transitions, tunneling effect, avalanche multiplication, switching in many p–n junction structures, and others (Kostylev and Shkut, 1978). Consider, as an example, the hypothesis of double carrier injection.

When voltage is applied across the device, initially the electric field is distributed uniformly in the semiconductor film. Consequently, the slope of the semiconductor energy bands is constant at this moment (Fig. 8.5a). The injection of electrons and holes from the electrodes into the semiconductor occurs under the electric field. However, due to the high density of traps (localized states) in the band gap of the semiconductor, the injected carriers are trapped. Filling the traps, the charge carriers are moving deeper into the semiconductor film and reach its central part. In the central part of the film, both electron and hole traps are filled and compensation of electrical charges takes place. This leads to a redistribution of the electric field in the film (Fig. 8.5b). Over time, the central region of the film with high mobility of charge carriers increases and fills almost the whole of the volume between the electrodes. The electric field is concentrated in narrow near-electrode regions (Fig. 8.5c), which leads to tunneling of carriers from the electrodes into the semiconductor and the device switches to the low-resistance state.

The model explains the presence of delay time, weak dependence of the voltage across the device in the low-resistance state on the film's thickness, and a number of other experimental results. However, some quantitative estimates on the basis of this model do not correspond to the experimental results (quantitative estimate of the delay time in thin films, for example).

Mixed electron–thermal hypotheses consider the contribution of both electronic and thermal processes to the threshold switching. In general terms, their idea is the following. Free charge carriers obtain additional energy from the electric field. Part of this energy is transferred to the atoms due to the thermal scattering of the charge carriers, which leads to an increase in the temperature of the structural network. The temperature increase leads to a rise in the concentration of the free carriers. Avalanche growth of this process switches the device into the low-resistance state. In other words, electronic hypothesis considers the electrical conductivity dependences on the electric field and on the temperature independently of each other ($\sigma = f[F, T]$). But in the case of electron–thermal hypotheses, temperature depends on the electric field as well ($\sigma = f[F, T(F)]$).

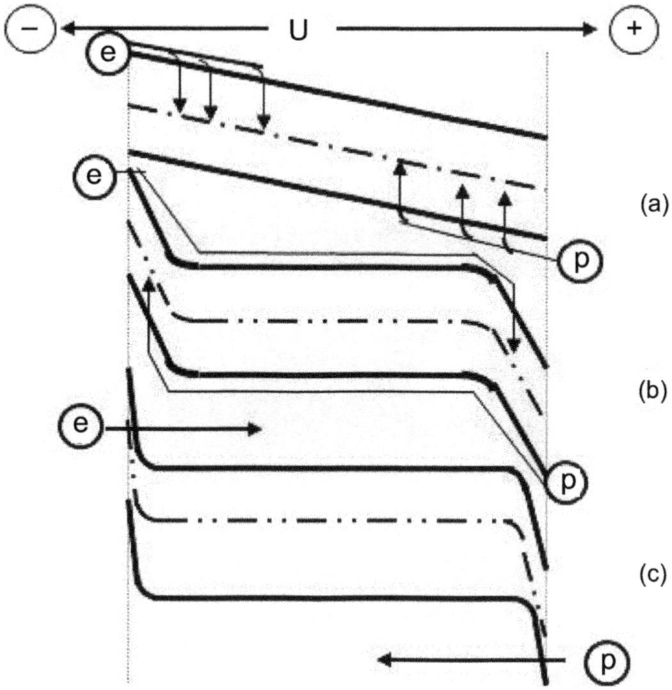

Figure 8.5 Switching process after the double injection model (see text).

Let us consider the percolation model of threshold switching. It is known that at the same width of the gap and, consequently, for the same carrier concentration, the electrical conductivity of amorphous semiconductors is many orders of magnitude less than the electrical conductivity of the crystal. The reason for this is that most of the time, charge carriers are in the localized states and do not take part in conduction. At equilibrium conditions, the generation of charge carriers from the localized states and the recombination of free charge carriers have the same rate. The electric field increases the rate of carrier generation and the concentration of free carriers. Regions with higher concentration of free carriers appear as a result. The electric field is redistributed in the space between the electrodes. Its value increases between the regions with higher carrier concentration. It stimulates the local emission of charge carriers and leads to the formation of percolation current filament between the electrodes. This percolation filament may be destroyed or transformed into a stable current channel. In the latter case, due to the additional heat effect, the area of channel increases and switching into the low-resistance state takes place.

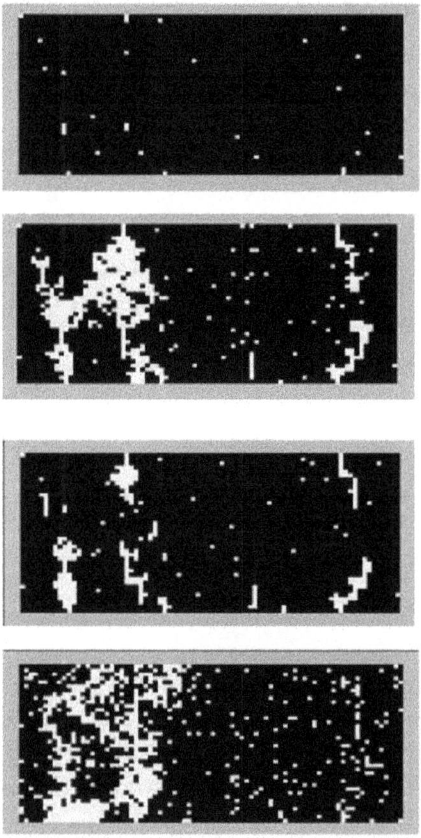

(a)
(b)
(c)
(d)

Figure 8.6 Simulation of current filament formation (light spots indicate the regions with higher carrier concentration): (a) corresponds to the moment of voltage application; (b), (c), and (d) are the model after 60, 71, and 200 iterations. Reprinted from Voronkov et al. (2006), with permission from Elsevier.

The simulation results of threshold switching in the frame of the percolation model are shown in Fig. 8.6 (Voronkov et al., 2006). Light spots in the figure indicate the regions with higher carrier concentration. After 60 iterations, the percolation current filament is formed between the electrodes (Fig. 8.6b). However, the filament is destroyed after 71 iterations (Fig. 8.6c), and the stable current channel is formed due to the increase in channel area only after 200 iterations (Fig. 8.6d).

8.1.2 Bistable (Memory) Switching in Chalcogenide Glassy Semiconductors

The memory switching effect in chalcogenide glassy semiconductors was discovered simultaneously with the effect of the threshold switching. The structure of the memory switch is completely the same as the structure of the threshold switch (Fig. 8.1a).

Initially, the current–voltage characteristic of the memory switch repeats the current–voltage characteristic of the threshold switch. The initial resistance of the device is high. The device switches into the low-resistance state under the action of the threshold voltage. Reducing the current through the device below the holding current value immediately after switching leads to its return to the high-resistance state, just as in the case of the threshold switcher. However, the low-resistance state is retained even when the external voltage is turned off, if the device was held in the low-resistance state for about 10^{-3} s (Fig. 8.7a). Once locked in the low-resistance state, the device can be returned to the original high-resistance state by applying a short intense "erase" current pulse of either polarity.

The time response of memory switches to voltage pulse is shown in Fig. 8.7b. In comparison with the threshold switches, there is one additional parameter in this case. It is the time of memorizing the low-resistance state (lock-on time, t_{mem}). This time is about 10^{-3} s. The lock-on low-resistance state is accompanied by a slight increase in the current through the device (Fig. 8.7b). Thus, this type of switch allows storing information without energy expenditure.

The difference between the threshold and bistable switch characteristics is determined by the different chemical composition of chalcogenide glassy semiconductors. Materials with high values of crystallization activation energy are used in threshold switches. Due to this, the semiconductor film remains in the glassy state during device operation. Composition of these materials is in the middle of the glass formation regions of the chemical systems (see Figs. 1.14 and 1.15). In contrast, memory switches are prepared on the basis of chalcogenide glasses with low activation energy of crystallization. The composition of these materials is near the borders of glass formation regions.

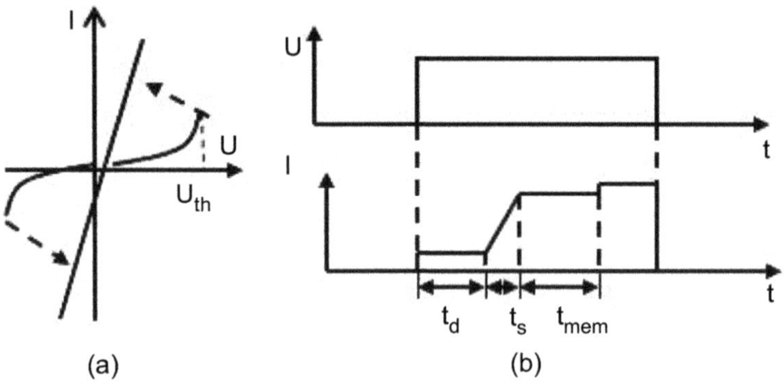

Figure 8.7 The current–voltage characteristic (a) and the time response to voltage pulse (b) of the memory switcher.

The low-resistance state in memory switches is realized due to the phase separation and crystallization of the glassy material in the current channel under the influence of Joule heating. Crystallization of the current channel between the electrodes leads to lock-on of the low-resistance state, since resistivity of the crystalline phase is less than the resistivity of glass by orders of magnitude. Two different mechanisms of material crystallization process in the current channel are possible depending on the parameters of current pulses and glass composition: glass–crystal transition and glass–melt–crystal transition. In the first case, the temperature of the material in the current channel increases under the current pulse action to a value above the glass transition temperature but below the melting temperature of the crystalline phase (curve 1 in Fig. 8.8). The result is crystallization of the material in the current channel between the electrodes.

In the second case, the temperature of the channel increases to a value higher than the melting temperature of the crystalline phase and the material turns into liquid state. The pulse quench must be sufficiently slow to ensure material crystallization during cooling (curve 2 in Fig. 8.8). The comparison of two versions shows that the glass–crystal transition is easier from the kinetic point of view and, consequently, more reproducible. The lock-on time is equal to about 10^{-3} s in this case. The advantage of the glass–melt–crystal transition is lower lock-on time of the low-resistance state.

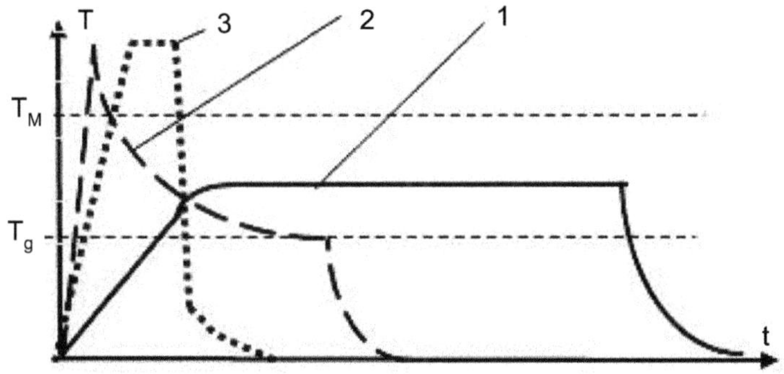

Figure 8.8 Dependences of current channel temperature on time of the low-resistance state recording in the case of glass–crystal transition (curve 1), glass–melt–crystal transition (curve 2), and on the time of the high-resistance state recording (curve 3).

Back switching into the high-resistance state can be achieved only through the crystal–melt–glass transition. This transition is implemented by a short current pulse with sharp quenching. The current value should provide increase in channel temperature above the melting point of the crystalline phase. The sharp quenching of the pulse leads to rapid cooling of the channel and vitrification of the melt (curve 3 in Fig. 8.8) and the device switches into the high-resistance state.

The choice of material is a key issue in designing any device. As noted above, materials with a sufficiently rigid and stable structural network (with high crystallization activation energy) are required for threshold switches. In contrast, materials with a fairly easily changeable atomic structure are used in bistable switches. The first industrially produced threshold switches were created on the basis of $Si_{12}Ge_{10}As_{30}Te_{48}$ (the criterion of structural modification efficiency is equal to 0.95, see Section 4.4.2) and bistable switches were created on the basis of $Ge_{15}Te_{81}As_4$ (the criterion of structural modification efficiency is equal to 1.68). However, these requirements are necessary but not sufficient, since all switch parameters depend on the chemical composition of the chalcogenide glassy semiconductor. For example, the threshold electric field varies from 10^4 to 10^6 V/cm depending on the chemical composition and on the band gap value of the material. In summary, the following requirements to materials for memory switches may be formulated:

- The chemical compositions of the material has to be near the boundary of glass formation region.
- The material has to have a minimum value of difference between the temperature of crystallization and the glass transition temperature.
- The material has to have a maximum value of difference between the temperature of melting and the temperature of crystallization.
- The band gap of the material should not exceed 1 eV.

Along with absolute values of switching characteristics, the stability of the characteristics during device operation is very essential. There are two types of changes in switch characteristics: consistent change in the parameters (threshold voltage, in the first place) during the first cycles of switching (called the forming process) and statistical variation in parameters during device operation (Fig. 8.9).

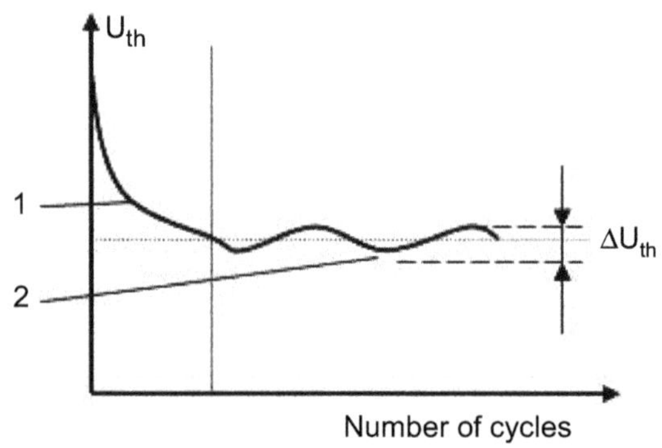

Figure 8.9 Schematic illustration of threshold voltage changing during device operation (1 is range of forming process, 2 is range of statistical variation).

The presence of the forming process is explained by the strongly non-equilibrium process of film formation. The atomic structure of the as-prepared films corresponds to the non-stable state, located on the slope of the metastable state minimum on the configuration diagram (point N_2 in Fig. 1.8). Therefore, the threshold voltage of as-prepared devices is generally about 30% larger than the threshold voltage of the formed devices. During the first 10–20 switching cycles, the system moves to the metastable state due to structural relaxation. After this, the threshold voltage is stable within the limits of statistical variations. In order to eliminate the need for forming process, it is necessary to obtain films with the structure corresponding to the metastable state. Some technological methods, consisting either in vapor–liquid–glass mechanism of film growth (Kostylev and Shkut, 1978), or in special heat treatment after device fabrication, have been developed to solve this problem.

In contrast to the forming process, the statistical variation of electrical parameters cannot be avoided and is determined by the material's atomic structure. It has been shown in Section 4.4.5 that the statistical variations of threshold voltage depend on the value of the criterion of structural modification efficiency (see Fig. 4.12). This dependence can be written as follows:

$$\Delta U_{\mathrm{th}} = \alpha(CSME)^{\beta} + \Delta U_0 \qquad (8.1)$$

where α and β are coefficients, $CSME$ is the criterion of structural modification efficiency, and ΔU_0 = 1–2%. Thus, it is necessary to choose chalcogenide glasses with a smaller value of this criterion in order to minimize statistical variation.

Integrated circuits of nonvolatile memory based on memory switches were developed and manufactured in the USA and the USSR in the 1970s. A diagram of these circuits is shown in Fig. 8.10. There is a matrix of mutually perpendicular buses, each cell of which contains a serially connected memory switch described above and a silicon diode.

However, as it was soon discovered, these integrated circuits were not sufficiently reliable. In the process of information storage and reading, some of the cells switched from one state to another spontaneously. The reasons for such behavior were related to both operational conditions and material properties. After switching from the low-resistance to the high-resistance state, the crystalline channel in the device did not disappear completely. Crystalline inclusions remained in the glassy matrix. They had small influence on the device resistance, but could play the role of spontaneous crystallization centers in the process of operation. As a result, spontaneous switching to the low-resistance state occurred. In the case of the reverse situation, the crystalline channel connecting the electrodes in a low-resistance state could destroy in a certain “weak” spot during the operation process. As a result, spontaneous switching to the high-resistance state occurred. For this reason, nonvolatile memory integrated circuits based on chalcogenide glassy semiconductors were not developed further and their production was discontinued. But 30 years later, these principles of information recording have found application in new devices, which will be discussed in the next section.

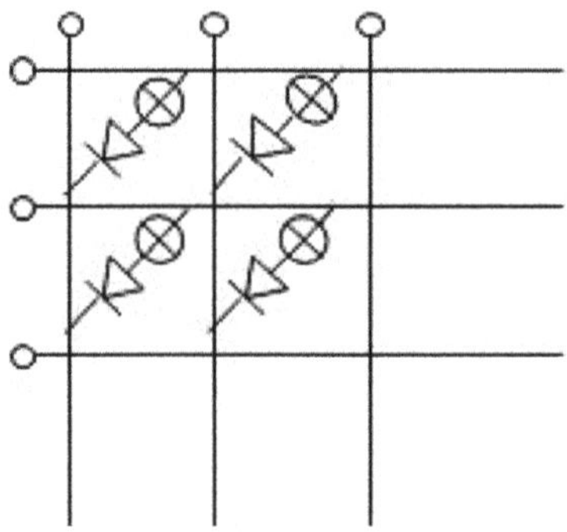

Figure 8.10 Diagram of nonvolatile memory integrated circuit on the basis of chalcogenide glasses.

8.1.3 Second Generation of Phase-Change Memory Devices

As noted in Chapter 1, the number of electronic devices connected to the network exceeded the earth’s human population in 2009. Rapid development of information technology has caused an increasing demand for basic electronic components and, in particular, for nonvolatile memory systems. Of course the dominant type of nonvolatile memory on the market today is the flash-NAND memory. However, this type of memory came to its theoretical limit in many important characteristics. In this regard, since the 1990s, nonvolatile memory devices based on new principles are developing intensively. These devices include resistive memory (RRAM), ferroelectric memory (FRAM), magneto-resistive memory (MRAM), and phase-transition memory (PCM or PCRAM). According to the forecast (Yole Développement, 2013), the market of new types of nonvolatile memory will increase from $209 million in 2012 to $2 billion in 2018.

In the early 2000s, new memory devices based on materials of the Ge–Sb–Te (GST) system were developed by Ovonyx, Inc. This memory device was first called ovonic unified memory (OUM), and later phase-change memory (PCM). The companies involved in the development of PCM integrated circuits include Intel and IBM (USA), BAE Systems and STMicroelectronics (Europe), Samsung (Korea), Elpida Memory (Japan), and others (Weiss, 2005). The developers believe that PCM can replace the traditional flash memory, static random access memory (SRAM), and dynamic random access memory (DRAM). Patenting related to PCM increased rapidly between 2002 and 2009 with more than 450 patent families published in 2009. After that, the number of published patents decreased, but more than 200 patents were published every year during 2010–13 (Bastard and Baron, 2014).

The advantages of this type of memory are numerous. There is a possibility to significantly increase the density of information recording because the bit is recorded in the volume with linear dimensions less than 20 nm. The ability to rewrite data without first having to erase it provides recording acceleration approximately by a factor of 30 compared with flash memory. Information storage time at room temperature is at least 10 years, and the number of rewrite cycles is greater than 10^8. PCM production has fewer processing steps than flash memory production, and the absence of high-temperature operations allows to produce a matrix of cells directly on silicon integrated circuits.

BAE Systems has created 2 Mb and 4 Mb radiation-resistant phase-change RAM integrated circuits. Samsung Electronics has demonstrated 64 Mb phase-change RAM integrated circuits based on 0.18 μm technology with the following parameters: cell size—0.56 μm × 0.9 μm = 0.504 μm^2; chip size—6.34 mm × 11.04 mm = 70 mm^2; supply voltage—3.0 V; reset current—0.8 mA; operation time—60–100 ns; endurance—10^9 cycles; and retention time—2 years at 85°C (Yeung et al., 2005). And in June 2009, the company began mass production of 512 Mb phase-change RAM chips. In 2012, Micron Technology (formerly Numonyx) launched serial production of PCM memory modules with a memory of 1 Gb on 45 nm technology for Nokia's smartphones (cell size 0.011 μm^2; Krishnamurthy, 2013).

Let us compare the new memory elements and memory elements based on chalcogenide glassy semiconductors, described in Section 8.1.2. The principle of recording in both cases is the same: reversible phase transitions glass ↔ crystal and associated changes in electrical conductivity of material. Therefore, new memory elements can be called the second generation of PCM. There are no significant differences in the matrix architecture as well. The decoupling diode (see Fig. 8.10) is replaced by a thin-film transistor (TFT). The latter is not a prerequisite but allows realization of end-to-end thin-film technology in the manufacture of memory integrated circuits. The structure of the switch has not changed either. It is a sandwich: conductive electrode–chalcogenide alloy–conductive electrode (Fig. 8.11).

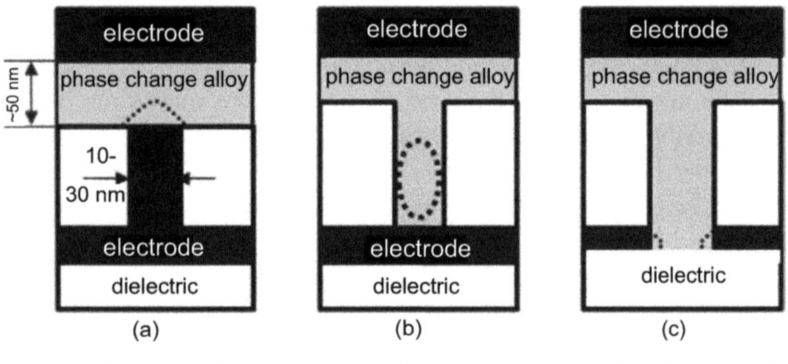

Figure 8.11 Phase-change memory cell structure: contact-minimized cell (a), volume minimized ("pore") cell (b), and "pore" cell with edge contact type (c). Active region is marked by a dotted line.

Along with the common features described above, new memory elements have a number of significant differences. The first difference lies in the fact that the chalcogenide alloy is in the crystalline state, and not in the glassy state. Hence, initially the device has the low-resistance state. The second difference lies in the structural dimensions of the memory element. As seen from Fig. 8.11a, the dimensions of element parts are in the range of tens of nanometers, which is an order of magnitude smaller than the first-generation devices. The smaller size comes out of application requirements (high density) but has an impact on the physics of device operation.

The next difference is in the design of the bottom electrode. The top and bottom electrodes in the first generation of memory elements had, as a rule, the same area. In contrast, in the case of second-generation memory elements, the area of contact of the bottom electrode with the chalcogenide alloy is several times smaller than that of the top electrode (see Fig. 8.11a). Yet another difference lies in the chemical composition of chalcogenide alloy. The same materials of Ge–Sb–Te and Ag–In–Sb–Te systems are used for both the electrical memory and for the optical disk storage (see Section 6.2.3). These materials ensure the stability of the non-crystalline phase of up to 10 years on one hand, and crystallization under the recording pulse action for 1–10 ns on the other hand.

As mentioned above, the device is initially in the low-resistance state. Switching to the high-resistance state is carried out by applying a current pulse with a sharp rear edge. Phase change takes place due to Joule heating up to the melting temperature of the crystalline phase and following vitrification of the melt. The highest heating of the material occurs in the area adjacent to the back electrode (active region in Fig. 8.11a), because the current density is greatest in this area. As a result, the semispherical volume of the non-crystalline phase is formed over the back electrode and the resistance of the device increases sharply. The radius of the non-crystalline region, as a rule (defined by application requirements), does not exceed 50 nm. In reverse switching, a longer current pulse, but with a smaller absolute value, is applied to the device. Under the current pulse, threshold switching, current channel formation, and the subsequent crystallization of glassy drop take place. The device switches to the low-resistance state (Popov, Savinov, and Voronkov, 2006). Figure 8.12 shows the dependence of device resistance on the recording current (on the basis of experimental data from Lowrey et al., 2004). Thus, instead of crystallization and melting of filaments between the electrodes, which occurred in the first generation of PCM devices (Section 8.1.2), in these devices the phase transitions occur in the bulk active region (Fig. 8.11a). This design provides a substantial increase in the reliability of information storage because the phase transition occurs in the three-dimensional mushroom-like area instead of the one-dimensional filament.

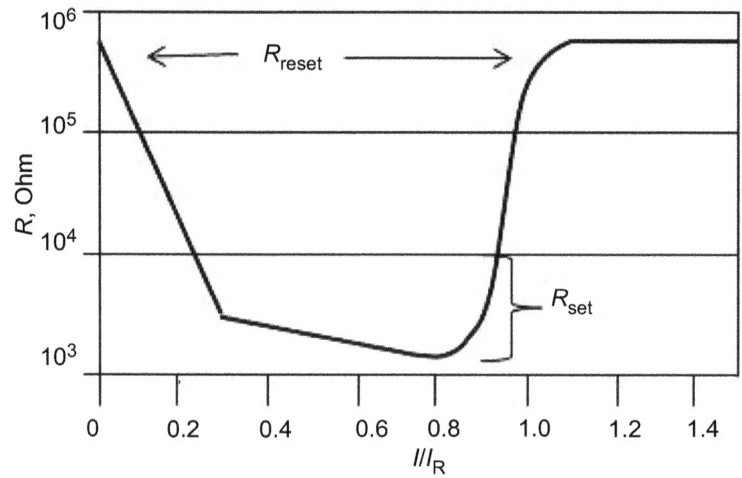

Figure 8.12 Dependence of device resistance on the recording current (I_R is the recording current of the high-resistance state; R_{set} and R_{reset} are resistances in the low-resistance and high-resistance states).

It should be noted that the change from phase transitions in the channel to phase transitions in the bulk region that is in contact with the electrode, while increasing the reliability of information storage, also leads to the necessity of strict control of the thermodynamic processes during the recording cycle. The necessary condition of reverse switching to the high-resistance state is the melting and amorphization of the crystalline phase. For this purpose, the parameters of the switching pulse and the conditions of heat removal should provide an increase in the temperature of the crystalline phase above the melting temperature. This condition is fulfilled reasonably easily in the case of memory cells of the first generation. There is a great difference in the thermal conductivity of metal electrodes and chalcogenide glass (two orders of magnitude). Due to this difference, almost all the heat (about 99%) dissipates through the electrodes. This leads to cooling mainly in the contact regions. Under such conditions, the distribution of temperature along the filament length is highly nonuniform. Its value in the middle of the filament can exceed the melting temperature of the crystalline phase, which results in transition of the central part of the filament to melt. In this case, the short rear front of the switching pulse provides fast cooling and amorphization of the melt. In the case of the second generation of memory cells, the greatest heating and melting of the crystalline phase, as well as its subsequent amorphization, occur in the area directly adjacent to the small contact (active region in Fig. 8.11a). Direct contact between the phase-transition region and the metal electrode with high heat conductivity causes stricter requirements for the design of cells in order for the phase transitions to be implemented at reasonable pulse powers because the increase in their power leads to a sharp decrease in the lifetime of the cells (Bogoslovskii and Tsandin, 2012) and to an increase in the time it takes to switch to the high-resistance state (Kostylev, 2009).

The investigation of the influence of the second-generation memory cell design on the parameters of switching pulses showed (Popov, Salnicov, and Anufriev, 2015) that their operability and stability of switching are defined by the balance between the amount of the heat released in the active region and the heat removed from it. The released heat is determined by the parameters of the switching pulse, and the removed heat is dependent on the heat sink conditions. As the heat sink from the active region in the second-generation cells is carried out mainly through the lower electrode (Fig. 8.11), the absolute values of removed heat are determined by the area of this electrode and its thermal conductivity.

Figure 8.13 shows the calculated dependence of heat quantity released in the cell's active region during the period of amorphizing pulse (20 ns, 5 V) and the dependence of heat amount that can be removed from the cell's active region in the same period on the size of the platinum contact area. It can be seen that the balance between the released and removed heat is achieved at a contact area of 0.001 μm^2. When the size of the electrode is more than the indicated value, the amount of removed heat exceeds the amount of heat released in the active region, which prevents an increase in the temperature of the active region to a value exceeding the melting temperature of the crystalline phase. In the above-mentioned work (Popov, Salnicov, and Anufriev, 2015), memory cells with the size of lower platinum electrode, 0.038, 0.023, and 0.001 μm^2, were prepared and studied.

As expected, a stable reverse switching was observed only in the cells with the smallest size of bottom electrode corresponding to the heat balance point.

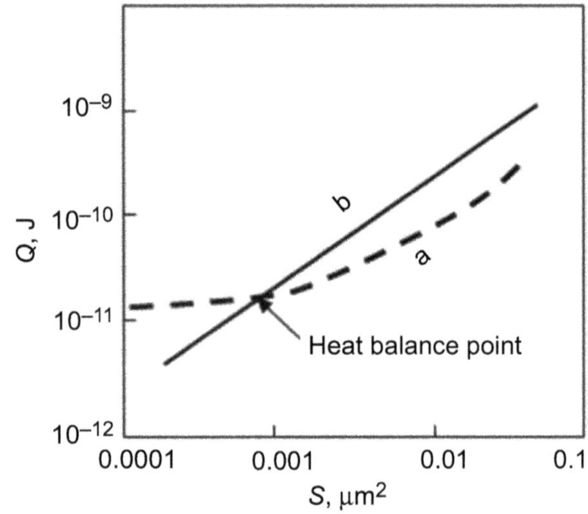

Figure 8.13 Calculated dependences of heat quantity released in the cell's active region (curve a) and of the heat amount that can be removed from the cell's active region (curve b) on the size of contact area between the electrode and the cell's active region.

Thus, reducing the contact area of the electrode with the active region of chalcogenide alloy is an obvious way to reduce heat sink from the active region and, consequently, improve the characteristics of memory cells. There are two ways to solve this (Burr et al., 2010): either by reducing the size of the electrode contacting with the chalcogenide alloy (contact-minimized cell, Fig. 8.11a) or by reducing the volume of chalcogenide alloy that is in contact with the electrode (volume minimized cell or "pore" cell, Fig. 8.11b). Minimizing the size of the electrode can be achieved by a combination of the above two approaches, namely, creation of edge contact in the "pore" cell (Ha et al., 2003). A possible implementation of such a design is shown in Fig. 8.11c. In this case, the lower electrode is etched up to a substrate. Then the chalcogenide alloy film is deposited. As a result, the film has a contact with an electrode in the form of a ring with a height equal to the thickness of the electrode. Since the electrode thickness can be a few nanometers, the contact area is significantly reduced.

The geometrical dimensions of the active region's contact with an electrode are not the only criterion of efficiency of heat sink from the active region. The efficiency of heat sink can be decreased by reducing the thermal conductivity of the electrodes. The dependence of the contact area, corresponding to the heat balance, on the thermal conductivity of the electrode material is represented in Fig. 8.14. As can be seen from the figure, the replacement of the platinum electrode with the tantalum carbide electrode increases the allowable area of the lower electrode to a value of 3.8×10^{-3} μm^2, which is about five times the difference.

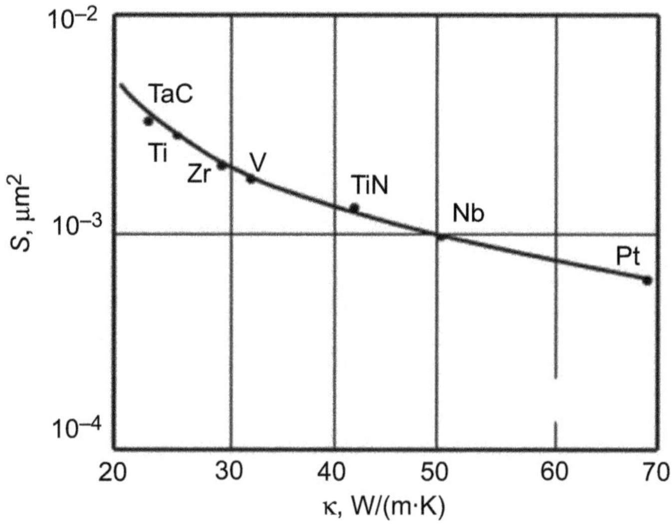

Figure 8.14 Dependence of the contact area, corresponding to the heat balance point, on the thermal conductivity coefficient of the electrode material.

Usually two groups of chalcogenide alloys are used in PCM elements, namely, the materials of Ge–Sb–Te system and the materials of Me–Sb–Te system (where Me is a metal). The first are pseudo-binary GeTe–Sb_2Te_3 compositions. These include the $Ge_2Sb_2Te_5$ (GST 225 is the most widely used material) and some other materials as $Ge_1Sb_2Te_4$ (GST 124) and $Ge_1Sb_4Te_7$ (GST 147). The governing process of glass–crystal phase transition in all these materials is nucleation, since the speed of this process considerably exceeds the rate of crystal growth. Materials close to GeTe in chemical composition possess high thermal stability (high crystallization temperature) but low rate of crystallization. On the contrary, materials close to Sb_2Te_3 in chemical composition have high crystallization rate but low thermal stability. Improved properties can be achieved by chemical modification of GST materials. For example, the addition of a few atomic percent of nitrogen in the material increases crystallization temperature and resistivity of the crystalline phase. The addition of oxygen leads to an increase in the resistance ratio of high- and low-resistance states of the memory element (Lacaita and Wouters, 2008). The effects of chemical composition of Ge–Sb–Te materials on phase-transition temperature and characteristics of memory cells are summarized in Table 8.1.

Table 8.1 Effects of chemical composition on the characteristics of memory cells

Chemical composition	% GeTe	% Sb_2Te_3	Crystallization temperature	Operation time	Stability
$Ge_2Sb_2Te_5$ (GST 225)	66	33	decreases ↓	decreases ↑	decreases ↓
$Ge_1Sb_2Te_4$ (GST 124)	50	50			
$Ge_1Sb_4Te_7$ (GST 147)	33	66			
GST + N			increases	resistance of SET state increases	
GST + O_2			resistance ratio R_{reset}/R_{set} increases		

The second group of chalcogenide alloys is metal-modified $Sb_{70}Te_{30}$ compound. Additions of indium $[In_x(Sb_{70}Te_{30})_{1-x}]$ and silver and indium $[Ag_xIn_y(Sb_{70}Te_{30})_{1-x-y}]$ (called AIST) are generally used. In contrast with GST, the glass–crystal phase transition in these materials is due to the growth of crystalline phase from the boundaries of the active region, rather than nucleation.

The ability to reduce sizes of PCM cells and increase the density of information recording (scalability) is a very important factor in terms of competition with existing flash and other types of memory.

The scaling of PCM includes (Lacaita and Wouters, 2008):

- Scaling of the thermal profiles in order to avoid thermal disturbances while shrinking cell separation
- Scaling of the program current and voltage
- Preservation of the material's characteristics with miniaturization

This analysis shows that the scaling potential of PCM is significantly superior to other types of memory.

Moreover, integrated circuits of PCM are quite easily produced by thin-film technology. This allows to place several integrated circuits one above the other (3D technology) and thus to increase the density of information recording by several times.

Further development of nonvolatile memory will include decreasing the energy of information recording and increasing the number of bits stored in one memory cell. An example of the first direction is the paper of Takaura et al. (2014). The authors, instead of the phase transition "crystal–amorphous phase" implemented in the traditional PCM in $Ge_2Sb_2Te_5$, have used the changes in the short-range order of the arrangement of atoms in the superlattice, made by alternately stacking Sb_2Te_3 and GeTe films (Fig. 8.15). This type of memory is named topological-switching RAM (TRAM). Changes in the short-range order are caused by the shift in germanium atoms as a result of injection of charge carriers in the material. As a result, the material becomes either high resistance or low resistance depending on the sign of the injected charge carriers. The resistance ratio in this case may be in the range of 10–1000 depending on the value of the programming current. Superlattice properties depend largely on the amount of germanium atoms therein. The authors of the aforementioned paper assumed that germanium atoms will move easily in the case of germanium deficit in the layers of GeTe and, therefore, investigated superlattices Ge_xTe_{1-x} / Sb_2Te_3 with $x << 0.5$. As a result, they got TRAM cells with recording current for both low-resistance and high-resistance states of 55 μA and a resistance ratio of 200. The number of endurance set and reset cycles of this memory cell was about 100 million.

Phase-change random access memory		Topological-switching random access memory	
PRAM		TRAM	
$Ge_2Sb_2Te_5$ (GST)		$GeTe/Sb_2Te_3$ (GT/ST)	
Alloy		Superlattice	
Joule heat		Charge injection	
Melting switch		Non-melting switch	
Order–disorder transition		Short-range movement of atoms	
Low-resistance state (crystal)	High-resistance state (amorphous)	Low-resistance state	High-resistance state
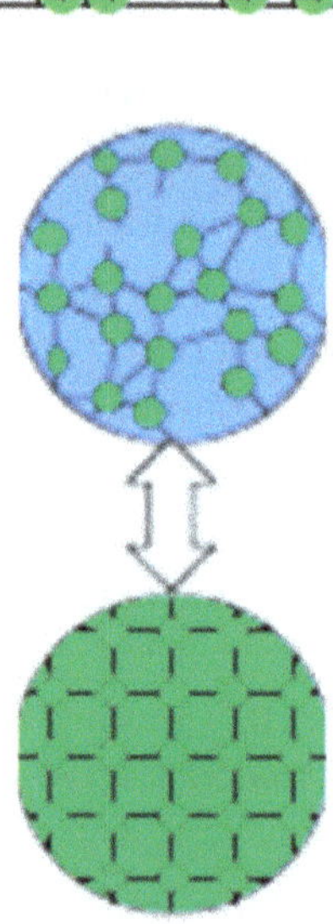		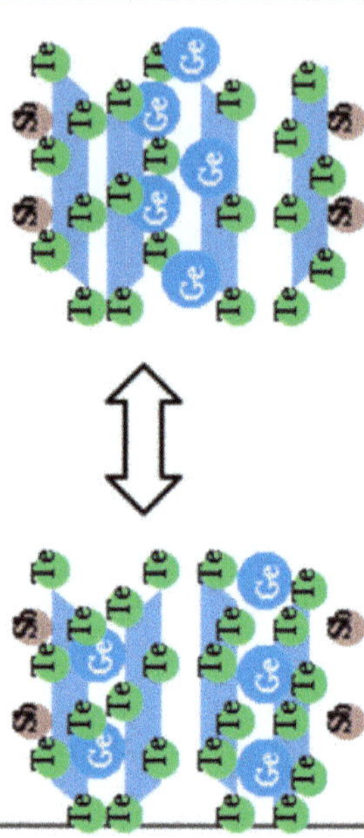	

Figure 8.15 Comparison of phase-change and topological-switching memory. Reprinted from Takaura et al. (2014), with permission from IEEE.

The second direction of nonvolatile PCM development is to store more than two levels (zero and one) per cell. PCM is ideally suited for this task. As mentioned above, switching from the low-resistance to the high-resistance state leads to the formation of glassy phase in the active region of the cell. The thickness of the formed glassy region depends on the programming current. In turn, the resistance of the cells in the high-resistance state is determined by the thickness of the glassy region. Consequently, varying the program current should lead to a change in cell resistance. More than 15 years ago, Ovshinsky (2000) demonstrated the possibility of "multiple bit storage in one memory cell" (Fig. 8.16). As seen from the figure, instead of two states (high-resistance and low-resistance), more than a dozen states with different resistances have been recorded.

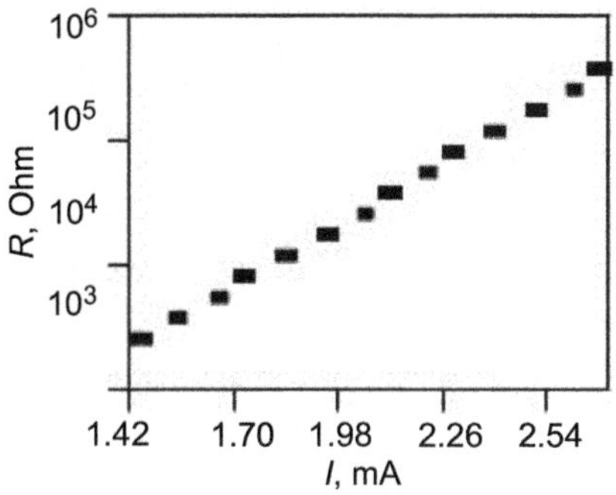

Figure 8.16 Cell resistance dependence on the magnitude of programming current.

However, in order for any memory to be considered a viable technology, its reliability should be verified at the array level. The task of creating nonvolatile PCM with recording and storage of 2 bits per cell has been solved initially for 32 kb cell arrays (Nirschel et al., 2007) and then for 64 kb cell arrays (Athmanathan et al., 2016). Dependence of cell resistance on the recording current in the case of four levels per cell storage is shown in Fig. 8.17.

The main problem arising in the development of multilevel cell array was the drift and variability of PCM cell characteristics, namely: resistance drift, which is caused by the nature of the amorphous phase (see Chapter 1); cell variability, including inter-cell variability in the array and intra-cell variability for any given cell; intrinsic array noise and noise from write/read circuitry.

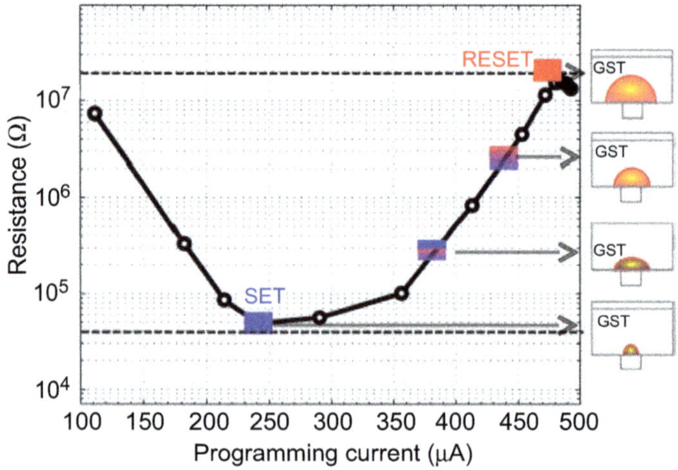

Figure 8.17 Dependence of cell resistance on the recording current in the case of four levels per cell storage. Reprinted from Athmanathan et al. (2016), with permission from IEEE.

In the aforementioned papers, these problems have been solved through the development of drift-immune cell-state metric schemes and employing iterative programming, where a sequence of "program and verify" steps is used in feedback until the cell is programmed at the target resistance level. This allowed to provide reliable storage of four levels per cell (2 bits/cell), with adequate data retention at elevated temperatures, even after endurance cycles. Moreover, Athmanathan et al. (2016) showed that by using these techniques, it is possible to reliably store and retrieve 3 bits/cell in PCM arrays, provided that temperature variations can be controlled.

8.2 Silicon Thin-Film Transistors

Metal-oxide semiconductor field-effect transistors (MOSFETs) based on single-crystal semiconductors are the most common type of semiconductor devices at present. The question is: if this type of devices has been successfully produced on the basis of single crystals for a long time, why is it necessary to develop similar devices based on disordered semiconductors? As it turns out, along with the general advantages of devices based on disordered semiconductors (lack of fundamental size limitations, low cost, etc.), there are tasks that cannot be fulfilled by using single-crystal devices. These tasks include creation of flat liquid-crystal displays (LCDs).

The market of flat displays is developing dynamically and is projected to be $120–150 billion by 2020. There are three main types of flat displays:

1. Plasma displays, which are based on the luminescence of phosphor under the action of radiation arising in the plasma of gas-filled cells
2. LCDs in which the optical transmission of cells changes under the influence of an electric field
3. Displays in which the cells are organic light-emitting diodes (OLEDs)

At present, LCDs are the most widely used type of flat displays (more than 50% of the market).

It is interesting to note that liquid crystals were discovered in the nineteenth century, even before the creation of cathode ray tubes. The first description of these materials dates back to 1888. But for a long time, they did not find any practical application. And only after almost 80 years, in 1966, the Radio Corporation of America (RCA) demonstrated a prototype of liquid-crystal monitor—a digital clock. And 10 years later, in 1976, Sharp produced monochrome TV with a screen diagonal of 5.5 inch, built on the basis of a liquid-crystal matrix with resolution of 160 × 120 pixels.

The operation principle of LCDs is based on the dependence of the plane of polarization of light that passes through the liquid crystal on the orientation of the liquid-crystal molecules. The light, after the polarizer acquires certain polarization, passes through the liquid crystal and is incident on the second polarizer (Fig. 8.18). If the planes of polarization of the first and second polarizers are perpendicular, and the polarization of light passing through the liquid crystal is not changed, the light does not pass through the second polarizer (Fig. 8.19a). If the electric field changes the orientation of liquid-crystal molecules with the result that the plane of light polarization changes, the light passes through the second polarizer (Fig. 8.19b). Thus, by changing the orientation of liquid-crystal molecules at various points of the screen using electrical signals, an image can be obtained.

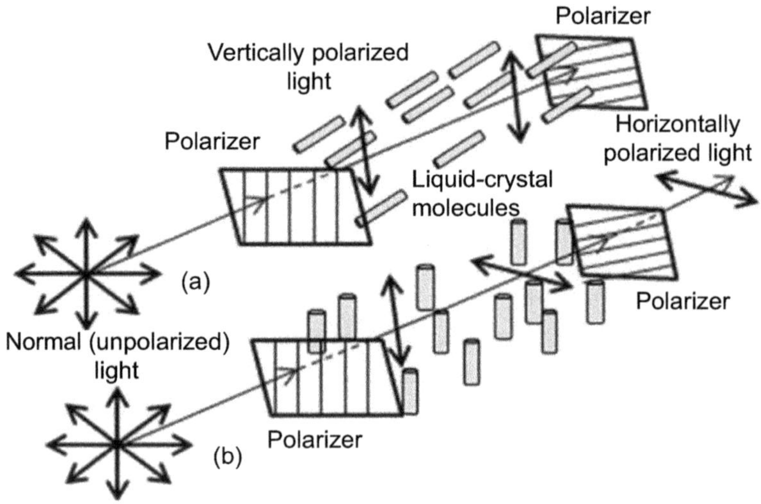

Figure 8.18 The operating principle of liquid-crystal display (see text).

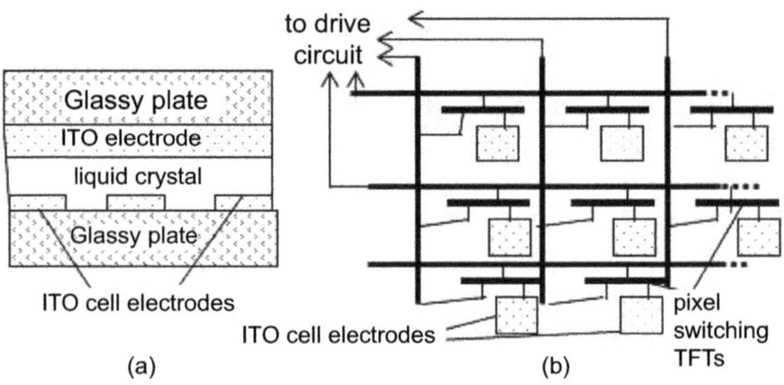

Figure 8.19 Structure (a) and TFT matrix diagram (b) of liquid-crystal display (TFTs are thin-film transistors and ITO is $SnO_2 + In_2O_3$).

Realization of the aforementioned principle is rather simple. There are two glass plates, one of which is covered with a transparent electrode (ITO) completely, and a similar electrode is formed as cells on the other plate. The gap between the plates is filled with liquid crystal (Fig. 8.19a).

The optical properties of liquid crystals are changed by the electric field in the cells, where voltage is applied. The problem lies in the control of the screen. In the case of resolution of 10 dots per millimeter, a screen of area 10 × 10 cm should have 10^6 cells, each of which must be independently controlled. This means that there has to be a million connections between the control circuit driving screen cells and the screen itself. The number of connections can be reduced only by placing the driving circuit directly on the glass substrate of the liquid-crystal screen. But this solution determines the technology of device preparation. It has to be thin-film technology. TFTs can be made of hydrogenated amorphous (nanocrystalline or microcrystalline) silicon, small-molecule or polymer organic semiconductors, or some polycrystalline compound semiconductors (CdSe, for example). Hydrogenated amorphous silicon TFTs have advantages both in terms of technological base and in terms of their characteristics. Since the deposition temperature of hydrogenated amorphous silicon films is in the range of 250–350°C (see Section 5.2), TFTs can be prepared on a glass substrate. Therefore, these transistors are widely used both as pixel-switching devices in active matrix LCDs and as peripheral driver circuits. The electrical circuit of the TFT matrix LCD is shown in Fig. 8.19b. It should be noted that the development of TFT active matrix arrays technology on glass substrates has also found application in medical electro-radiography in creating digital flat panel X-ray image detectors (see Section 6.1.4).

The technological manufacturing process of the TFT matrix LCD includes the following operations (Fig. 8.20):

1. Deposition of transparent ITO film on a glassy substrate
2. The first photolithography process to form ITO cell electrodes
3. Deposition of chromium film
4. The second photolithography process to form chromium gate electrodes and matrix horizontal buses
5. Deposition of silicon nitride (Si_3N_4) film
6. Deposition of undoped hydrogenated amorphous silicon film
7. The third photolithography process to form transistor active regions
8. Deposition of n^+-type hydrogenated amorphous silicon film
9. The fourth photolithography process to form transistor source and drain regions and windows on the silicon nitride
10. Deposition of aluminum film

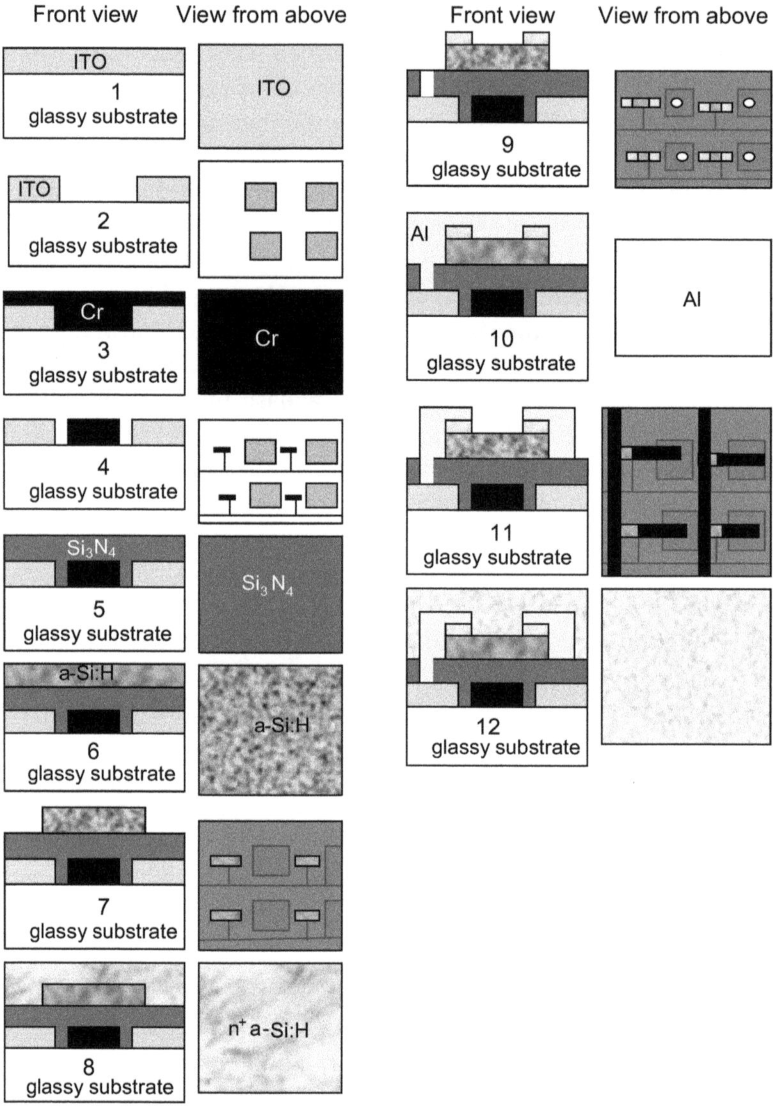

Figure 8.20 Basic steps of TFT active matrix LCD fabrication (see text).

11. The fifth photolithography process to form aluminum contacts to the drain and source of transistors and to cell ITO electrodes and to form matrix vertical buses
12. Deposition of protective layer

13. The sixth photolithography process to form windows on protective layer to create the contacts to matrix buses (not shown in the figure).

TFTs based on hydrogenated amorphous silicon have the following distinctive features:

- Current in the off state is very small (about 10^{-12} A) due to the very high resistance of undoped hydrogenated amorphous silicon (see Section 4.1). In the on state, current increases by six to eight orders of magnitude. This provides a large on–off ratio (more than six orders of magnitude).
- All fabrication processes are carried out by conventional photolithographic technique that provides reproducible and cheap production of large-scale integrated LCDs.
- The presence of localized states in the band gap of hydrogenated amorphous silicon limits the field-effect mobility of charge carriers in the transistor channel ($\mu_n = 0.5–1.0$ cm^2/Vs and $\mu_p \approx 0.01$ cm^2/Vs). This leads to limitation of the switching speed.

There are two ways to increase the operation speed of transistors: reducing the channel length between the source and the drain, or increasing the field-effect mobility of the material. However, reducing the channel length means decreasing the minimum feature size of the photolithographic process and, consequently, a rise in the probability of defects. This is a serious problem in the case of active matrix LCDs, since any technological defect causes visual defect. Therefore, reducing the channel length of transistors by traditional methods is not an acceptable way to solve the problem.

One possible solution to the problem noted above is a TFT with a vertical structure. The channel length of this transistor is determined by the thickness of the dielectric film and is equal to 0.5–1.0 μm (Aivasov et al., 1995). The structure of the transistor is shown in Fig. 8.21. Films of metal, amorphous (or microcrystalline) hydrogenated n^+-type silicon, silicon nitride (thickness of about 1 μm), amorphous (or microcrystalline) hydrogenated n^+-type silicon and metal are deposited onto the substrate sequentially. Then photolithography and etching of films are carried out. The upper and lower metal films are the source and drain electrodes. Films of amorphous (or microcrystalline) hydrogenated n^+-type silicon are the source and drain regions. Then films of undoped hydrogenated amorphous silicon, silicon nitride (gate insulator), and metal (gate electrode) are deposited. The channel is formed in the undoped a–Si:H film between two n⁺-type layers, and its length (*L*) is approximately equal to the thickness of the silicon nitride layer. This allows to increase the operation speed of transistors by more than an order of magnitude at the same technological minimum features size.

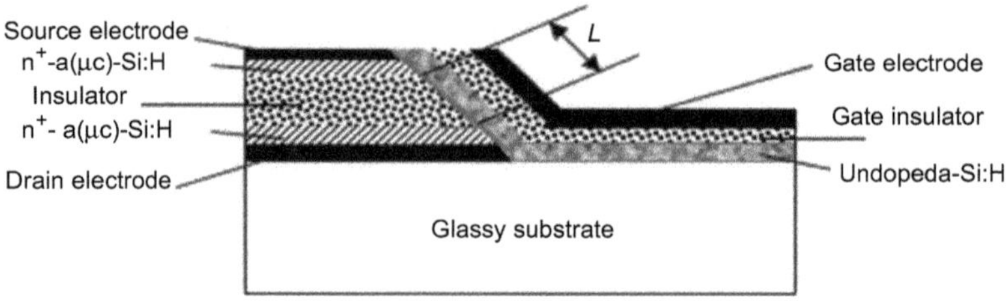

Figure 8.21 Thin-film transistor with vertical structure (*L* is channel length).

Another way to increase the transistor operation speed is to obtain a semiconductor material with a higher mobility of charge carriers. The increased mobility of electrons and holes in silicon TFTs can be achieved by replacing the amorphous to nanocrystalline or microcrystalline hydrogenated silicon (Wagner et al., 2003). The mobility of charge carriers increases with the size of crystal grains in the material (Fig. 8.22).

The nanocrystalline or microcrystalline silicon film can be obtained either by crystallization of the initial a–Si:H film or from direct deposition. As a result, thin-film n-type channel transistors with mobility of several tens of cm^2/Vs (Matsumara et al., 2003) and even with mobility above 100 cm^2/Vs (Stannowski, Rath, and Schropp, 2003) were manufactured. Moreover, thin-film p-type channel transistors have been created. This cannot be achieved in amorphous silicon due to low hole mobility.

The next goal of the flat panel industry is displays fabricated on flexible plastic foil substrates. Flexible displays are attractive due to light weight and ruggedness. There are desirable features for portable applications today and for large-area displays, electronic textiles, and mechatronics tomorrow (Wagner et al., 2003). According to the forecast (Kang, 2016), shipments of flexible displays will grow in 8 years (during 2015–23) from 50 to 560 million units. There are two main technologies for producing flexible displays: LCDs and OLEDs. While the LCD technology confidently takes the first place in the total production of flat displays, the OLED technology is dominant at present in the case of flexible displays, and LCDs have not yet come to the level of commercial production. The main reason for this is that plastic substrate foils may be processed up to only 150–200°C. On the other hand, the optimal range of substrate temperature for device-quality hydrogenated amorphous silicon deposited by glow discharge decomposition method is 250–350°C. Decreasing the deposition temperature leads to deterioration of a–Si:H film quality (see Section 5.2). To overcome this problem, dilution of silane with hydrogen in the glow discharge decomposition method (Wagner et al., 2003) or the method of hot wire chemical vapor deposition (see Section 5.4) is used. This allows to obtain device-quality films of hydrogenated silicon at a lower substrate temperature. Example of the a–Si:H TFT structure made on SiN_x passivated polyimide foil is shown in Fig. 8.23 (Wagner et al., 2003).

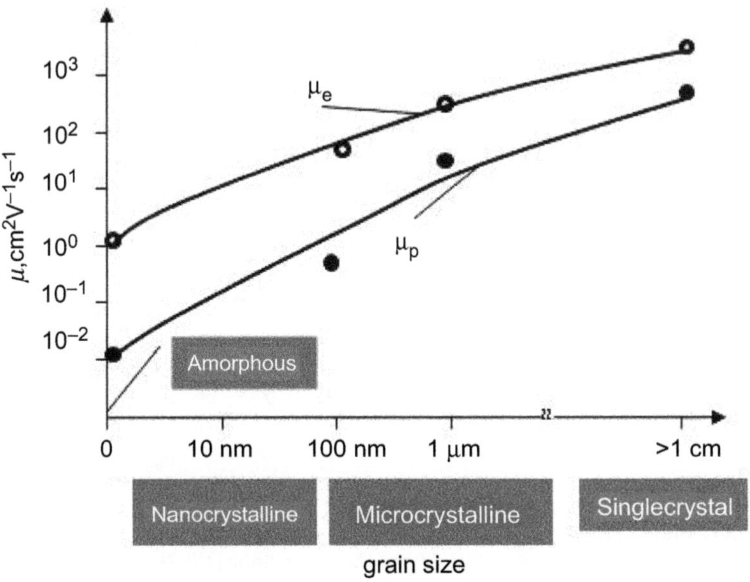

Figure 8.22 Dependence of electron (μ_e) and hole (μ_p) mobility in silicon on crystal grain size.

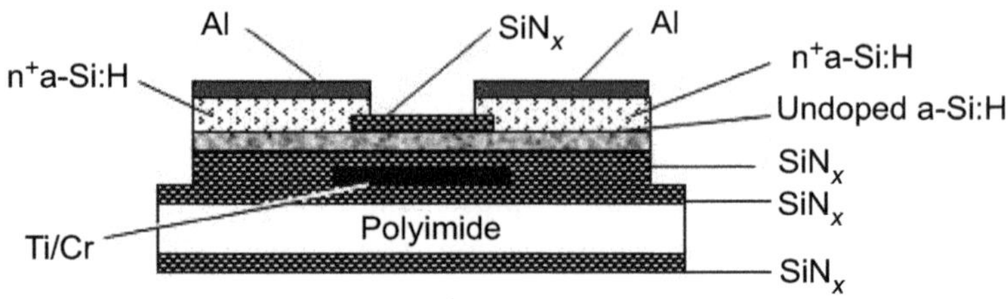

Figure 8.23 Cross section through a–Si:H thin-film transistor made on SiN_x passivated polyimide foil. Reprinted from Wagner et al. (2003), with permission from Elsevier.

For some applications (e.g., in certain types of electrophotographic printers), TFTs with a high-voltage (hundreds of volts) source–drain and a small gate voltage (a few volts) are needed. High-voltage TFTs based on hydrogenated amorphous silicon (Fig. 8.24a) have been created to solve this problem. As one can see from the figure, the gate electrode is located under the source area, and the inter-electrode space source–drain is large enough (about 20 μm). The part of the device that is located above the gate electrode operates as an ordinary field-effect transistor. Electrons from this region are injected into the inter-electrode source–drain space, and the most part of the applied voltage drops there. The current–voltage characteristics of the transistor are shown in Fig. 8.24b (Hack and Shur, 1987). The operating voltage of the transistor reaches 400 V at gate voltages of 0–10 V.

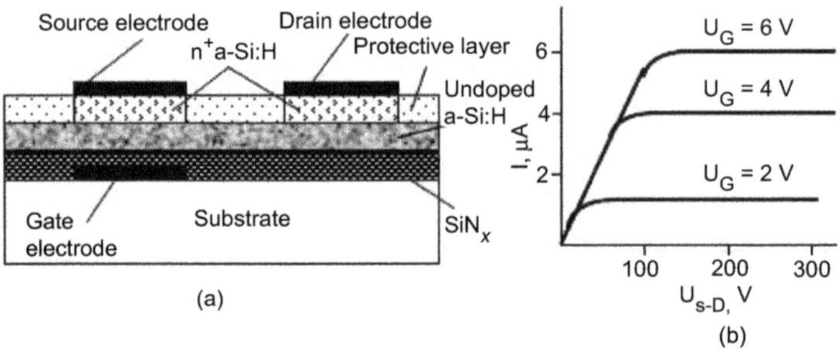

Figure 8.24 Structure (a) and current–voltage characteristics of high-voltage thin-film transistor (b). Reprinted from Hack and Shur (1987), with permission from Elsevier.

Bibliography

Adler, D. (1978) Density of states in the gap of tetrahedrally bonded amorphous semiconductors, *Phys. Rev. Lett.*, **41**, pp. 1755–1758.

Aitken, B. G. (2001) Structure–property relationships in Ge–As selenide and sulphide glasses. In: Proceedings of the International Congress on Glass, Edinburgh, Invited papers, pp. 135–140.

Aivasov, A. A., Budagyn, B. G., Vihrov, S. P., and Popov, A. I. (1995) *Disordered Semiconductors*, "Vishaya shcola," Moscow, p. 352 (in Russian).

Anderson, P. W. (1975) Model for the electronic structure of amorphous semiconductors, *Phys. Rev. Lett.*, **34**, pp. 953–955.

Andrievsky, A. I. and Nabitovich, I. D. (1960) Structure of β-cubic modification of selenium, *Sov. Phys. Crystallogr.*, **5**, pp. 442–443.

Andrievsky, A. I., Nabitovich, I. D., and Krinykovicn, P. K. (1959) Structure of selenium in thin layers, *Sov. Phys. Dokl.*, **4**, pp. 16–18.

Appleyard, D. (2015a) UK cut red tape for commercial rooftop solar. *Renewable Energy World*, March 27. URL: http://www.renewableenergyworld.com

Appleyard, D. (2015b) France mandates solar roofs for commercial zones. *Renewable Energy World*, March 30. URL: http://www.renewableenergyworld.com

Athmanathan, A., Stanisavljevic, M., Papandreou, N., Pozidis, H., and Eleftheriou, E. (2016) Multilevel-cell phase-change memory: A viable technology, *IEEE J. Emerg. Sel. Topics Circuits Syst.*, **6**, pp. 87–100.

Averynov, V. L., Kolobov, A. V., Kolomiets, B. T., Lyubin, V. M., and Fedorov, V. A. (1980) Photo-darkening and photo-bleaching in chalcogenide glassy semiconductor films As–Se and As–S systems: Physical phenomena in non-crystalline semiconductors. In: Proceedings of the International

Conference "Amorphous Semiconductors-80", Kishinev, AN SSSR, pp. 213–215 (in Russian).

Baratov, A. G. and Popov, A. I. (1990) Molecular Structure and Properties of Chalcogens (Sulphur, Selenium and Tellurium), "Aiastan," Erevan, p. 88 (in Russian).

Barinov, A. D. (2014) Physical modification of the silicon–carbon films, Proc. VII All-Russian School Seminar in the Direction of "Diagnostics of Nanomaterials and Nanostructures," 3, pp. 268–271 (in Russian).

Barinov, A. D. , Kutuzov, L. V., Popov, A. I., and Presnyakov, M. Yu. (2014) Effect of nanocrystalline phase on the electrical conductivity of amorphous tantalum-containing silicon–carbon films, Inorg. Mater., 50, pp. 939–944.

Barinov, A. D., Popov, A. I., Presniakov, M. Y., and Shupegin, M. L. (2015a) Influence of heat treatment on the structure, chemical composition and electrical conductivity of diamond-like carbon–silicon nanocomposites, Vestnik MPEI, N1, pp. 85–90 (in Russian).

Barinov, A. D., Vorontsov, V. A., Popov, A. I., Presnyakov, M. Y., and Shupegin, M. L. (2015b) Effect of metals on the electrical conductivity of silicon–carbon films. In: Fluctuation and Degradation Processes in Semiconductors Devices, MPEI, Moscow, pp. 92–96 (in Russian).

Barinov, A. D., Popov, A. I., Presniakov, M. Y., and Shupegin, M. L. (2016) Formation of nanophase in amorphous silicon–carbon matrix. In: Proceedings of the International Conference on Amorphous and Microcrystalline Semiconductors, St. Petersburg, pp. 32–33 (in Russian).

Bastard, F. and Baron, N. Emerging non-volatile memories patent landscape. KnowMade Web site. http://www.knowmade.com/downloads/emerging-non-volatile-memories. Published February 2014.

Belev, G. and Kasap, S. O. (2009) The influence of substrate temperature on charge transport in vacuum deposited amorphous selenium, *J. Optoelectron. Adv. Mater.*, **11**, pp. 1053–1059.

Bogoslovskii, N. A. and Tsandin, K. D. (2012) Physics of switching and memory effects in chalcogenide glassy semiconductors, *Semiconductors*, **46**, pp. 559–590.

Bowman, D. L. and Schottmiller, J. C. (1968) Vitreous Bi–Se layers as near-infrared photodetectors, *J. Appl. Phys.*, **39**, pp. 1659–1662.

BP (2015) Renewables section, *BP Statistical Review of World Energy*, June 2015.

Brenner, T. M., Egger, D. A., Kronik, L., Hodes, G., and Cahen, D. (2016) Hybrid organic—inorganic perovskites: Low-cost semiconductors with intriguing charge-transport properties, *Nat. Rev. Mater.*, **1**, Article number: 15007, doi:10.1038/natrevmats.2015.7.

Brutting, W. Organic semiconductors, pp. 1–11, University of Augsburg Web site. http://www.physik.uni-augsburg.de/organic/Dateien/OrganicSemiconductors.pdf. Published February 24, 2005.

Bugaev, S. P., Korotaev, A. D., Oskomov, K. V., and Sochugov, N. S. (1997) Properties of diamond-like films obtained in a barrier discharge at atmospheric pressure, *Tech. Phys.*, **42**, pp. 945–949.

Bunker, G. (2010) *Introduction to XAFS: A Practical Guide to X-ray Absorption Fine Structure Spectroscopy*, Cambridge University Press, Cambridge, p. 260.

Burr, G. W., Breitwisch, M. J., Franceschini, M., Garetto, D., Gopalakrishnan, K., Jackson, B., Kurdi, B., Lam, C., Lastras, L. A., Padilla, A., Rajendran, B., Raoux, S., and Shenoy, R. S. (2010) Phase change memory technology, *J. Vac. Sci. Technol., B*, **28**, pp. 223–262.

Busse, L. (1984) Temperature dependence of the structures of As2Se3 and AsxS1-x glasses near glass transition, *Phys. Rev. B*, 29, pp. 3639–3651.

Carlson, C. (1942) Electrophotography, U.S. Patent 2.297.691.

Carlson, D. (1984) Solar energy conversion. In: *The Physics of Hydrogenated Amorphous Silicon I. Structure, Preparation and Devices*, J. Joannopoulos and G. Lucovsky (ed.), Springer-Verlag, pp. 203–244.

Cherfi, R., Farhi, G., Aoucher, V., and Zellama, K. (2001) Hydrogenated amorphous silicon deposited by DC magnetron sputtering at high temperature, *Thin Solid Films*, **383**, pp. 192–195.

Cherkasov, Yu. A. and Kreitor, L. G. (1974) Exiton absorption and photoconductivity in amorphous analogs of various crystalline forms of selenium, *Fisica Tverdogo Tela*, **16**, pp. 2407–2410 (in Russian).

Chiang, C. K., Fincher, C. R., Park, Jr. Y. W., Heeger, A. J., Shirakawa, H., Louis, E. J., Gau, S. C., and MacDiarmid, A. G. (1977) Electrical conductivity in doped polyacetylene, *Phys. Rev. Lett.*, **39**, 1098.

CISION. German solar breaks three records within two weeks. PR-Newswire Web site. http://www.prnewswire.com/news-releases/german-solar-breaks-three-records-within-two-weeks-263664861.html. Published June 18, 2014.

Cohen, M. H., Fritzsche, H., and Ovshinsky, S. R. (1969) Simple band model for amorphous semiconducting alloys, *Phys. Rev. Lett.*, **22**, pp. 1065–1068.

Coropceanu, V., Cornil, J., Filho, D., Olivier, Y., Silbey, R., and Bredas, J. (2007) Charge transport in organic semiconductors, *Chem. Rev.*, **107**, pp. 926–952.

Cormack, A. N., Du, J., and Zeitler, T. (2001) Molecular modeling of glass. In: Proceedings of the International Congress on Glass, Edinburgh, Invited papers, pp. 170–174.

Davis, E. A. and Greaves, G. N. (1976) Amorphous arsenic: Properties and structural model. In: Proceedings of the 6th International Conference on Amorphous and Liquid Semiconductors, Leningrad, pp. 212–220.

Davis, E. A. and Mott, N. F. (1970) Condition in non-crystalline systems, *Phil. Mag.*, **22**, pp. 903–922.

Dembovsky, S. A. (1977) Empirical glass forming theory in chalcogenide systems, *J. Neorganicheskoy Chimie*, **22**, pp. 3187–3199 (in Russian).

Dembovsky, S. A. (1981) The connection of quasi-defects with glass formation in the substances with high long-pair electron concentration, *Mat. Res. Bul.*, **16**, pp. 1331–1338.

Dembovsky, S. A. (2007) New sight into the structure of selenium: Four-valence Se in glass, *J. Non-Cryst. Solids*, **353**, pp. 2944–2948.

Dembovsky, S. A. and Chechetkina, E. A. (1981a) The molecular-kinetic model for critical cooling rate of glass forming melts, *Mat. Res. Bul.*, **16**, pp. 723–727.

Dembovsky, S. A. and Chechetkina, E. A. (1981b) Kinetico-thermodynamic aspect of glass formation and critical cooling rates in chalcogenide systems, *Mat. Res. Bul.*, **16**, pp. 505–511.

Dembovsky, S. A., Chechetkina, E. A., and Kozukhin, S. A. (1985) Anomalous influence of weak magnetic fields on diamagnetic glassy semiconductors, *JETP Lett.*, **41**, pp. 88–90.

Dembovsky, S. A., Zyubin, A. S., and Grigor'ev, F. V. (1998) Modeling of hypervalent configurations, valence alternation pairs, deformed structure and properties of a–S and a–As_2S_s, *Semiconductors*, **32**, pp. 899–916.

Descoeudres, A., Holman, Z., Barraud, L., Morel, S., De Wolf, S., and Ballif, C. (2013) >21% Efficient silicon heterojunction solar cells on n- and p-type wafers compared, *IEEE J. Photovolt.*, **3**, pp. 83–89.

Directorate for Fuel and Energy Sector. Development of solar technologies in the world. Information Reference (2013), Moscow, p. 10 (in Russian).

Doe, P. (2015) More change for the chip industry, *The Pulse*. Solid State Technology Web site. http://electroiq.com. Published June 12, 2015.

Dorfman, V. F. (1992) Diamond-like nanocomposites (DLN), *Thin Solid Films*, **212**, pp. 267–273.

Dvorkovich, V. P. and Dvorkovich, A. V. (2012) *Digital Video Information Systems (Theory and Practice)*, Technosphera, Moscow, p. 1008 (in Russian).

Elliott, S. R. (1984) *Physics of Amorphous Materials*, 2nd edn., Longman, London, p. 386.

Elliott, S. R. (1987) Medium-range order in amorphous materials: Documented cases, *J. Non-Cryst. Solids*, **97&98**, pp. 159–162.

Elliott, S. R. (1991) Origin of the first sharp diffraction peak in the structure factor of covalent glasses, *Phys. Rev. Lett.*, **67**, pp. 711–714.

Elliott, S. R. and Davis, E. A. (1979) Defect states in Group-V amorphous semiconductors, *J. Phys. C*, **12**, pp. 2577–2587.

Flasck, R., Izu, M., Supru, K., Anderson, T., Ovshinsky, S. R., and Fritzsche, H. (1977) Optical and electronic properties of modified amorphous materials. In: Proceedings of the 7th International Conference on Amorphous and Liquid Semiconductors, Edinburgh, pp. 524–528.

Fujishima, D., Inoue, H., Tsunomura, Y., Asaumi, T., Taira, S., Kinoshita, T., Taquchi, M., Sakata, H., and Maruyama, E. (2010) High-performance HIT solar cells for thinner silicon wafers. In: Photovoltaic Specialists Conference (PVSC), 2010 35th IEEE, pp. 3137–3140, doi: 10.1109/PVSC.2010.5614460.

Golikova, O. A. and Kazanin, M. M. (2000) Effect of nanocrystalline inclusions on the photosensitivity of amorphous hydrogenated silicon films, *Semiconductors*, **34**, pp. 737–740.

Golikova, O. A. (2001) Medium-range order and optoelectronic properties of a tetrahedrally coordinated hydrogenated amorphous semiconductor, *Semiconductors*, **35**, pp. 1314–1319.

Greaves, G. N., Elliott, S. R., and Davis, E. A. (1979) Amorphous arsenic, *Adv. Phys.*, **28**, pp. 49–141.

Grigorovici, R. (1974) Structure of amorphous semiconductors. In: *Amorphous and Liquid Semiconductors*, J. Tauc (ed.), Plenum Press, pp. 45–99.

Golikova, O. A. (1991) Doping and pseudodoping of amorphous silicon, *Sov. Phys. Semicond.*, **25**, pp. 915–938.

Gubanov, A. I. (1963) *Quantum Electronics Theory of Amorphous Semiconductors*, Academy of Science, Moscow-Leningrad, p. 250 (in Russian).

Gurevich, S. B., Ilyshenko, N. N., Kolomiets, B. T., Lyubin, V. M., Fedorov, V. A., and Shilo, V. P. (1976a) Holographic media on the base of chalcogenide glassy semiconductor films of the As–Se system. In: *Holography and Information Transmission*, Nauka, Leningrad, pp. 146–157 (in Russian).

Gurevich, S. B., Ilyshenko, N. N., Kolomiets, B. T., Lyubin, V. M., and Shilo, V. P. (1976b) Photo-induced changes of optical properties of some chalcogenide glassy semiconductors. In: Structure and Properties of Non-Crystalline Semiconductors, Proceedings of the 6th International Conference on Amorphous and Liquid Semiconductors, FTI, Leningrad, pp. 451–455 (in Russian).

Ha, Y. H., Yi, J. H., Horii, H., Park, J. H., Joo, S. H., Park, S. O., Chung, U. I., and Moon, J. T. (2003) An edge contact type cell for phase change RAM featuring very low power consumption. *Symposium on VLSI Technology Digest of Technical Papers*, 4-89114-033-X, pp. 175–176.

Hack, M. and Shur, M. (1987) Analysis of amorphous silicon thin-film transistors, *J. Non-Cryst. Solids*, **97&98**, pp. 1291–1294.

Hamakawa, Y. (1993) Physics and technology of amorphous semiconductors: Introduction. In: *Current Topics in Amorphous Materials: Physics and Technology*, Y. Sakurai, Y. Hamakawa, T. Masumoto, K. Shirae, and K. Suzuki (eds), Elsevier, pp. 301–311.

Hata, N., Yamasaka, S., Oheda, H., Matsuda, A., Okushi, H., and Tanaka, K. (1981) A photoluminescence study of amorphous-microcrystalline mixed-phase Si:H films, *Jpn. J. Appl. Phys.*, **20**, pp. L793–L797.

Hirai, T. and Maruyama, E. (1982) Integrated potosensors and imaging devices, *JARECT*, **2**, pp. 264–276.

Hirata, A., Guan, P., Fujita, T., Hirotsu, Y., Inoue, A., Yavari, A., Sakurai, T., and Chen, M. (2011) Direct observation of local atomic order in a metallic glass, *Nat. Mater.*, **10**, pp. 28–33.

Hirotsune, A., Mukoh, M., and Terao, M. (2006) Layer-selection-type recordable optical disk with inorganic electrochromic film, *Jpn. J. Appl. Phys.*, **45**, pp. 1235–1238.

Hoheisel, M. and Batz, L. (2001) Requirements on amorphous semiconductors for medical X-ray detectors, *Thin Solid Films*, **383**, pp. 132–136.

Jandieri, K., Rubel, O., Baranovskii, S. D., Reznik, A., Rowlands, J. A., and Kasap, S. O. (2008) Scattering and movement asymmetry in the one-dimensional lucky-drift simulation of the avalanche processes in disordered semiconductors, *J. Non-Cryst. Solids*, **354**, pp. 2657–2661.

Juska, G., Arlauskas, K., and Montrimas, E. (1987) Features of carriers at very high electric fields in a–Se and a–Si:H, *J. Non-Cryst. Solids*, **97&98**, pp. 559–562.

Kang, J. (2016) Flexible display is ready to jump up. INS Technology Abstract, Flexible Display Market Tracker. https://technology.ihs.com/569926

Kaplan, D. (1984) CVD material. In: *The Physics of Hydrogenated Amorphous Silicon I. Structure, Preparation and Devices*, J. Joannopoulos and G. Lucovsky (eds.), Springer-Verlag, pp. 177–202.

Kasap, S., Fray, J., Belev, G., Tousignant, O., Mani, H., Laperriere, L., Reznik, A., and Rowlands, J. (2009) Amorphous selenium and its alloys from early xeroradiography to high resolution X-ray image detectors and ultrasensitive imaging tubes, *Phys. Status Solidi*, **B246**, pp. 1794–1805.

Kastner, M. (1977) Radiative and non-radiative processes at valence alternation and other defects in lone-pair semiconductors. In: Proceedings of the 7th International Conference on Amorphous and Liquid Semiconductors, Edinburgh, pp. 504–508.

Kastner, M., Adler, D., and Fritzsche, H. (1976) Valence-alternation model for localized gap states in lone-pair semiconductors, *Phys. Rev. Lett.*, **37**, pp. 1504–1507.

Kastner, M. and Hudgens, S. L. (1978) Evidence for neutrality of luminescence centers in chalcogenide glasses, *Phil. Mag.*, **B37**, pp. 665–681.

Kawamura, T., Yamamoto, N., and Nakayama, Y. (1983) Electrophotographic application of amorphous semiconductors, amorphous semiconductor technologies and devices, *Jap. Annu. Rev. Electron., Comput. Telecommun.*, **6**, pp. 325–336.

Keating, P. N. (1966) Effect of invariance requirements on the elastic strain energy of crystals with application to the diamond structure, *Phys. Rev.*, **145**, pp. 637–645.

Khokhlov, A. F., Mashin, A. I., and Khokhlov, D. A. (1998) New allotropic form of silicon, *JETF Lett.*, **67**, pp. 675–679.

Knights, J. S. and Lujan, R. A. (1979) Microstructure of plasma-deposited a–Si:H films, *Appl. Phys. Lett.*, **35**, pp. 214–216.

Kolobov, A. V. and Shimakava, K. (1996) Photoinduced metastability in CGS. In: *Electronic Phenomena in Chalcogenide Glassy Semiconductors*, K. Tsendin (ed.), Nauka, St. Peterburg, pp. 365–395 (in Russian).

Kolobov, A. V., Oyanagi, H., Tanaka, K., and Tanaka, Ke. (1997) Structural study of amorphous selenium by in situ EXAFS: Observation of photoinduced bond alternation, *Phys. Rev. B*, **55**, pp. 726–734.

Kolobov, A. V., Fons, P., Frenkel, A. I., Ankudinov, A. L., Tominaga, J., and Uruga, T. (2004) Understanding of phase-change mechanism of rewritable optical media, *Nat. Mater.*, **3**, pp. 703–708.

Kolobov, A. V. and Tominaga, J. (2012) *Chalcogenides: Metastability and Phase Change Phenomena*, Springer-Verlag Berlin Heidelberg, p. 284.

Kolomietz, B. T. (1960) *Glassy State* (*Stekloobraznoe Sostoynie*), Moscow, Leningrad (in Russian).

Kolomiets, B. T., Lebedev, E. A., and Taksami, I. A. (1969) Mechanism of the breakdown in films of glassy chalcogenide semiconductors, *Sov. Phys. Semicond.*, **3**, pp. 267–269.

Kolomiets, B. T. (1981) Studies of chalcogenide vitreous semiconductors in Ioffe physico-technical institute, *J. Phys. Colloque C4*, **42**, pp. C4-887–C4-897. doi: 10.1051/jphyscol:19814194.

Kolomietz, B. T., Lyubin, V. M., and Shilo, V. P. (1978) Photo-induced changes in the etching rate of chalcogenide glasses, *Phys. Chem. Glasses*, **4**, pp. 351–357 (in Russian).

Kostylev, S. (2009) Programming speed, cycle-life and failure mechanisms in binary and multistate ovonic (phase-change) chalcogenide memory devices, *J. Optoelectron. Adv. Mater.*, **11**, pp. 1972–1982.

Kostylev, C. A. and Shkut, V. A. (1978) Electronic Switching in Amorphous Semiconductors, Kiev, "Naukova dumka," p. 203 (in Russian).

Kostyshin, M. T., Mikhaylovskaya, Ye. V., and Romanenko, P. F. (1966) On the effect on photosensitivity of thin semiconductor layers on metallic substrates, *Sov. Phys. Solid State*, **8**, pp. 451–452.

Krishnamurthy, R. (2013). First volume production phase change memory by Micron. [Blog] *Chipworks*. Available at: http://www.chipworks.com/en/technical-competitive-analysis/resources/blog/first-volume-production-phase-change-memory-by-micron/.

Kurashige, M., Tanioka, K., Ooka, M., and Hirai, T. (1993) 1-inch all electrostatic super HARD tube and camera for HDTV: A new possibility for very dark images with super-fine resolution, *Proc. SPEI*, **1987**, pp. 1–10.

Kuwano, Y. and Ohnishi, M. (1983) Integrated amorphous silicon solar modules. In: *Japan Annual Reviews in Electronics, Computers and Telecommunications, Amorphous Semiconductors: Technology and Devices*, Y. Hamakawa (ed.), OHM, Norton-Holland, pp. 204–227.

Lacaita, A. L. and Wounters, D. J. (2008) Phase-change memories, *Phys. Stat. Sol.* (a), **205**, pp. 2281–2297.

Lacourse, W. C., Twaddell, V. A., and Mackenzie, J. D. (1970) Effects of impurities on the electrical conductivity of glassy selenium. *J. Non-Cryst. Sol.*, **3**, pp. 234–236.

Leadbetter, A. J. and Apling, A. J. (1974) Diffraction studies of glass structure (V). The structure of some arsenic chalcogenide glasses, *J. Non-Cryst. Solids*, **15**, pp. 250–268.

Lebedev, A. A. (1921) On polymorphism and annealing in glass, *Proc. State Optical Institute*, St. Petersburg, **2(10)**, pp. 1–21 (in Russian).

Long, M., Galison, P., and Alben, R. (1976) Structural model of amorphous selenium and tellutium, *Phys. Rev. B*, **13**, pp. 1821–1829.

Lowrey, T. A., Hudgens, S. J., Czubatyj, W., Dennison, C. H., Kostylev, S. A., and Wicker, G. C. (2004) Characteristics of OUN phase change materials and devices for high density nonvolatile commodity and embedded memory applications, *Mat. Res. Soc. Symp. Proc.*, **803**, pp. HH2.1.1–2.1.12.

Lukovsky, G. (1987) Specification of medium-range order in amorphous materials, *J. Non-Cryst. Solids*, **97&98**, pp. 155–158.

Lukovsky, G. and Galeener, F. L. (1980) Intermediate range order in amorphous solids, *J. Non-Cryst. Solids*, **35&36**, pp. 1209–1214.

Lukovsky, G. and Pollard, W. (1984) Vibrational properties of amorphous alloys. In: *The Physics of Hydrogenated Amorphous Silicon II.*, J. Joannopoulos and G. Lucovsky (eds.), Springer-Verlag, pp. 377–431.

Madan, A. (1984) Devices using fluorinated material. In: *The Physics of Hydrogenated Amorphous Silicon I. Structure, Preparation and Devices*, J. Joannopoulos and G. Lucovsky (eds.), Springer-Verlag, pp. 245–283.

Madan, A. (2006) Amorphous silicon: From doping to multi-billion dollar applications, *J. Non-Cryst. Solids*, **352**, pp. 881–886.

Madan, A. and Ovshinsky, S. R. (1980) Properties of amorphous Si:F:H alloys, *J. Non-Cryst. Solids*, **35&36**, pp. 171–181.

Madan, A., Le Comber, P. G., and Spear, W. E. (1976) Investigation of the density of localized states in a–Si using the field effect technique, *J. Non-Cryst. Solids*, **20**, pp. 239–257.

Malaurent, J. C. and Dixmier, J. (1977) *A Random Chain Model for Amorphous Selenium. The Structure of Non-Crystalline Materials*, Cambridge University Press, Cambridge, UK, pp. 49–51.

Maruyama, M. (1992) Development of the underwater super-HARD color TV camera. In: Proceedings of ITEC'92, **N22-6**, pp. 425–426.

Mashin, A. I. and Khokhlov, A. F. (1999) Multiple bonds in hydrogen-free amorphous silicon, *Semiconductors*, **33**, pp. 911–914.

Masters, J. I. and Goldberg, G. M. (1981) Process of doping silver image in chalcogenide layer, US Patent 4,269,935, Class 430/323.

Matsuda, A. (1993) Mechanism of plasma chemical-vapor deposition of hydrogenated amorphous silicon. In: *Current Topics in Amorphous Materials: Physics and Technology*, Y. Sakurai, Y. Hamakawa, T. Masumoto, K. Shirae, and K. Suzuki (eds), Elsever, pp. 349–350.

Matsumura, H., Umemoto, H., Izumi, A., and Masuda, A. (2003) Recent progress of Cat-CVD research in Japan: Bridging between the first and second Cat-CVD conferences, *Thin Solid Films*, **430**, pp. 7–14.

Meškinis, Š. and Tamulevičienė, A. (2011) Structure, properties and applications of diamond-like nanocomposite, *Mater. Sci. (Medžiagotyra)*, **17**, pp. 358–370.

Minaev, V. S. (1991) *Glass-Forming Semiconductor Alloys*, "Metalurgija," Moscow, p. 406 (in Russian).

Mishima, T., Taguchi, M., Sakata, H., and Maruyama, E. (2011) Development status of high-efficiency HIT solar calls, *Sol. Energy Mater. Sol. Cells*, **95**, pp. 18–21.

Miyauchi, K., Qui, J., Shojiya, M., and Kawamura, N. (2001) Structural study of GeS_2 glasses permanently densified under high pressures up to 9 GPa, *J. Non-Cryst. Solids*, **279**, pp. 186–195.

Mizushima, Y. and Yoshikawa, A. (1982) Photoprocessing and lithographic applications, *Jpn. Annu. Rev. Electron. Comput. Telecommun.*, **2**, pp. 277–295.

Moller, S., Perlov, C., Jakson, W., Taussing, C., and Forrest, S. R. (2003) A polymer/semiconductor write-once-read-many-times memory, *Nature*, **426**, pp. 161–169.

Montrimas, E. and Petretis, B. (1974) The relation between deep trapping levels and the structure of Se layers, *J. Non-Cryst. Solids*, **15**, pp. 96–106.

Morigaki, K. (1999) *Physics of Amorphous Semiconductors*, Imperial College Press and World Scientific Publishing, p. 418.

Mott, N. F. and Davis, E. A. (1979) *Electron Process in Non-Crystalline Materials*, Clarendon Press, Oxford.

Muller, R. D. (1960) *Glassy State*, "Nauka," Moscow-Leningrad, pp. 61–71 (in Russian).

Nakmanson, S. M., Voyles, P. M., Mousseau, N., Barkema, G. T., and Drabolt, D. A. (2001) Realistic models of paracrystalline silicon, *Phys. Rev. B*, **63**, pp. 235207-1 –235297-6.

Nirschel, T., Philipp, J. B., Happ, T. D., Burr, G. W., Rajendran, B., Lee, M.-H., Schrott, A., Yang, M., Breitwisch, M., Chen, C.-F., Joseph, E., Lamorey, M., Cheek, R., Chen, S.-H., Zaidi, S., Raoux, S., Chen, Y. C., Zhu, Y., Bergmann, R., Lung, H.-L., and Lam, C. (2007) Write strategies for 2 and 4-bit multi-level phase-change memory. In: *2007 IEEE International Electron Devices Meeting*, Washington, DC, pp. 461–464.

NPD Solarbuzz. Marketbuzz 2013. HIS Web site. https://technology.ihs.com/Research-by-Market/450473/power-energy.

Okamoto, H., Hattori, K., and Hamakawa, Y. (1993) Hall effect near the mobility edge, *J. Non-Cryst. Solids*, **164–166**, pp. 445–448.

Ovshinsky, S. R. (1968) Reversible electrical switching phenomena in disordered structures, *Phys. Rev. Lett.*, **21**, pp. 1450–1453.

Ovshinsky, S. R. (1977) Chemical modification of amorphous chalcogenides: Amorphous and liquid semiconductors. In: Proceedings of the 7th International Conference on Amorphous and Liquid Semiconductors, Edinburgh, pp. 519–523.

Ovshinsky, S. R. (1991) Chemistry and structure in amorphous materials: The shapes of thinks to came. In: *Disordered Materials. Science and Technology. Selected Papers*, Plenum Press, pp. 333–342.

Ovshinsky, S. (2000) The letter to the stockholders of ECD company. Troy, Michigan, p. 28.

Park, W.-D. and Tanioka, K. (2003) Avalanche-type high sensitive image pickup tube using an a–Se photoconductive target, *Jpn. J. Appl. Phys.*, **42**, pp. L209–L211.

Pazin, A. V. and Borisova, Z. U. (1969) Glass-forming region in bismuth-germanium-selenium system, *Vestnik LGU, Phisika i himia*, **4**, 22, pp. 140–144 (in Russian).

Petkov, K. and Fedorov, B. A. (1982) Photo-induced changes of chalcogenide glassy semiconductor films As–S system, In: Proceedings of the International Conference "Amorphous Semiconductors-82", Bucharest, CIP-AP, pp. 205–207 (in Russian).

Pfister, G. and Scher, H. (1978) Dispersive (non-Gaussian) transient transport in disordered solids, *Adv. Phys.*, **27**, pp. 747–798.

Phillips, J. C. (1979a) Topology of covalent non-crystalline solids I: Short-range order in chalcogenide alloys, *J. Non-Cryst. Solids*, **34**, pp. 153–181.

Phillips, J. C. (1979b) Hollow claster models of atomic structure of chalcogenide glasses, *Inst. Phys. Conf. Ser.*, **43**, pp. 705–708.

Phillips, J. C. (1980) Chemical bonding, internal surfaces, and the topology of non-crystalline solids, *Phys. Status Solidi*, **B101**, pp. 473–479.

Phillips, J. C. (1981) Topology of covalent non-crystalline solids II: Medium-range order in chalcogenide alloys and a–Si(Ge), *J. Non-Cryst. Solids*, **43**, pp. 37–77.

Pimentel, G. C. and Spratley, R. D. (1970) *Chemical Bonding Clarified through Quantum Mechanics*, Holden-Day Inc.

Polk, D. E. (1971) Structural model for amorphous silicon and germanium, *J. Non-Cryst. Solids*, **5**, pp. 365–376.

Poltavzev, Yu. G. (1984) *Structure of Semiconductor Melts*, Metallulgiya, Moscow (in Russian).

Popescu, M. and Bradaczek, H. (2001) Microparacrystalline model for medium-range order in non-crystalline chalcogenides, *J. Optoelectron. Adv. Mater.*, **3**, pp. 249–254.

Popov, A. I. (1980) The influence of the structural changes on viscosity of selenium, *Fisika i Khimia Stecla*, **6**, pp. 307–311 (in Russian).

Popov, A. I. (1994) Determination of the rigidity threshold of the structural network in covalent glasses, *Glass Phys. Chem.*, **20**, pp. 553–554.

Popov, A. I. (1997) Electro-photographic materials, *Mater. Sci. Trans. (Materialovedenie)*, **2**, pp. 27–33 (in Russian).

Popov, A. I. (2002) Structural characteristics and structural modifications of non-crystalline semiconductors, *J. Optoelectron. Adv. Mater.*, **4**, pp. 481–488.

Popov, A. I. (2004) Atomic structure and structural modification of glass, *Semiconduct. Semimet.*, **78**, pp. 51–95.

Popov, A. I. (2008) *Physics and Technology of Disordered Semiconductors*, Publishing House MPEI, Moscow, p. 272 (in Russian).

Popov, A. I. (2009) Two generation of phase-change memory devices: Differences and common problems, *Phys. Status Solidi (b)*, **246**, pp. 1837–1840.

Popov, A. I., Geller, I., Karalunets, A., and Ipatova, N. (1980) Effect of doping on the molecular structure and VAP states of amorphous selenium, *J. Non-Cryst. Solids*, **35&36**, pp. 871–876.

Popov, A. I., Ligachev, V. A., Vasil'eva, N. D., and Stuokacn, S. N. (1995) Preparation of thin hydrogenated carbon films by HF sputtering method, *Proc. Ukr. Vac. Soc.*, **1**, pp. 248–251 (in Russian).

Popov, A. I., Michalev, N. I., and Shemetova, V. K. (1983) Structural modification of some glassy chalcogenudes, *Philos. Mag. B*, **47**, pp. 73–81.

Popov, A., Mikhalev, N., Karalynets, A., and Vasilyeva, N. (1991) Photosensitivity of selenium–bismuth films with varigap structure. In: Proceedings of the International Conference on Thin Film Physics and Applications, Shanghai, China (SPIE, vol. 1519), pp. 457–462.

Popov, A. I., Vorontsov, V. A., and Popov, I. A. (2001) Levels of structural modification of non-crystalline semiconductors and limits of their applicability, *Semiconductors*, **35**, pp. 637–642.

Popov, A. I., Presnyakov, M. Yu., Shupegin, M. L., and Vasil'ev, A. L. (2014a) Nanostructuring of metal-containing silicon–carbon composite films, *Nanotechnol. Russ.*, **9**, pp. 448–452.

Popov, A. I., Salnikov, S. M., and Anufriev, Y. V. (2015) Conditions of stable switch in the phase transition memory cells, *Semiconductors*, **49**, pp. 498–503.

Popov, A. I. and Shupegin, M. L. (2012) Diagnostics of the silicon–carbon nanocomposite film structure, *Proceedings of the V All-Russian School-Seminar "Diagnostics of Nanomaterials and Nanostructures,"* Ryazan, pp. 154–170 (in Russian).

Popov, A. I., Vorontsov, V. A., and Popov, I. A. (2001) Levels of structural modification of non-crystalline semiconductors and limits of their applicability, *Semiconductors*, **35**, pp. 637–642.

Popov, A. I., Savinov, I. S., and Voronkov, E. N. (2006) Simulation of phase-change processes in non-volatile memory cells, *J. Non-Cryst. Solids*, **352**, pp. 1624–1627.

Popov, I. A. (1996) Structural modification of amorphous hydrogenated silicon with help of ultra-violet radiation, *Semiconductors*, **30**, pp. 258–265.

Popov, I. A., Van Doorselaer, G., Van Calster, A., De Smet, H., Callens, F., and Boesman, E. (1998a) Prototype of 2D direct X-ray a–SIN:H sensor array. In: Proceedings of SPIE, Volume 3301, Solid State Sensor Arrays: Development and Applications II, pp. 126–130.

Popov, I. A., Van Doorselaer, G., Van Calster, A., De Smet, H., Callens, F., and Boesman, E. (1998b). a–SiN:H TFD sensor array for digital radiography. In: Proceedings of the International Congress on Imaging Science—ICPS`98, Antwerp, Belgium, pp. 315–319.

Popov, N. A. (1980) New model of defects in chalcogenide glassy semiconductors, *Pisma v GETF*, **31**, pp. 437–440 (in Russian).

Popov, N. A. (1981) Quasimolecular defects in glassy chalcogenide semiconductors, *Soviet Phys. Semicond.*, **15**, pp. 212–215.

Presniakov, M., Popov, A., and Vasiliev, A. (2013) Electron microscopy of the effect of heat treatment on the structure of metal-containing nanocomposites with silicon–carbon matrix, *J. Phys. Conf. Series*, **471**, 012036, pp. 1–4.

Presnyakov, M. Y., Popov, A. I., Vasil'eva, N. D., and Shupegin, M. L. (2013) Electron-microscopy study of the distribution of dopants (Ta) in Si–C nanocomposite films, *J. Surf. Invest.*, **7**, pp. 724–728.

Presniakov, M. Yu., Popov, A. I., Usol'tseva, D. S., Shupegin, M. L., and Vasiliev, A. L. (2014) Thermal stability of metal–silicon–carbon nanocomposites, *Nanotechnol. Russ.*, **9**, pp. 518–532.

Resnik, A., Lui, B. J. M., Lubin, V., Klebanov, M., Ohkawa, Y., Matsubara, T., Miyakawa, K., Kubota, M., Tanioka, K., Kawai, T., and Rowlands, J. A. (2006) The effect of temperature on photoinduced metastability in avalanche a–Se layers, *J. Non-Cryst. Solids*, **352**, pp. 1595–1598.

Rhee, J. H., Chun Chung, C., and Eric Wei-Guang, E. D. (2013) A perspective of mesoscopic solar cells based on metal chalcogenide quantum dots and organometal-halide perovskites, *NPG Asia Mater.*, **5**, e68; doi:10.1038/am.2013.53.

Richer, Y. and Breiting, G. (1971) Verschiedene formen von amorphen selen, *Z. Naturforsch*, **26a**, pp. 1699–1708 (in German).

Robertson, J. (2002) Diamond-like amorphous carbon, *Mater. Sci. Eng. R*, **37**, pp. 129–281.

Robertson, J. (2003) Improving the properties of diamond-like carbon, *Diamond Relat. Mater.*, **12**, pp. 79–84.

Sadoc, J. F. and Mosseri, R. (1982) Order and disorder in amorphous tetrahedrally coordinated semiconductors: A curved-space description, *Phil. Mag. B*, **45**, pp. 467–483.

Sakaguchi, Y., Asaoka, H., Uozumi, Y., Kawakita, Y., Ito, T., Kubota, M., Yamazaki, D., Soyama, K., Sheoran, G., and Mitkova, M. (2016) Processes of silver photodiffusion into Ge-chalcogenide probed by neutron reflectivity technique, *Phys. Stat. Sol. A*, **213**, pp. 1894–1903.

Sakaguchi, Y., Asaoka, H., Uozumi, Y., Kawakita, Y., Ito, T., Kubota, M., Yamazaki, D., Soyama, K., Ailavajhala, M., Latif, M. R., and Mitkova, M. (2014) Studies of silver photodiffusion dynamics in Ag/Ge_xS_{1-x} (x = 0.2 and 0.4) films using neutron reflectometry, *Can. J. Phys.*, **92**, pp. 654–658.

Sarsembinov, Sh. Sh., Prikhodko, O. Yu., Ryaguzov, A. P., and Maksimova, S. Ya. (2001a) Correlation between medium-range order in atomic structure and optical properties of amorphous AsSe films, *Semicond. Sci. Technol.*, **16**, pp. L44– L46.

Sarsembinov, Sh. Sh., Prikhodko, O. Yu., Ryaguzov, A. P., and Maksimova, S. Ya. (2001b) Comparison of atomic structures and electronic properties of amorphous AsSe films, *Semicond. Sci. Technol.*, **16**, pp. 872–876.

Sharber, M. C. and Sariciftci, N. S. (2013) Efficiency of bulk-heterojunction organic solar cells, *Prog. Polym. Sci.*, **38**, pp. 1929–1940.

Schottmiller, J. C., Bowman, D. L., and Wood, C. (1968) New vitreous semiconductors, *J. Appl. Phys.*, **39**, pp. 1663–1669.

Schroeder, B. (2003) Status report: Solar cell related research and development using amorphous and microcrystalline silicon deposited by HW(Cat)CVD, *Thin Solid Films*, **430**, pp. 1–6.

Schropp, R. E. I. (2009) Hot wire chemical vapor deposition: Resent progress, present state of the art and competitive opportunities, *Electrochem. Soc. Trans.*, **25**, pp. 3–14.

Sherchenkov, A. A. and Budagyn, B. G. (2007) *Physics and Technology of Semiconductor Power Converters, part 2*, MIET, Moscow, p. 280 (in Russian).

Shirai, S. and Kajita, I. (1993) a–Si photoreceptors. In: *Current Topics in Amorphous Materials: Physics and Technology*, Y. Sakurai, Y. Hamakawa, T. Masumoto, K. Shirae, and K. Suzuki (eds.), Elsever, pp. 410–414.

Shupegin, M. L. (2013) Deposition of metal-containing nanocomposite films with silicon-carbon matrix, *Zavodskaya Laboratory N*, **02** (79), pp. 28–33 (in Russian).

Shustin, E. G., Isaev, N. V., Klykov, I. L., Peskov, V. V., Polyakov, V. I., Rukovishnikov, A. I., and Temiryazeva, M. P. (2013) Synthesis of carbon films in a plasma-chemical reactor based on beam-plasma discharge, *Technical Physics*, **58**, pp. 245–250.

Sladkov, A. M., Korshak, V. V., Kudryavtsev, Yu. P., and Kasatochkin, V. I. (1972) USSR Discoverer's Certificate No 107, *Byull. Izobret.*, **6**(3).

Smorgonskaya, E. A. and Tsendin, K. D. (1996) Atomic and electronic structure of chalcogenide glassy semiconductors. In: *Electronic Phenomena in Cnalcogenide Glassy Semiconductors*, "Nauka," Saint Petersburg, pp. 9–33 (in Russian).

Sony and Panasonic. (2015) *White Paper: Archival Disc Technology*, July 2015, p. 16.

Spear, W. E. and Le Comber, P. G. (1975) Substitutional doping of amorphous silicon, *Solid State Commun.*, **17**, pp. 1193–1196.

Spear, W. E. and Le Comber, P. G. (1984) Fundamental and applied work on glow discharge material. In: *The Physics of Hydrogenated Amorphous Silicon I. Topics in Applied Physics*, vol 55, Springer, Berlin, Heidelberg, pp. 63–118.

Staebler, D. L. and Wronski, C. R. (1977) Reversible conductivity changes in discharge-produced amorphous Si, *Appl. Phys. Lett.*, **31**, pp. 292–294.

Stannowski, B., Rath, J. K., and Schropp, R. E. I. (2003) Thin-film transistors deposited by hot-wire chemical vapor deposition, *Thin Solid Films*, **430**, pp. 220–225.

Street, R. A. (1977). Non-radiative recombination, photostructural changes and Urbach's rule in chalcogenides. In: Proceedings of the 7th International Conference on Amorphous and Liquid Semiconductors, Edinburgh, pp. 509–513.

Street, R. A. and Mott, N. F. (1975) States in the gap in glassy semiconductors, *Phys. Rev. Lett.*, **35**, pp. 1293–1296.

Stutzmann, M., Biegelsen, D. K., and Street, R. A. (1987) Detailed investigation of doping in hydrogenated amorphous silicon and germanium, *Phys. Rev. B*, **35**, pp. 5666–5701.

Sunayama, H., Yamada, K., Karasawa, M., and Ishibashi, K. (2003) Preparation of poly-Si films by Cat-CVD for thin transistor, *Thin Solid Films*, **430**, pp. 226–229.

Takaura, N., Ohyanagi, T., Tai, M., Kinoshita, M., Akita, K., Morikawa, T., Shirakawa, H., Araidai, M., Shiraishi, K., Saito, Y., and Tominaga, J. (2014) 55-μA Ge_xTe_{1-x}/Sb_2Te_3 superlattice topological-switching random access memory (TRAM) and study of atomic arrangement in Ge–Te and Sb–Te structures, *IEEE International Electron Device Meeting (IEDM 2014)*, paper number 29.2.

Tanaka, K. (1975) Reversible photoinduced change in intermolecular distance in amorphous As_2S_3 network, *Appl. Phys. Lett.*, **26**, pp. 243–245.

Tanaka, K. (1980) Reversible photostructural change: Mechanism, properties and application, *J. Non-Cryst. Solids*, **35&36**, pp. 1023–1034.

Tanaka, K. (1981) Photo-induced phenomena in amorphous semiconductors. In: *Fundamental Physics of Amorphous Semiconductors*, F. Yonezawa (ed.), Springer-Verlag, Berlin, pp. 104–118.

Tanaka, K. (1983) Mechanism of photodarkening in amorphous chalcogenides, *J. Non-Cryst. Solids*, **59&60**, pp. 925–928.

Tanaka, M., Taouchi, M., Matsuyawa, T., Sawada, T., Tsuda, S., Nakano, S., Hanapusa, H., and Kuwano, Y. (1992) Development of new a-Si/c-Si heterojunction solar cells: ACJ-HIT (Artificially Constructed Junction–Heterojunction with Intrinsic Thin-Layer), *Jpn. J. Appl. Phys.*, **31**, pp. 3518–3522.

Terukov, E. I. (2016) Public–private partnership in the field of solar energy: An example of successful cooperation between science and business. In: Proceedings of the International Conference on Amorphous and Microcrystalline Semiconductors, St. Petersburg, pp. 19–20 (in Russian).

Thompson, M. (1984) Sputtered material. In: *The Physics of Hydrogenated Amorphous Silicon I. Structure, Preparation and Devices*, J. Joannopoulos and G. Lucovsky (eds), Springer-Verlag, pp. 119–175.

Tiedje, T., Moustakas, T. D., and Cebulka, J. M. (1981) Effect of hydrogen on the density of gap states in reactively sputtered amorphous silicon, *Phys. Rev. B*, **23**, pp. 5634–5437(R).

Uchino, D., Clary, D. C., and Elliott, S. R. (2000) Mechanism of photoinduced changes in the structure and optical properties of amorphous As_2S_3, *Phys. Rev. Lett.*, **85**, pp. 3305–3308.

Uvarova, R. M., Vannikov A. V., and Churkin, A. V. (2005) *Basics of Digital Printing*, MGUP, Moscow, p. 448 (in Russian).

Vasil'eva, N. D. and Popov, A. I. (1995) Effect of deposition conditions on the atomic structure of hydrogenated amorphous carbon films, *Inorg. Mater.*, **31**, pp. 417–422.

Vasil'eva, N. D., Popov, A. I., and Khalturin, V. P. (1982) Application of subgradient method to the non-crystalline semiconductor structure simulation. In: Proceedings of the International Conference "Amorphous Semiconductors-82", Bucharest, pp. 50–52 (in Russian).

Venkatraman, C., Goel, A., Lei, R., Kester, D., and Outten, C. (1997) Electrical properties of diamond-like nanocomposite coatings, *Thin Solid Films*, **308–309**, pp. 173–177.

Vikhrov, S. P. and Bodygin, N. V. (1994) *New Approach to Technology System Design (Preparation of a–Si:H Layers)*, RSRTA, Ryzan, p. 108 (in Russian).

Vinogradova, G. Z. (1984) *Glass Formation and Phase Equilibria in Chalcogenide Systems*, Nauka, Moscow, p. 176 (in Russian).

Voronkov, E. N., Popov, A. I., Savinov, I. S., and Fairushin, A. R. (2006) Percolation breakdown of amorphous semiconductors, *J. Non-Cryst. Solids*, **352**, pp. 1578–1581.

Vorotnikov, V. (2015) Russia turns to sun, wind to improve electricity supply for contry's Far East, *Renewable Energy World Magazine*, October 9, URL: http://www.renewableenergyworld.com

Voyles, P. M., Zotov, N., Nakhmanson, S. M., Drabold, D. A., Gibson, J. M., Treacy, M. M. J., and Keblinsky, P. (2001) Structural and physical properties of paracrystalline atomic models of amorphous silicon, *J. Appl. Phys.*, **90**, pp. 4437–4451.

Wagner, S., Gleskova, H., Cheng, I.-C., and Wu, M. (2003) Silicon for thin-film transistors, *Thin Solid Films*, **430**, pp. 15–19.

Wan, J. Z., Pollak, F. H., and Dorfman, B. F. J. (1997) Micro-Raman study of diamond-like atomic-scale composite films modified by continuous wave laser annealing, *Appl. Phys.*, **81**, pp. 6407–6414.

Watabe (2010) Next-generation optical disk technologies. URL: www.semicon.toshiba.co.jp/product/storage/pdf/ToshibaReview_vol66n8_06.pdf

Weiss, P. (2005) Superfast atom shuffling inspires data-storage alternatives, *Sci. News*, **167**, pp. 363–364.

Wright, V. and Uddin, A. (2012) Organic–inorganic solar cells: A comparative review, *Sol. Energy Mater. Sol. Cells*, **107**, pp. 87–111.

Wuttig, M. and Yamada, N. (2007) Phase-change materials for rewriteable data storage, *Nat. Mater.*, **6**, pp. 824–832.

Yamada, N. (2002) Phase-change material for use in rewritable dual-layer optical disk, *Proc. SPIE*, **4352**, pp. 55–63.

Yang, W. J., Choab, Y.-H., Sekinoc, T., Shima, K. B., Niiharac, K., and Auha, K. H. (2003a) Thermal stability evaluation of diamond-like nanocomposite coatings, *Thin Solid Films*, **434**, pp. 49–54.

Yang, W. J., Choa, Y.-H., Sekino, T., Shim, K. B., Niihara, K., and Auh, K. H. (2003b) Structural characteristics of diamond-like nanocomposite films grown by PECVD, *Mater. Lett.*, **57**, pp. 3305–3310.

Yeh, N. and Yeh, P. (2013) Organic solar cells: Their developments and potentials, *Renew. Sustain. Energy Rev.*, **21**, pp. 421 – 431.

Yeung, F., Ahn, S.-J., Hwang, Y.-N., Jeong, C.-W., Song, Y.-J., Lee, S.-J., Lee, S.-H., Ryoo, K.-C., Park, J.-H., Shin, J.-M., Jeong, W.-C., Kim, Y.-T., Koh, G.-H., Jeong, G.-T., Jeong, H.-S., and Kim, K. (2005) $Ge_2Sb_2Te_5$ confined structures and integration of 64 Mb phase-change random access memory, *Jpn. J. Appl. Phys.*, **44**, pp. 2691–2695.

Yole Développement (2013) By 2018, MRAM/STTMRAM and PCM will represent a $1.6B business. 19 February 2013. URL: http://www.i-micronews.com/news/archives/158-archives-2013/manufacturing.

Yukimoto, Y. (1983) Multi-layered and stacked solar cells: Amorphous semiconductor technologies and devices, *Jpn. Annu. Rev. Electron. Comput. Telecommun.*, **6**, pp. 228–238.

Yun, S. S., Li, H., Cappelletti, R. L., Enzweiler, R. N., and Boolchand, P. (1989) Onset of rigidity in $Se_{1-x}Ge_x$ glasses: Ultrasonic elastic module, *Phys. Rev. B*, **39**, pp. 8702–8706.

Zakery, A. and Elliott, S. R. (2003) Optical properties and applications of chalcogenide glasses: A review, *J. Non-Cryst. Solids*, **330**, pp. 1–12.

Zhou, H., Yang, L., and You, W. (2012) Rational design of high performance conjugated polymers for organic solar cells, *Macromolecules*, **45**, pp. 607–632.

Zolotukhin, I. V., Kalinin, Y., and Sitnikov, A. V. (2006) Nanocomposite structures on the way to nanoelectronics, *Priroda*, **N1**, pp. 11–19 (in Russian).

Index

For Product Safety Concerns and Information please contact our EU
representative GPSR@taylorandfrancis.com
Taylor & Francis Verlag GmbH, Kaufingerstraße 24, 80331 München, Germany

www.ingramcontent.com/pod-product-compliance
Ingram Content Group UK Ltd.
Pitfield, Milton Keynes, MK11 3LW, UK
UKHW021501240926
475702UK00035B/108

* 9 7 8 9 8 1 4 7 7 4 3 7 6 *